Ernest Lawrence Rossi, Ph.D.

# A DISCOURSE WITH OUR GENES

*The Psychosocial and Cultural Genomics of Therapeutic Hypnosis and Psychotherapy*

Edited by
*Salvatore Iannotti, M.D.*

EDIZIONI Editris sas

First Edition: July 2004

Rossi Ernest Lawrence, Ph.D.
A Discourse with our Genes
*The Psychosocial and Cultural Genomics*
*of Therapeutic Hypnosis and Psychotherapy*

**ISBN 88-89396-01-6**



*Printed in Italy*

Cover design by *Veneziano Scocca*

Special thanks to Professor ***Hermona Soreq*** - Jerusalem University -
for allowing us to use, and modify her figure.

Special thanks to Professor ***Eduardo Munoz*** - Psychobiology Unit,
Universidad Diego Portales, Santiago - for permitting us to modify
and utilize the DNA image he created,
to announce Dr Ernest Lawrence Rossi's 2003 workshop
in Chile in association with his hands hypnotic inductions.

Many sincere thanks to all who, directly or indirectly,
have contributed to this book

# Contents

# Introduction

Salvatore Iannotti, M.D.

This book is a necessity. In year 2000 Dr Rossi inducted me into a hypnotherapeutic state in a demonstration workshop for training students and professional psychotherapists in Rome, Italy. That was my first hypnotic experience. What a remarkable experience it was! I had been acutely ill with one of the most serious cases of Guillain-Barré Syndrome (GBS), a stress-related acute auto-immune polyradiculo-neuropathy with damage of my peripheral nerves that kept me in a complete flaccid paralysis for over twelve months. For nine and half months I was on a respirator.

It was only after 15 months that I, with the help of four strong men holding me, could stand on my feet for five seconds. They have certainly been the longest 5 seconds of my life. All my body, tired to the extreme for the great effort, was soaked wet with perspiration. Each day, though, I couldn't wait for that moment to stand up enjoying the feeling and working as hard as I could to beat the previous record. My muscles were almost completely atrophied and my nerves not well regenerated. The only displeasure I used to have was the fact that I could not do more exercises; I was pushing for it, but colleagues and physiotherapists would not allow me to go to far because it is counter productive in GBS. Six months of intensive physical and occupational therapy allowed me to start to walk with a four wheeled walker and 2 physiotherapists holding me on each side by handles attached laterally, front and back to a special, huge lather "training belt". Several times I was rescued by these well trained physiotherapists who held me up avoiding more damage to my body. Few other times I was not as lucky (above all when I started to do more on my own) and my knees hit the floor with all my weight. Thank

God my weight was not much at that time: I had lost about 70 lbs. Many muscles never came back fully and most had disappeared, completely atrophied.

During that first hypnotherapeutic session with Dr. Rossi I experienced something very unusual: muscles all over my body, from my head to my toes, started to contract and move involuntarily. As I learned later, these apparently involuntary movements were a classical experience of the "*ideodynamic action hypothesis*" about the fundamental nature of therapeutic hypnosis: *an idea can generate involuntary muscle activity on an unconscious or implicate level.* At first I could not stop these involuntary movements, which Rossi did not suggest in any way. Then, with growing delight, I did not want them to stop! I was experiencing and rediscovered the vestiges of muscles I did not know I had anymore. Too much atrophy due to the GBS. The moment I came out of the hypnotherapeutic state, the first thing that came to my mind and I immediately expressed was: "I wish I had the benefit of this experience during the acute phase of my illness!" I felt I now knew exactly when this hypnotherapeutic treatment should be started with the GBS patient.

My first insight was that this type of naturalistic hypnotherapy would have accelerated my nervous regeneration while offering many benefits – shorter hospital stay – less cost to society, and above all, more effective recovery with less sequela and fewer long lasting deficits. At this point Rossi exclaimed: "We are doing pioneering work here!" From then on I did not leave Rossi alone and I continued to follow his therapeutic suggestions to work on rehabilitating my entire body, mind and emotions with appropriate activity. At that time Dr Rossi was writing *The Psychobiology of Gene Expression: Neuroscience and Neurogenesis in Hypnosis and the Healing Arts*" (2002, W.W. Norton Professional Books). After I learned more about modern scientific theories of therapeutic hypnosis I organized a symposium at the 9th European Congress of Hypnotherapy in Rome in 2002 to discuss his work in integrating activity on the mental and genomic levels in psychology and medicine. In 2003 I legally incorporated in the state of

Connecticut, USA, the non-profit Ernest Lawrence Rossi Foundation (ELRF) with this mission statement.

"The Ernest Lawrence Rossi Foundation exists to facilitate research in consciousness studies via research in Psychosocial Genomics. Psychosocial Genomics explores how psychological states of consciousness modulate gene expression, brain plasticity, and mind-body healing. Whereas traditional behavioral genetics and evolutionary psychology focus on how genes modulate behavior, psychological traits, and psychosocial processes, Psychosocial Genomics focuses on the reverse - how consciousness, social processes, and cultural rituals can modulate gene expression and its sequelae for good or ill."

Currently the scope of the ELRF's mission is articulated in Rossi's *The Psychobiology of Gene Expression: Neuroscience and Neurogenesis in Hypnosis and the Healing Arts.* In 2004 I also incorporated the Iannotti-Rossi Foundation (F.I.R.) in San Lorenzo Maggiore, Benevento, Italy with similar and broader purposes. The major focus of both foundations at this time is to facilitate and support neuroscience research for the prevention and healing of mind-body and psychosomatic disorders. A *Discourse with Our Genes* is being published simultaneously in an English and Italian version (*Discorso tra Geni*). This book sets the stage for the great research adventure of psychosocial genomics with many innovative scientific projects proposed throughout.

Each chapter of this book is an update of one of Rossi's scientific papers that was peer-reviewed and published in a variety of research journals. Rossi's central themes are concerned with healing by integrating mind, body, and soul. Many of his original images integrate fundamental concepts in biology, medicine, and psychology in a new and intriguing manner that is updated with modern neuroscience. Without realizing it, Rossi has been introducing a new philosophy of life by exploring the deep meaning and surprising implications of the psychosocial genomics of creativity in everyday life as well as the arts, culture, spiritual practices, and psychotherapy in general.

Ernest Lawrence Rossi's Grandfather Ernest and his wife

Raffaela came from tiny village of San Lorenzo Maggiore, 1200 feet above sea level in the Benevento province of Italy, where the family had lived for many generations as small farmers subsisting by raising vegetable, grapes, olives, and figs. Grandfather Ernest, his wife, and their first born 6 year old son, Angelo, who was later to become Ernest Lawrence's father, immigrated to America in the early 1900s. Life was very difficult economically for the young family. Young Angelo loved to read but his education ended without graduation from high school because he had to go to work to help support the growing family. He married Mary De Libero when they were in their late teens and Ernest Lawrence Rossi was born a year later in the great depression of 1933.

One of Ernest's earliest memories was when grandfather suffered a very serious stroke at the age of about 45 that left him permanently crippled. The family could not afford a wheelchair so grandfather had to sit all day in a stout chair with cushions specially built for him by his eldest son, Angelo. With grandfather incapacitated, the whole family, now with 5 children, all had to go to work. Grandfather, handicapped as he was, became Ernest's first baby sitter. Ernest remembers playing happily alone about the house and fetching grandfather a glass of water or following other such simple request. Grandfather told the family when they all returned from work at the end of the day that Ernest usually behaved well but every now and then he "escaped" when he ran out of the house to explore the mysterious and forbidden far away territory across the street in the little town of Shelton, Connecticut.

A therapeutic ritual soon developed around grandfather's primitive efforts at self-rehabilitation. On many weekends the family gathered together to assess the value of yet another patent medicine that was reputed to possess miraculous healing powers. Little Ernest, about 4 or 5 years old, watched with fascination as grandfather was first administered a few huge tablespoonfuls of medicine. Then there was a dramatic pause as grandfather smacked his lips from the usually huge alcoholic content of the "*medicine.*" Grandfather would then make a heroic effort to stand-up with the use of his cane, which was actually a broomstick with a handle made by his son Angelo. With the excited and reverently praying family standing in a close circle around him, grandfather would then dramatically drop his cane and take a few lumbering steps unaided. He would pause momentarily as if suddenly stricken by a dark angel; the family would gasp audibly and move in to embrace him as he inevitably collapsed in awkward, uncoordinated, despair. Once again, the family thought, the "medicine" had not worked.

From our current neuroscience perspective on rehabilitative medicine, however, we now know that grandfather had not failed entirely. Rather, 70 years ago there was not enough understanding of how to facilitate stroke rehabilitation with persistent physical exercise and motivated emotional effort. If grandfather had the constant support of a modern physical rehabilitation team that understood the principles of activity-dependent gene expression and brain plasticity, grandfather undoubtedly would have recovered his ability to walk. I, myself, have been lucky to have survived one of the most debilitating cases of Guillain-Barré Syndrome only because modern medical knowledge, care, and technology are so much more advanced. I am convinced, however, that we can do much more to improve the quality of our health and life by learning how to better utilize our own inner psychosocial genomic resources as Rossi proposes.

In chapter one of this book, Ernest Lawrence Rossi tells his personal story of how he facilitated his rehabilitation from a stroke at the age of 69 by using his innovative psychosocial genomic approaches for evoking activity-dependent gene expression, brain plasticity, and mind-body healing that were inspired by a remarkable series of his own healing dreams. Rossi's basic insight is that the classical *ideodynamic action theory* of therapeutic hypnosis actually engages *activity-dependent and behavior-state related gene expression* to facilitate brain plasticity (synaptogenesis and possibly neurogenesis) at the molecular and cellular levels to optimise physical and mental rehabilitation from trauma and stress-related illnesses. This deep psychobiological theory about the fundamental nature of therapeutic hypnosis, psychotherapy, occupational therapy, and the creative process in the arts and sciences is entirely consistent with modern neuroscience research that documents how novelty, enrichment, as well as physical and mental exercise can generate gene transcription, the synthesis of new proteins, and brain plasticity.

Many of Rossi's most profound ideas are only hypotheses at this time, however. These hypotheses need to be tested scientifically with the new DNA microarray technology ("*gene chip*") to explore the details of the psychobiological mechanisms by which psychological, social, cultural, and spiritual practices can modulate gene expression, brain plasticity, and mind-body healing. What are the possibilities and limitations of Rossi's innovative approaches to facilitating mind-body communication and healing? To exactly which dysfunctions and psychosomatic illnesses do they apply? To what extent can these new approaches to self-care and rehabilitation be taught? What are the best methods of training health professionals to administer these new approaches to what populations of patients? These are some of the fundamental questions that the Ernest Lawrence Rossi Foundation and the Iannotti Rossi Foundation seek to answer by sponsoring modern neuroscientific and psychosocial genomics research and education programs throughout the world wherever we can organize and facilitate like-minded creative workers.

**In the Creative Tradition of Leonardo da Vinci**

Leonardo da Vinci had a profound influence on the development of the arts and sciences from his day to the present. His inventions have been ahead of his time: he drew plans for parachutes, helicopters, various machines, the bicycle, etc. Leonardo da Vinci, besides engineering a large number of practical objects, tools and machines to the point of gave birth to the Industrial Era. Leonardo has been recognized even more for his new ideas and theories about creativity, which have revolutionized design.

He pioneered of the concept of automation and the discipline of modern comparative anatomy in his drawings of the fetus, the heart and of the brain. He is a pioneer of modern botanical science with his concepts of eliotropism, geotropism, determination of trees' age and leaf arrangement in plants. Fossilization and soil erosion are well acknowledged phenomena discovered by Leonardo da Vinci. He could also very well be considered the father of the modern fields of mechanics, hydrostatics and optics.

Michael J Gelb (p. 45, 1998) rightly and very correctly has stated: "Leonardo's investigations led him to anticipate many great scientific discoveries including many great breakthroughs by Copernicus, Galileo, Newton, and Darwin."

*40 years before Copernicus*—- da Vinci noted, in large letters for emphasis, "IL SOLE NON SI MUOVE," "The sun does not move." He added, "The earth is not in the center of the circle of the sun, nor in the center of the universe."

*60 years before Galileo*—-He suggested that "a large magnifying lens" should be employed to study the surface of the moon and other heavenly bodies.

*200 years before Newton*—- Anticipating the theory of gravitation, Leonardo wrote, "Every weight tends to fall towards the center by the shortest possible way." And elsewhere he added that because "every heavy substance presses downward, and cannot be upheld perpetually, the whole earth must become spherical."

*400 years before Darwin*—-He placed man in the same broad category as monkeys and apes and wrote, "Man does not vary from the animals except in what is accidental."

Despite all of the above Leonardo did not have as much influence on the integration of the arts and science he should have had. Some say this is because he did not organize his scientific thoughts and did not publish them in our modern fashion. Others argue he was so far ahead of his time that his work would not have been appreciated even if it was formulated in comprehensible general theories. In all of this I can see similarities between the creativity of Leonardo da Vinci and Ernest Lawrence Rossi who has developed many theories during the last forty years about the nature of consciousness, the creative process, and the relationship between mind and body in psychology and medicine. Many of Rossi's ideas have not been easily to understand because they are so far ahead of our time.

The most striking similarity between Leonardo da Vinci and Ernest Lawrence Rossi is their focus, analysis, application, and teaching about the creative process in the integration of the arts and sciences. Leonardo da Vinci originally described 7 characteristics of the creative process: *Curiosità—Dimostrazione—Sensazione—Sfumato—Arte/Scienza—Corporalità—Connessione.* Centuries of introspection by creative workers as well as research in psychology and modern neuroscience has simplified and condensed these 7 characteristics into a 4-stage creative process that is easy for most people to learn and practice. Rossi has made many important contributions to the theory and application this 4-stage creative process in psychotherapy and psychosomatic medicine that are described in the chapters of this book. A brief summary of Rossi's mind-body perspective of the integration of these 4 stages of the creative process in the arts and sciences may be introduced here as an initial orientation for the reader.

**Stage 1: Preparation, Data Gathering**. It is now known that any intense psychological states of arousal —- such as trauma, pain, stress, novelty, dreaming (REM sleep), and creative moments in everyday life as well as all the arts and sciences can initiate the activity of Immediate Early Genes, Activity (or Experience) Dependent, and Behavior-State Related Genes in our brain and body. Our genes are not always in an active state; genes have to be stimulated in everyday life by internal and external environmental and psychosocial signals to generate the proteins that are the molecular machines of life that do creative work. Stage one of the creative process includes Leonardo da Vinci's principles of *curiosità* and *sensazione.* Sensations stimulate neural activity and curiosity, the desire to learn more, sets us forth on deeply motivating outer and inner journeys of discovery and self-creation.

**Stage Two: Incubation.** In his teaching demonstrations of psychotherapy for students and professionals Rossi hypothesizes how Immediate Early, Activity Dependent and Behaviour State-Related Gene Expression may be taking place during creative problem solving. Often the incubation stage is characterized by a mild state of confusion, stress and even psychosomatic symptoms. The involuntary *ideodynamic* muscle movements I experienced so dramatically in my first hypnotherapeutic experience with Rossi may be an example of how *ideas* give rise to *dynamic action* at the *genomic level.* Rossi demonstrates how emotional conflicts and symptoms experienced in this stage are mind-body language about unresolved problems at an unconscious level that require review, re-synthesis and therapeutic reframing on the conscious level. This stage often corresponds to Leonardo's principle of *dimostrazione.* We have to discover for ourselves what we believe in. We need to look at things from different perspectives and to learn from our mistakes. Leonardo's principle of *sfumato* has a gradation of meanings from the literal translation of smoke gradually thinning out, to soft, shaded, and, the more popularly expressed, "come to nothing" - to progress or "go up in smoke." The concept of *sfumato* characterizes the often difficult transition from stage two to stage three of Rossi's creative healing cycle.

**Stage Three: Illumination.** This is the very rewarding creative moment experienced in the arts and sciences as well as psychotherapy. This stage is the initial experience of what Leonardo describes as *arte/scienza, corporalità,* and *connessione.* Rossi hypothesizes that gene expression and new protein synthesis at this numinous stage may generate brain plasticity - the actual synthesis of new synapses and connections between brain cells that encode human experience and the creative transformations of consciousness. It is of essence that people learn how to recognize and support these new developments in their consciousness. The main job of the psychotherapist at this stage is to help people recognize and appreciate the new. Often a person's psychosomatic symptoms disappear dramatically as personal problems are resolved with the new perspectives that develop.

**Stage Four: Verification.** In this final stage of the creative cycle the person must verify the value of the new experiences of stage three by practicing them in real world. The new experiences and realizations are often fragile and can be easily lost in smoke – Leonardo's precious *arte/scienza, corporalità,* and *connessione* can be lost in *sfumato* again! It is ironic that our family and friends, who wish us well, often do not recognize the new that develops within us. Because they do not recognize the potential value of the new, the people who are closest to us often do not know how to support us in the realization of our new reality. Thus adolescents have difficulties with their family and friends. Falling in love can be fragile and fickle. Innovators and creative worker have been perpetually misunderstood and persecuted throughout history for daring to assert their new consciousness. Rossi has demonstrated many practical methods for supporting people through this difficult labour of verifying and realizing the new.

Each chapter of this book explores the 4-stage of creative process in the arts and sciences in one way or another by integrating their manifestations from mind to gene. In this way Ernest Lawrence Rossi, like Leonardo da Vinci, corrects Rene Descartes' error in separating mind and body in psychology and medicine. The appendix outlines many types of research programs that the Ernest Lawrence Rossi Foundation and the Iannotti Rossi Foundation seek to support so that students and professional scientists will delight in deepening our understanding of the fundamental nature of art, creativity, imagination, drama, dreams, and mind-body healing in science and medicine. Let's go to work! *

* F.I.R. will became operative with a congress headed by Rossi himself on "***Rossi's Innovative Hypnopsychotherapeutic Approach***", in the Benevento province, Italy.

# Preface

There are profound turning points in the history of the human mind that are little recognized when they are first taking place. Consider how little the general public valued Socrates' appreciation of the self-examined life, Copernicus' realization that the earth goes around the sun [*Galileo Galilei was jailed by the "Sant'Uffizio" for following this theory*], Darwin's evidence for the evolution of life, Mendel's discovery of the first laws of genetics, and Einstein's equation, $E=MC^2$.

Humanity is now at another profound but little understood transition: we are making epochal discoveries about how everyone can learn to turn their genes on and off to create a better brain, health, and well-being throughout their lifetime. Art, science, history, philosophy, and consciousness itself gain deeper meaning in the light of this fundamentally new understanding of how we co-create ourselves in our daily dialogue with our genes, our *Discorso Tra Geni*, for good or for ill.

Genes are *inner resources* that we can all learn to use in a creative manner in daily life to construct and reconstruct our brain and mind for optimum health and well being. We are not the puppets dancing to the tune of our genes as many popular but superficial theories maintain. The reverse is true. Genes are an active society within every cell of our brain and body that responds in a cooperative and adaptive manner to signals from our environment to make life possible. Our hopes and fears, our thinking, passions, and striving all can turn families of genes on and off to make the proteins that can generate new brain cells and the connections between them to recreate our mind and personality. Matt Ridley (2003) pioneered this understanding of genes by reviewing the scientific literature and interviewing many of the researchers that made the original scientific discoveries.

> *The first and most general moral is that genes are enablers, not constrainers. They create new possibilities for the organism; they do not reduce its options.* Oxytocin receptor genes allow pair bonding; without them the prairie vole would not have the option of forming a pair bond [finding a mate]. CREB genes allow memory; without these genes, it would be impossible to learn and recall. BDNF allows the calibration of binocular vision through experience; without it you could not so easily judge depth and see the world as three-dimensional. FOXP2 mysteriously allows human beings to acquire the language of their people; without it, you cannot learn to speak. And so on. These new possibilities are open to experience, not scripted in advance. Gene no more constrain human nature than the extra programs constrain a computer. A computer with Word, PowerPoint, Acrobat, Internet Explorer, Photoshop, and the like not only can do more that a computer without these programs but can also get more from the outside world. It can open more files, find more Websites, and accept more e-mail... *Genes, unlike gods, are conditional. They are exquisitely good at simple if-then logic: if in a certain environment, then develop in a certain way... I suspect that science has so far greatly underestimated the number of gene sets which act in this way – conditioning their output to external conditions.* (p. 250)... Genes themselves are implacable little determinists, churning out utterly predictable messages. *But because of the way their promoters switch on and off in response to external instruction, genes are very far from being fixed in their actions. Instead, they are devices for extracting information from the environment. Every minute, every second, the pattern of genes being expressed in your brain changes, often in direct or indirect response to events outside the body. Genes are the mechanisms of experience.* (p. 248, italics added)

In this book I explore, extend, and apply this creative relationship between our genes, our psychosocial and cultural environment, and our conscious sense of free will and spiritual seeking. The truth, however, is that we are all in kindergarten in this profound playground of understanding human nature. At best I can only hope to inspire you with the wonder of this new perspective about the co-creative relationship between our mind, genes, and nature – and propel to explore further on your own.

***Ernest Lawrence Rossi***
Los Osos, California,
2004

# I
# MY STORY: GENE EXPRESSION & BRAIN PLASTICITY IN STROKE REHABILITATION

I was brutishly clubbed on the head in my sleep. I felt heavy and unable to move out of a cramped fetal position in the nightmarish darkness. I wanted to groan but could not. I did not know whether I was asleep or awake. But I must have opened one eye at least momentarily to glance at the dim luminous glow of a clock by my bed that registered about 2:30 a.m.

I awakened at about 4:30 a.m. with a strange sluggishness, stiffness and awkwardness of movement. I could hardly roll out of bed. I thought I was suffering from a sudden attack of arthritis or post-polio syndrome. I staggered to the medicine cabinet to try out some new anti-inflammatory medication my primary physician had given me a few days before. I wanted to record the exact time I took the medicine to see how long it took to give me relief. There was no paper handy so I tried to write "4:30" on the medicine package. But something was radically wrong with my writing; it was incredibly tiny and the numbers were written on top of each other! I thought it must be the unevenness of the writing surface on the medicine package. I finally found a sheet of paper but my handwriting was no better – the numbers remained tiny and I was almost amused to realize I probably was experiencing micro-graphia for some unknown reason. But I was incredibly uncoordinated trying to take a shower and almost fell. It was a struggle to dress; my right leg could not find its way properly into my trousers. It wasn't till my wife awakened about two hours later that I realized I had badly slurred speech. A quick check of my face in the mirror revealed the truth: the right side of my face was pulled down completely out of its normal symmetry. I now knew I was experiencing a stroke. My wife, Kathryn, called the doctor and rushed me to the hospital in our van.

The CD player in our van automatically turned on playing my favorite contemporary composer, Kitaro. One passage of the music played a haunting drumbeat reminiscent of American Indian music. Suddenly I was seized with a paroxysm of hot tears as I glanced over the pristine countryside of the San Louis Obispo area through which we were driving on the way to the hospital. I was experiencing the truth of the deep tragedy of the Indian Nation with what seemed to be a profound clarity - a cleansing of perception. Did not anyone else recognize the truth of this exquisite tragedy of the Indian Nation? Why wasn't something more being done to correct it! In reality I saw the mortified fear in my wife's face – she never saw me crying that way. She thought I was crying about myself and tried to reassure me. I had great difficulty with my slurred speech in trying to explain about the Indian Nation. I saw her growing alarm as she thought I was surely going daft. I now realized I might be experiencing "affect labiality," which is typical in stroke. I tried to calm down and smile to reassure her but all that came out was a strange, strangled laughing along with the tears that I still could not control.

The music of Kitaro continues to move me to a mystical sense of hyper-reality even today a few months after that fateful ride. It is the kernel of a personality change that is still taking place within me as I carefully cultivate a greater accessibility and expression of my emotions that my wife now greatly enjoys. Here is the full record of a series of dreams of recovery I had during this period that seem to reflect Jung's (1966) synthetic or constructive method of psychotherapy as well as Erickson's (1959/1980) utilization or activity-dependent approach to healing and rehabilitation.

### The Body Out of Control

(*First week after the stroke*)

I am driving on a freeway overpass but there is an accident and pile-up of cars ahead. Some people got out of their cars and went to the railing to get out of the way. But I fear I cannot stop my car because of some weakness in my leg and may run into them. I greatly fear that a tragic accident is about to take place.

This dream relates to the real weakness I feel throughout my body – especially on the right side. It really is a good idea not to drive at this time. I tell this dream to my physical therapist explaining how it probably related to my primary physician having to report my stroke to the DMV (Department of Motor Vehicles) by law and having my drivers license restricted. She looked at me softly and sympathetically but quietly questioned, "Have you considered the dream also could be reflecting your cerebral-vascular accident where the freeway was a blood vessel in your brain accidentally piling-up blood corpuscles?"

"Well, yes," I say with a crooked little smile doing my best to make my face more symmetrical just to please her, "how come I did not think of that?"

## Emergency Self-Observation of Lonely Radar

*(Second week of recovery)*

A small, lonely, emergency radar on an infinite, desolate, and dusty plain seemingly coughs, sputters a bit and then begins sweeping the sky rapidly in a circular pattern - frantically but futilely seeking a response. Gradually, in the far distance, I see a response beginning to develop as an output image on a computer screen. The image is a 3-D wire mesh outline of a human head circulating so one can get a complete 360-degree view of the interior of the head. With growing excitement I realize that it is a MRI image of my own head that may enable me to see the exact area of my brain that was damaged by the stroke. I start to go lucid in the dream: I realize that I am dreaming and if I can slow down the rotating image enough I will be able to see exactly where the damage is to better guide my rehabilitation program. The rotating image does indeed slow down and I am just about to see the damaged area in my left cerebral hemisphere when I realize, "Oh, no, I am waking up!" I lose the image.

I am at first greatly disappointed when fully awakened because I could not exactly locate the damaged area. But gradually tears of gratitude come as I realize the most important significance of this dream. The lonely radar is a metaphor for a signaling system becoming activated in the self-reflective capacity of my mind-brain. My brain realizes that it is still in an emergency situation but enough recovery has taken place over the past two weeks so that at least a small radar is now operating seeking the information it needs to further direct its own self-repair.

It is well known that all cells of the brain and body are continually sending each other molecular messengers (neurotransmitters, hormones, growth factors etc.) to direct development, adaptation, and repair. The desolate plain is the damaged area of my brain where molecular signals from dead and still struggling neurons are thrashing about seeking contact and connection to reestablish lines of communication for my recovery. The dream inspires me to redouble my daily rehabilitation efforts in speech, memory, eye-hand coordination, balance and mildly strengthening physical exercises. I begin to go on little unsupervised walks around the block but I am not allowed to try to cross the street by myself yet.

My many rehabilitation therapists (occupational, speech, physical, neurologists, etc.) congratulate me on my positive attitude. They really don't know half the story. My body is still very weak and uncoordinated. I have to hold the banister with two hands to get up and down stairs. My wife winces with worry as she watches me use both my arms to lift and haul my somewhat limp right leg into and out of our van. Paradoxically her worried wince evokes a grim satisfaction in me as I suddenly recall the analogous situation about 30 years when I lifted the famous Milton H. Erickson, M.D. (founder of The American Society of Clinical Hypnosis) out of his wheel chair into his family's old station wagon to take him for a drive through the deserts around Phoenix to visit rock shops, which we both loved. On those occasions the situation was reversed since I was the one who winced with worry least I hurt

Erickson as I hauled him about.

Erickson on the other hand, ever the healing mentor, tried to reassure me with his jaw firmly set in grim satisfaction as he told me yet another story about his efforts at self-rehabilitation from his lifetime of coping with polio. Life, he would explain, actually is a continuous process of rehabilitation. Every day and every moment when you *consciously choose* to work cheerfully and creatively with your handicap rather than complain - literally gives you a leg up. Getting in and out of an auto is *utilized as another opportunity* to gain yet another increment of muscle coordination and strength. He called this the essence of his *naturalistic or utilization approach* (Erickson, 1958/1980, 1959/1980). I now call it, "The Symptom Path to Enlightenment" (Rossi, 1996) whereby the *process of actively of coping with one's symptoms and problems* guides and facilitates real physical healing as well as further psychological development. This is the source of my positive attitude toward rehabilitation – it is the source of Erickson's greatest legacy of healing to me although I could not appreciate its significance 30 years ago. Into my current desperate situation comes the following compensatory dream of the gathering of the archetypal forces of healing at the deepest level.

**Gathering the Archetypal Resources for Healing**

(*Third week of recovery*)

Real life circumstances of this dream: A neurologist filling out a routine medical form asks me if my illness has resulted in a loss of physical strength so that I can no longer do my job. I grimly grin at him with my best Ericksonian attitude and humorously respond, "Well, I'm not exactly an iron foundry worker you know." That night I have the following dream.

A huge Paul Bunyan type man is using gigantic iron pliers and tongs to manipulate small metal objects. He is going to teach me how to do it skillfully. I am experiencing great awe that he notices me and I feel very grateful about the prospect his help.

I interpret this dream figure to be analogous to my occupational therapist who is, in reality, assessing and facilitating recovery of my damaged hand-eye coordination by giving me many tasks involving puzzles, picking up small metal objects with tweezers etc. I tell him this dream and explain my interpretation: Paul Bunyan is a metaphor of an inner implicit healing process operating via activity-dependent gene expression and brain plasticity (the growth and reorganization) of synapses and neurons in the brain) that hopefully are now being activated by all this occupational therapy to repair my brain. He has never heard of this neuroscience concept of rehabilitation but I assure him that my new book (Rossi, 2002) is coming out soon so he can read about it. I'm just too exhausted to try to explain it all right now.

As I continue to replay this dream in active imagination the Paul Bunyan figure becomes evocative of the archetype of Vulcan. My Vulcan is somewhat like Mr. Spock of Star Trek fame. But he is Chinese blood red-orange with a thunderous body stretching from the center of the earth to the sky. He has huge massive muscles and an impassive mien. He does not speak and he hardly notices me but I am given to understand he actually is a genial gene genii ready, available, and fully capable of firing the sources of life should he be called.

Ok, so I'm calling, now! After a week or so of watching him do nothing but stand there silently poised with his huge iron tools, the Vulcan figure finally becomes activated in my imagination. One morning I awaken with grateful tears when I see Vulcan pounding a huge glowing gold ingot on a mighty anvil with flashing lightening leaping about with every stroke. Even now, several months later, I gain strengthening reassurance as I witness his continuing methodical pounding whenever I call him forth in my active imagination.

Jung (1959) describes Vulcan as follows:

> " ...we learn that the lowest sphere corresponds to Vulcan, the earthy fire; the second to Mercurius, the vegetative life-spirit; the third to the moon, the female, psychic principle; the fourth to the sun, the male, spiritual principle. It is evident from Maier's commentary that he is concerned on the

> one hand with the four *elements* and on the other with the four kinds of *fire,* which are responsible for producing different states of aggregation. His *ignis elementalis re et nomine* would, according to its place in the sequence, correspond to Vulcan... The idea is... the elements or states of aggregation can be reduced to a common denominator. Today we know that the common factor to antagonistic elements is *molecular movement,* and that the states of aggregation correspond to different degrees of this movement. Molecular movement corresponds to a certain quantum of energy. One of the stepping-stones to the modern concept of energy is Stahl's phlogiston theory, which is based on the alchemical premises discussed above. We can see in them, therefore, the earliest beginnings of a theory of energy." (Pp. 249-251)

How would we update Jung's concept of the "molecular movement... the elements or states of aggregation [that] can be reduced to a common denominator?" A way of understanding the common denominator of these elements of molecular aggregation that are fundamental to the sources of life is to regard them as genes. How does this relate to my concern with rehabilitation? This healing dream implies that rehabilitation will be facilitated by my being *actively engaged in exercises* like manipulating the tongs as illustrated with the Paul Bunyan figure representing the archetypal Vulcan.

This, ironically, is the subject of my new book (Rossi, 2002) that was completed a few months before my stroke. Therein I discuss the possible molecular mechanism of rehabilitation *via activity-dependent gene expression*. It is now believed that facilitating gene expression, brain plasticity, and stem cell differentiation and maturation into newly functioning tissues via *activity-dependent cognitive-emotional-behavioral experiences* is a basic mechanism of healing that makes rehabilitation possible. Patients with severe trauma resulting in loss or paralysis of sensory-motor functions due to physical injury, cardio-vascular accidents, stroke, etc., for example, can recover their abilities via occupational and physical therapy that works primarily by *activating their behavior*. Until recently this molecular-genomic mechanism of rehabilitative healing by behavioral activation was not understood. *The new neuroscience hypothesis is that behavior activation initiates activity-dependent gene expression. This activity-dependent gene expression initiates healing by the generation of proteins that facilitate brain plasticity and stem cell differentiation into new tissues.*

We do not yet know the full range and limitations of this new neuroscience approach to healing and rehabilitation. Intense research efforts are now underway to discover the optimal conditions whereby cognitive-behavioral stimulation can evoke *activity-dependent gene expression, brain plasticity and mind-body healing* (Kempermann et al., 1997, 1999). It is now known, for example, that when experimental animals experience *novelty, environmental enrichment and physical exercise* the *zif-268 gene* is expressed during their REM (dream) sleep (Ribeiro et al., 1999, 2004). *Zif-268* gene expression is associated with the generation of proteins and growth factors that facilitate synaptogenesis and neurogenesis, literally brain growth. I generalize this to the human level by describing it as the *novelty-numinosum-neurogenesis effect*, which I propose (Rossi, 2000a, 2000b, 2002) as an emerging neuroscience foundation for Jung's (1966) synthetic or constructive method as well as Erickson's (1958/1980) naturalistic and utilization approach to healing and rehabilitation. In my most recent book (Rossi, 2002) I generalize this idea of healing to the cultural and spiritual arts as follows.

"The *Creative Replay of the Novelty-Numinosum-Neurogenesis Effect in the Arts, Humanities, and Cultural Rituals:* Enriching life experiences that evoke the novelty-numinosum-neurogenesis effect during creative moments of art, music, dance, drama, humor, literature, spirituality, awe, joy, and cultural rituals can optimize the psychosocial genomics of consciousness, personal relationships, and healing.

"The entire history of human approaches to healing that evoke the novelty-numinosum-neurogenesis effect — from ancient spiritual rituals of exorcism, shamanism, fire walking, to the still "mysterious" methods of acupuncture, energy medicine, and neurofeedback is the data base for this hypothesis. *Psychobiological healing during ecstatic religious experiences of the numinosum involving a combined sense of fascination, the mysterious, and the tremendous has much in common with modern rituals of healing associated with the self-help groups, twelve-step programs, and the so-called "miracle cures" of therapeutic hypnosis*. I hypothesize that just as negative states of emotional arousal can evoke the psychogenomic network to initiate gene expression cascades leading to the overproduction of stress proteins and illness, so can positive psychological experiences initiate the novelty-numinosum-neurogenesis effect to facilitate gene expression, brain plasticity, problem solving, and healing."

### A Dream of Numinous Beauty and Clarity

(*Fourth week of recovery*)

I enjoy the numinous beauty and wonderment of looking through a new clear crystal cover on our swim spa seeing the delightful light blue, clean water in the sparkling sunlight.

We actually don't have such a new crystal cover over our swim spa but one does not have to be a rocket scientist to realize this dream is a metaphor for some sunlight clarity coming into the waters of my brain. A battery of psychological tests administered to me at this time tells the story of my mental status in a stark manner that clearly outlines my assets and deficits. The good news is that my *abstract reasoning* is at the 99th percentile level and my capacity for *mental organization* is at the 97th percentile. The bad news is that I am way below normal in *perception and discrimination* at the 45th percentile level and, even worst, is my *short-term memory*, which is down to the 37th percentile.

I spend several afternoons sitting silently in the sunlight gazing into the clear crystal water of our swim spa with a deep hunger to drink it all in somehow to assuage my stunned brain. I smile at the surrounding little meditation paths I built when I was strong years ago in our backyard garden with their statues of the Buddha and Quan Yin. For the longest time it seems I find myself struggling to recall the words of the Yeats poem *Vacillation*.

My fiftieth year has come and gone,
I sat, a solitary man,
In a crowded London shop,
An open book and empty cup
On the marble table-top.

While on the shop and street I gazed
My body of a sudden blazed;
And twenty minutes more or less
It seemed, so great my happiness,
That I was blessed and could bless.

After I repeatedly replay this dream in active imagination with wisps of poetry throughout the day for many days, it finally dawns on me that my hunger and preoccupation with the numinous experience of crystal clarity may be an example of the "novelty-numinosum-neurogenesis effect." The psychological experience of numinous beauty, wonderment, and crystal clarity may correspond to the activation of gene expression and neurogenesis to facilitate the healing of my *perception and discrimination,* which is way below par at the $45^{th}$ percentile level. Rudolph Otto (1923/1950) formulated the concept of the numinosum, consisting of the experiences of *fascination, mystery, and tremendousness*, to describe the common denominator of all original spiritual experience. I muse over these three facets characteristic of the numinosum in spiritual development and the three facets of *novelty, environmental enrichment and physical exercise* that neuroscience research now finds characteristic of the development of consciousness (memory, learning etc.) via activity-dependent gene expression and brain plasticity to build a better brain in daily life.

After replaying the numinous beauty of this dream for weeks I have a particularly vivid experience of it, an epiphany of sorts, while listening to Carlo Ponti, Jr. conducting The Russian National Orchestra's performance of Mussorgsky's "Pictures at an Exhibition" at California Polytechnic University on Saturday evening, August 17, 2002. The program notes quote Mussorgsky's own description of his creative fervor while composing this piece: "Ideas, melodies, come to me of their own accord... I gorge and overeat myself. I can hardly manage to put it all down on paper fast enough." Likewise, while listening to this performance, I have a similar experience by recognizing how music, art, poetry, philosophy, and science all come together as one with my Italian heritage in a new theory of aesthetics: *the numinous experience of beauty could generate gene expression, brain plasticity, and the actual reconstruction of the brain during creative moments described by John Keats as the equivalence of truth and beauty.* At this moment all these connections seem to be an astonishing reification of Carl Jung's (1916/1960) concept of "The psychological 'transcendent function' [that] arises from a union of conscious and unconscious contents as well as the real and imaginary" (p. 69).

Is my repeated replaying of the numinous beauty and clarity of this dream a *novel, enrichening, exercise* that can facilitate brain plasticity so that my *perception and discrimination* will really improve? Can the numinous experiences of mind really facilitate physical healing in this way? I cannot really tell, yet. The next dream, however, certainly suggests continuing rehabilitation via the very acute perceptions and fine discriminations that are now taking place between the various sides of my personality in creative confrontation with each other.

### A Little Bus Ride with Jung

(*Fifth week of recovery*)

Real life circumstances: Since my driver's license is restricted my wife is getting a bit burned out driving me to all my rehabilitation therapists every day. One of my physical therapists takes me on a few field trips walking through the city learning how to negotiate street curbs, how not to bump into people, police or telephone poles, and how to safely step on and off an escalator: "Pay full attention to getting your full foot squarely on the step!" I struggle haltingly with these amazingly complex techniques of daily living as if in a haze but I fall only once during this period. True, I fall in the street right in front of a bus but fortunately it was not moving at that moment. I vividly remember desperately shifting my body balance as if in slow motion so I could brake my fall with my hands and rump rather than risking more sever injury to my knees. As if from a great distance I heard a tremulous gasp of alarm and dismay from a crowd of people around me. Eventually my therapist decides I may be ready to try taking a bus by myself to at least some of my therapy appointments. And so I dream.

> My wife Kathryn, my daughter April, and I are talking in our living room. A patient (a diminutive, older, very shy woman) knocks on the door. I tell her to wait a moment as I seek for a room in the back of the house that we can use for her therapy session.
>
> The scene now changes: I am on a bus. In the seat across the aisle from where I am sitting is a large distinguished man, evidently a university professor type, but not really from the university. He is to be my new patient and we are taking this bus to the new location of my office. As we travel together we talk casually but with carefully studied purpose in preparation for our first session. He is very kind and listens compassionately as I try to give him the most adequate answers I can in response to his highly intelligent and penetrating questions regarding my professional qualifications to be his therapist and so forth. Gradually I become subtly and humbly aware that I am in the presence of an intellect and humanity that is vastly more profound than my own.
>
> Intruding on our conversation is the diminutive woman patient who, I now realize, is sitting a few rows ahead of us. She continues with a quiet litany as if to herself over and over: "Oh, yes, I was Dr. Rossi's patient *some time ago* and he was very good. But I remember a time when Dr Rossi was not well. His mind seemed a bit cloudy and his speech was not clear. Oh, no, his mind was not always there! He really was not well enough to be seeing patients."
>
> It seems as if all eyes and ears on the bus now are focused on me to pick up what my response will be to this intruding litany. Will I be irritated, try to ignore her, or deny what she is saying? I decide to be gracious with good humor and

publicly accept the truth she is divulging. I try to put a good spin on it, however, by emphasizing her own words by saying, "Yes, what you say is really true, I was not well back then, but that was '*some time ago*'." A sigh of relief seems to go through the bus as everyone realizes that I am not as defensive as I might have been. I am basically telling the truth even though it is difficult for me. The distinguished man immediately understands the situation completely on all levels and seems favorably impressed that I am probably okay.

The bus comes to what appears to be an exotic tropical garden on a college campus. This seems to be a good place where the large distinguished man and I can walk along well-manicured paths conducting our first therapy session together. But I have never been in this place before and I fear I may get lost in the meandering paths because I will be concentrating so intensely on our therapy work. I realize, however, that these paths are probably familiar to the distinguished man and all I need to do is carefully allow him to choose the way so I can follow along as he leads us correctly.

I spent many hours replaying this dream in *active imagination* with many variations over many days. The most numinous figure of the dream is the large distinguished man. But who is he? He is what Jung would call, "A Big Fish. He would not suffer fools gladly." Quietly, with a fresh flood of tears, the stately but haunting music from the film "A Matter of Heart" (a spiritual biography of Jung) comes to mind. The music identifies the large distinguished man as none other than Jung himself (or, rather, the Jung complex in my mind if you will)! It is profoundly reassuring to experience how I have such an inner mentor who can guide me in the guise of a patient checking me out. Perhaps I could come out of this okay after all.

### Caring for a Lost Inner Personality

(*Eighth Week of Recovery*)

I am in the back seat of a car crowded with people. Suddenly the police force our car to stop and surround us with drawn guns to capture a dangerous person in the front seat. It is a tense and frightening moment but I remain calm, cool, and competent carefully sizing up the situation. I surreptitiously open the back door a bit and signal the police not to shoot because I was confident I could talk the dangerous fellow into giving up peacefully.

I then turn my full attention to the white frightened face of the dangerous fellow in the front seat - he is clutching a canvas bag with a hole burned through it by some acid he is hiding inside. He murmurs in a pathetic far away voice, "I want some more acid." I think to myself, 'Oh, my poor dear fellow, what would you want with more acid?' But instead of confronting him with such a question I simply smile gently to support him with the words, "Of course, lets figure out together how we can get you the *kind of acid* you want so you can get the *kind of help* you really need." I seem to be going lucid in this last part of the dream wherein I am consciously using a therapeutic approach called "presupposition," or "reframing" (or "overlapping," or "hitchhiking one idea on another"). My plan is to carefully shift him away from the *kind of acid* he says he wants towards the *kind of psychological help* he really needs.

I am very comfortable with this dream, which I recognize as part of the continuing saga of a highly introverted part of me that seems to have had a hidden independent existence since my earliest childhood. In reality I was a rather clever child who played with chemistry sets and eventually contrived secretly to buy dangerous acids, manufacture my own gun powder, nitroglycerine, vacuum tubes, radios, batteries, a lathe to make my own chess pieces etc. Miraculously, I never did hurt others or myself although I was expelled from grammar school once on graduation day for exploding one of my bombs. The authorities did not take kindly to my way of celebrating. My mother never did connect the dots when she puzzled over the peculiar small holes in my clothing. She dutifully sewed them up never realizing they actually were acid burns. It was a different story the day I made and liquefied chlorine gas that eventually escaped and rusted all of my father's precision tools. I was really beaten up pretty bad for that one.

The hurt introvert gave up his dangerous pursuits when at 12 he discovered Kant's *Critique of Pure Reason,* Yoga, girls, the numinous mysteries of mind, and eventually psychology. That damn little introverted kid eventually did grow up a bit, but he remained a troublesome shadow who would occasionally verge on getting into stupid little scrapes in real life from which I would always have to rescue him. So, as I said, I feel very comfortable with this dream that restores me once again - well within my accustomed therapeutic role of taking care of my child self - before I was so rudely interrupted by the stroke.

It is probably no coincidence that this healing dream of the recovery of my more competent adult self comes after 2 months of rehabilitation. It is well known in many stroke rehabilitation programs that the amount of recovery within the first few months pretty much tells the story of how much healing the patient will experience. This clinical experience is now confirmed by neuroscience research on neurogenesis and healing that is well described by Henrietta van Praag et al. (2002).

> Our findings demonstrate that newly generated cells mature into functional neurons in the adult mammalian brain... We have identified one-month-old neurons with functional properties similar to those of mature dentate granule cells in the adult hippocampus. We also show that newly generated neurons are initially smaller and reach a more mature morphology after 4 months. (p. 1030)

Current neuroscience research such as this spells the death of a dogma we have had in neurology for over 100 years: the dogma that the brain does not grow after birth and the earliest period of infancy. The brain can continue to grow in competence throughout our lifetime if we learn how to develop and care for it. The well maintained brain can continue to reorganize, reconstruct, and resynthesize itself with carefully nurtured creative experiences of all sorts – particularly joyful, deeply meaningful, and spiritually numinous experiences that turn on gene expression, brain plasticity, and mind-body healing. This careful nurturing on all levels from mind to gene of others as well as ourselves and is a process of *co-creation* – cooperation between consciousness and nature whose possibilities we are only beginning to intimate.

### A Dream of Recovery

(*One year recovery*)

I was a mental patient in a gray, barren, and dark mental institution (as many of them are in reality). I walk along the main hall with a huge, disheveled man who is a mental patient and a friend of mine. I notice that my sensations are sharp, my perceptions are vibrant, and my mind is very clear. No one in the hospital seems aware that I have recovered - I feel that I am well now. On a stairway up to the left is a dear, sad, little girl patient (who reminds me of Anna Frank). I playfully bounce a pink rubber ball in her direction to try to cheer her up. On another stairway up on the right side is the office of the medical superintendent of the mental hospital. A nun standing outside his office confides to me that the medical superintendent has an interesting but little known personal history. It seems that he too was mentally ill but he was able to keep his illness under control with his sensible and rational attitude. In fact, all who knew this medical superintendent had a deep love and respect for him because of his humble and helpful attitude to everyone. At that point I turn to my disheveled friend and tell him in a louder and happy voice, "My mind is clear now and I am well! The doctors are going to let me go home now." In this dream I did feel a wonderful clarity that I actually lost for a while after my stroke last April. My friend turns to me and replies, "Well, of course they are going to let you go. You were able to explain yourself to them clearly so that they could understand who you are." The sad implication of his words was that he and the other patients would have to remain in the hospital diagnosed as mentally ill until they could explain themselves.

I believe this dream set the feeling tone of my 70th birthday. My basic feeling is Gratitude. Gratitude for my recovery, my good fortune of living in such a beautiful place with such a wonderful wife and gratitude for my returning mental faculties so that I can write a book like this one. My wife tells me that in fact she has noticed over the past few weeks that I really am acting clear with full recovery from my stroke. She can no longer see any lingering after effects of the stroke. She is almost completely right in this. A recent retest documents that after 15 months of rehabilitation both of my major stroke induced cognitive deficits improved dramatically: my *perception and discrimination* improved to the $90^{th}$ percentile from the $45^{th}$; my *short-term memory* improved to the $66^{th}$ percentile from the $37^{th}$.

* * * *

In an anecdotal and personal report such as this it is difficult to determine the healing factors that contributed to my mental and physical rehabilitation. Here is a short list of what was important from my personal perspective and what remains to be done from a scientific point of view.

Accessing and re-experiencing the numinosum in repeated creative replay is the essence of the psychological, cultural, and spiritual approaches to healing. Notice how often I report being preoccupied with certain numinous experiences in dreams and active imagination that continues over days, weeks, and months during my rehabilitation. Active imagination that generates healing usually requires repeated exercises of novel, enrichening, and positive, inner experience. Replaying numinous dramas of healing with deep emotions over and over with many creative variations allows a natural Darwinian variation and selection of optimal healing strategies to take place on all levels from mind to gene throughout the days, weeks, and months of rehabilitation.

I hypothesize that this creative replay evokes gene expression, brain plasticity, and mind-body healing via stem cell differentiation throughout the brain and body; this is the essence of what I call the "Novelty-Numinosum-Neurogenesis Effect."

While the subjective aspects of human consciousness, meaning, and myth making are often disparaged in current scientific literature, ***the creative subjective experiences of consciousness that replay the novelty-numinosum-neurogenesis effect are here regarded as the royal road to mind-body healing and rehabilitation.***

Although we can cite many scholarly sources from the history of therapeutic hypnosis, psychoanalytic traditions, and current neuroscience to support these ideas about mind-body communication and the psychosocial genomics of healing via the novelty-numinosum-neurogenesis effect, these conceptions remain speculative until they receive experimental confirmation. ***The most direct route to such experimental confirmation is to measure gene expression, brain plasticity and stem cell differentiation and healing in response to the creative replay of the novelty-numinosum-neurogenesis effect in psychotherapy, rehabilitation, and the creative arts.*** There is an emerging scientific technology to assess moment to moment changing profiles of gene expression from the blood, for example, that will allow us to make these measurements (Whitney et al., 2003).

Experimental confirmation of the value of creative replay of the novelty-numinosum-neurogenesis effect in the arts, humanities, and cultural rituals would have profound implications for an understanding of consciousness role in human nature. *Consciousness can turn on gene expression, synaptogenesis, neurogenesis, and mind-body healing!* It would mean that enriching life experiences that evoke the novelty-numinosum-neurogenesis effect during creative moments of art, music, dance, drama, humor, literature, spirituality, awe, joy, and cultural rituals could optimize the psychosocial genomics of consciousness, personal relationships, and healing. In essence it would mean that there is something more involved than "Art for the sake of art." Rather it would mean "Art for the sake of building a better brain" in the daily construction and re-construction of our lives.

This contains the seed of a new theory of aesthetics that reifies Keats' poetic and philosophical conundrum, "Beauty is truth, truth beauty, - that is all Ye know on earth, and all ye need to know." If beauty and truth are both numinous experiences that activate gene expression and brain plasticity, then beauty could actually reconstruct our brain to generate new experiences of truth and visa versa. This takes us quite beyond my personal story to a deepening exploration of the co-creation between consciousness and nature in human experience that is the main subject of this book.

# II
# THE DEEP PSYCHOBIOLOGY OF PSYCHOTHERAPY

**Definition**. - The Deep Psychobiology of Psychotherapy may be defined as the exploration of mind-body experience, communication and healing at all levels from the cultural and psychosocial to the cellular-genetic-molecular and the quantum. It is a highly integrative approach that greatly expands the traditional domains of phenomenological, analytical and cognitive-behavioral psychology to include new insights into creativity, consciousness and the human condition as they are continually updated by research in biology, physics and mathematics. It seeks to breakthrough the Cartesian dualism between mind and body by exploring questions such as these. How is it possible for thoughts, emotions, imagination and personal experience to influence physical health and visa versa? We know that our genes are expressed in our behavior, for example, but to what extent can we have a "psychobiological dialogue" with our genes to modulate how their information is expressed in self-creation and the process of physical healing? How do we facilitate our daily work of synthesizing the organic structure of our brain to optimize relationships with ourselves and our neighbors in harmony with the evolutionary informational dynamics of consciousness and cosmos?

**History**. - The history of The Deep Psychobiology of Psychotherapy began with the author's exploration of the implications of early research in the 1960's and 1970's that documented how the psychological experience of novelty and enriched environments was encoded as new memory and learning in the organic structure of the brain on a molecular level (Rossi 1972/1985/2000). This led to the formulation of the dream-protein hypothesis: "Dreaming is a process of psychophysiological growth that involves the synthesis and modification of protein structures in the brain that serves as the organic basis for new developments in personality."

Within this perspective the essence of psychotherapy becomes a process of facilitating "creative moments" that are encoded in new proteins and neural networks in the brain. But what is a creative moment? Such moments have been celebrated as the exciting 'hunch' by scientific workers and inspiration by people in the arts. The creative moment occurs when a habitual pattern of association is interrupted. There may be a spontaneous lapse or relaxation of ones associative process. There may be a psychological shock, an overwhelming sensory or emotional experience; a psychedelic drug, a toxic condition or sensory deprivation. Yoga, Zen, spiritual and meditative exercises may likewise interrupt habitual associations and introduce a momentary void in awareness. In that fraction of a second when the habitual contents of awareness are knocked out there is a chance for pure awareness and an experience of new awareness or heightened consciousness. This fraction of a second may be experienced as a mystic state, satori, a peak experience or an altered state of consciousness. It may be experienced as a moment of fascination or falling in love when the gap in one's awareness is suddenly filled with the new that has been created in a semi-autonomous manner within the deep psychobiology of our being.

From this perspective the new that appears in creative moments is the basic unit of original thought and insight as well as personality transformation. Experiencing creative moments is the phenomenological correlate of a critical change in the molecular structure of proteins within the brain associated with the creation of new cell assemblies, memory and learning. Molecular transformations in the brain in response to psychological shock, arousal and novelty was now recognized by the author as the deep psychobiological basis of psychopathology as well as the educational, constructive and synthetic approach to healing and psychotherapy that had its roots in the early work of pioneers such as James Braid in

Scotland, Pierre Janet in France, Maurice Bucke in Canada, Carl Jung in Switzerland, Rolando Assagioli in Italy and Milton Erickson in America. James Braid, one of the fathers of hypnosis, for example, attributed the therapeutic effects of trance induction to the "psychophysiology of fascination" which we now regard as a special state of psychobiological arousal and focused attention. Maurice Bucke described the most heightened state of psychobiological arousal as "cosmic consciousness." He attributed mental illness to a breakdown in the natural process of evolving consciousness. Borrowing from the work of the religious scholar, Rudolph Otto, Carl Jung described heightened states of consciousness in the creative breakthroughs of spiritual innovators, artists and scientists as an experience of the *"numinosum"* (an overwhelming experience of *fascination* with the *mysteriousness* and *tremendousness* of life and the universe).

Some of the most intriguing, innovative approaches to hypnosis and psychotherapy based upon psychological shock, emotional arousal and creative moments were pioneered by the late Milton Erickson and described as follows.

> "The induction and maintenance of a trance serve to provide a special psychological state in which the patient can reassociate and reorganize his inner psychological complexities and utilize his own capacities in a manner in accord with his own experiential life...therapy results from an inner resynthesis of the patient's behavior achieved by the patient himself. It is true that direct suggestion can effect an alteration in the patient's behavior and result in a symptomatic cure, at least temporarily. However, such a "cure" is simply a response to suggestion and does not entail that reassociation and reorganization of ideas, understandings and memories so essential for actual cure. It is this experience of reassociating and reorganizing his own experiential life that eventuates in a cure, not the manifestation or response behavior [to hypnotic suggestion] which can, at best, satisfy only the observer [and stage hypnotist]." (Erickson, 1948).

**Current Status.** - The fundamental question for a deep psychobiology of psychotherapy this, "How do we integrate the many levels of mind-body communication and healing from the psychosocial to the cellular-genetic?" Is it possible to use the concept of "information" and the "transformations of information" to do this? Is it possible to create a new science of "information transduction" that explores how information experienced as human cognitive behavior (thoughts, words, images, emotions, meaning etc.) is transformed into other forms of information expressed as the physical structure of our genes and proteins and visa versa? A major mission of this new science of information transduction is to trace the pathways by which human cognition and emotional experience modulates biological processes in health, sickness and the dynamics of mind-body healing.

An interesting step in the creation of this new science of information transduction was pioneered by the biologist, Thomas Stonier (1990). A visual summary of his ideas was illustrated by the author (Rossi, 1986/1993) in Figure 2.1. From the broadest perspective the informational dynamics of the evolution of the physical universe as well as living systems is currently described as "getting its from bits" (Wilczek, 1999). That is, how can "its" (the apparently physical, molecular and biological structures of the universe) evolve out of "bits" (binary information) from the quantum flux of the Big Bang origin of the universe? Information is regarded as the mathematical common denominator that brings together physics, biology, psychology, human consciousness, creativity, and culture. Is the *information* that organizes the *energy* and *physical* dynamics of nature the same as the information contained in thoughts, words, feelings, images and imagination? If so, how do we facilitate the *transduction of information* from one modality to another in creative experience and healing?

The French mathematician Henri Pöincare, the originator of *non-linear dynamics* (currently called "chaos theory" and "adaptive complexity theory"), first described the four stage creative process about 100 years ago at the same time that Sigmund Freud and Carl Jung were originating *psycho-dynamics*. This four stage creative process is illustrated in the inner circle of Figure Two. The middle circle of Figure 2.2 illustrates the

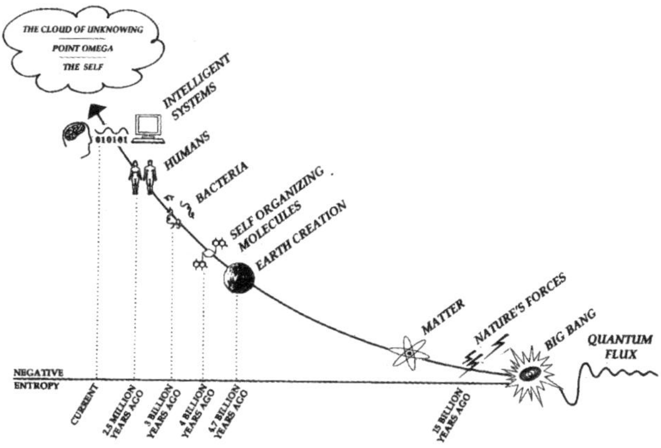


**Fig. 2.1.** - The evolution of information transduction from the Big Bang to the Self.

Breakout Heuristic as a model of psychotherapy (Rossi, 1972/1985/2000) that corresponds to Pöincare's creative cycle. The outer circle is a condensation of Joseph Campbell's monomyth of the cultural dynamics of the hero that corresponds to the same four stage creative process. The outer labels of Figure 2.2 illustrate how socio-political identity follows the same 4 stage paradigm. Let us now focus on the dynamics of mindbody communication (also called "information transduction") in this four stage creative process as it is manifest in current theory of the deep psychobiology of psychotherapy.

**Theory.** - Figure 2.3 illustrates a four stage outline of the complex field of mindbody communication and healing that corresponds to the creative cycle of Figure 2.2. Let us review these four stages as a foundation for a new understanding of what is actually going on between patient and therapist at the deepest levels of "mind-gene communication" (Rossi, 1972/1985/2000,1986/1993, 1996).

### Stage 1: MindBrain Information Transduction

The Limbic-Hypothalamic-Pituitary System is currently recognized as a major information transducer between the

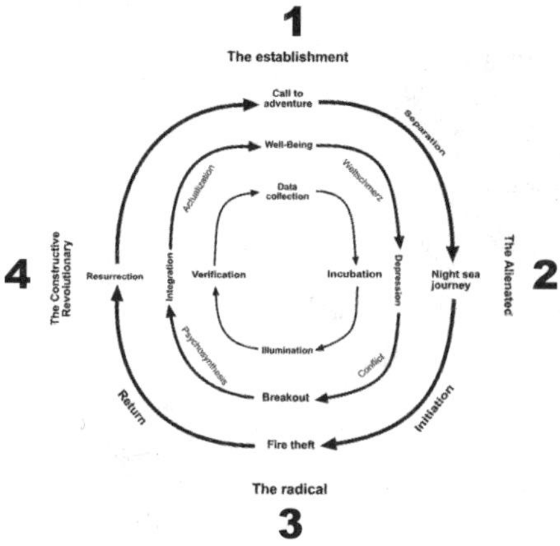


**Fig. 2.2.** - The breakout heuristic in psychoterapy.

brain and the body. Cells within the hypothalamus transduce (transformation information from one form into another) the essentially electrochemical neural impulses of the brain that encode the phenomenological experience of "mind" and emotions into the hormonal "messenger molecules" of the endocrine system. These messenger molecules then travel through the blood stream in a cybernetic loop of information transduction in Figure Three where the four major stages of interest for psychotherapy are numbered.

The complex loop of mindbody communication illustrated in Figure 2.3 modulates the action of neurons and cells of the body at all levels from the basic pathways of sensation and perception in the brain to the intracellular dynamics of gene transcription and translation. It has been proposed that the molecular messengers (also called "informational substances") of the endocrine, autonomic and immune systems mediate stress, emotions, memory, learning, personality, behavior and symptoms (Rossi, 1986/1993,1996). This communication loop is a two way street by which (1) mind can modulate physiology of the brain and body and (2) biology can modulate mind, emotions, learning and behavior.

### Stage 2: Immediate-Early Genes in Creative Adaptation

A generation ago it was believed that genes were simply the

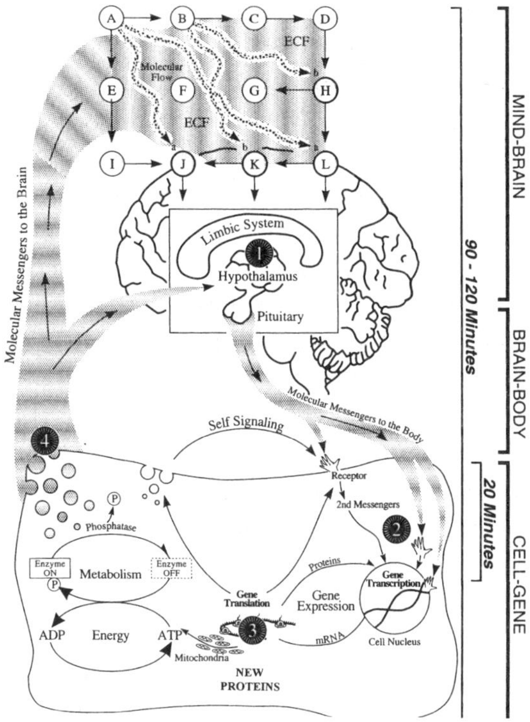


**Fig. 2.3.** - The mindbody communication loop.

units of physical heredity that were transmitted from one generation to another through sexual reproduction. Today we know that genes have a second major function: a major class of genes, sometimes called, "Immediate-Early Genes," (also called "Primary Response Genes" or "Third Messengers") are continuously active in responding to the hormonal messenger molecules signaling the need for creative adaptation to important changes in the environment. Everything from novelty, shock, surprise, touch, sexual stimuli, maternal behavior and psychosocial stress to temperature, food, physical trauma and toxins in the environment can be signaled to the genes via the hormonal messenger molecules that arrive from the limbic-hypothalamic-pituitary system (Merchant, 1965).

Immediate-early genes (IEGs) are the newly discovered mediators between nature and nurture. Immediate-early genes act as information transducers allowing signals from the external environment to regulate genes within the internal matrix of the nucleus of life itself. Many researchers now believe that memories along with new experiences are encoded in the central nervous system by changes in the structure and formation of new proteins within the synapses between neurons (Eriksson et al., 1998). IEGs function as transcription factors regulating the "housekeeping genes" that make the proteins within the neuron that encode new memory and learning (Tölle et al., 1995). More than 100 immediate-early genes have been reported at this time. While many of their functions still remain unknown, neuroscientists are exploring the interrelated biological and psychological functions that immediate-early genes such as *c-Fos* and *c-Jun* are already known to serve as illustrated in Figure Four. Most arousing psychosocial stimuli can induce immediate-early genes within minutes and their main effects are mediated within 20 minutes to an hour or two. A more detailed discussion and complete citation of all the references of Figure 4 have been published previously (Rossi, 1972/1985/2000).

**Stage 3: Protein Synthesis in Memory, Stress, and Healing**

The third stage in the mind-gene communication loop is the process of gene translation leading to the production of new proteins in Figure Three. The time required to make new proteins in response to psychological arousal and physical stress as illustrated on the right side of Figure 2.3 provides an important window into the informational dynamics of new approaches to psychotherapy. The domain of psychological time in minutes and hours as illustrated in Figure 2.3 relates the Basic Rest-Activity Cycle of human behavior to the processes of mindbody communication. This is in sharp contrast to the more recent evolutionary form of more rapid mindbody communication mediated by the central nervous system in small fractions of a second that are briefer than the usual phenomenological span of consciousness.

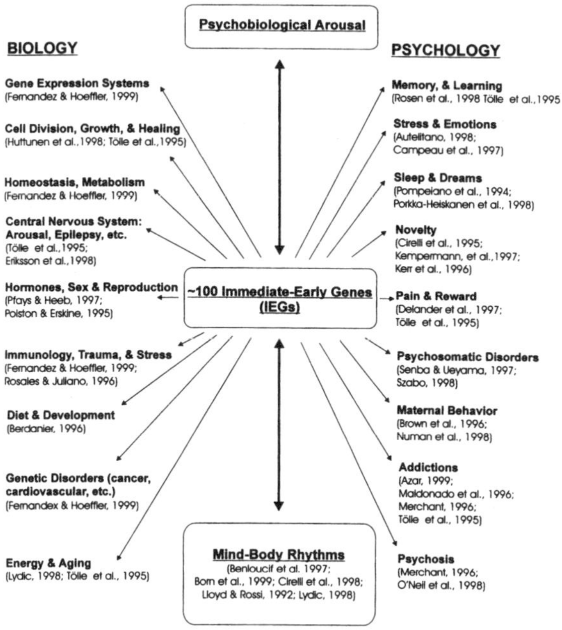


**Fig. 2.4.** - The central role of immediate-early genes (IEGs) in the deep psychobiology of psychotherapy.

Detailed research on the genetic, neuroendocrinal and psychosocial levels suggests that the 90-120 minute ultradian rhythm (Lloyd & Rossi, 1992), originally described as the Basic Rest-Activity Cycle by Kleitman, is a more fundamental "work cycle of life" than the circadian cycle (the 24 hour rhythm). *Ultradian* in this context means any rhythm faster than the 24 hour circadian cycle; in this chapter we emphasize only the major 90-120 ultradian rhythms on the genetic, endocrine and cognitive-behavioral levels during sleeping, dreaming, and creative waking states that have important implications for psychotherapy.

**Stage 4: Messenger Molecules and State Dependent Memory**

Stage four of Figure 2.3 illustrates how messenger molecules (such as hormones, neurotransmitters, growth factors, etc.)

that have their origin in the processing of the larger protein "mother-molecules" in stage three may be stored within the cells of the brain and body as a kind of "molecular memory." These molecular messengers are released into the blood stream where they can complete the complex cybernetic loop of information transduction by passing through the "blood-brain barrier" to modulate the brain's neural networks and psychological experience as illustrated by the block of letters A through L at the top of Figure 2.3. Such localized neuronal networks of the brain are modulated by a complex field of messenger molecules that can reach the limbic-hypothalamic-pituitary system as well as certain areas of the cerebral cortex. This is a new model of how the sexual hormones, stress hormones, immune system cytokines (messenger molecules like IL-1 and Il-2) and growth factors of the body can modulate mind and emotions and visa versa.

If we are willing to grant that communication within the neuronal networks of the brain is modulated changes in the strengths of synaptic connections, then we could say that *meaning* is to be found in the complex dynamic field of messenger molecules that continually bath and contextualize the information of the neuronal networks in ever changing patterns. It is truly amazing to learn, for example, that most of the sexual and stress hormones that have been adequately tested have state dependent effects on our mental and emotional states as well as memory and learning.

Recent research indicates that most forms of learning (Pavlovian, Skinnerian, imprinting, sensitization, etc.) are now known to involve these hormonal messenger molecules from the body that can reach the brain to modulate the neural networks that encode mind, memory, learning and behavior. Insofar as these classical forms of learning use messenger molecules, they *ipso facto* have a "state-dependent component" (Rossi, 1986/1993, 1996).

**Methodology. -** The methodology of the deep psychobiology of psychotherapy expands traditional psychological research to include all the new approaches of neuroscience. Current research on state-dependent memory, learning and behavior is a clear example. When subjects are given memory and learning tasks while under the influence of stress hormones such as adrenocorticotropic hormone (ACTH), epinephrine or sex hormones (or even psychoactive drugs that mimic these natural hormonal messenger molecules) there is a varying degree of amnesia when the stress hormone or drug has been metabolized out of the system. That is, when memory is encoded under conditions of high emotional arousal, novelty, sex, stress or trauma, it tends to become state-dependent or statebound to that psychobiological condition. Memory and learning is state dependent on the original psychobiological conditions when it was first encoded. This state-dependent memory becomes dissociated or apparently "lost" after the person apparently recovers when the stress or sexual hormones are metabolized and return to normal levels. Reactivating stress or sex in another context, however, has a tendency to reestablishes the original encoding condition with varying degrees of emotion and memory of the trauma. This is the psychobiological basis of much psychopathology related to early sexual and stressful life events as uncovered by classical psychoanalysis. This illustrates how state-dependent memory, learning, and behavior (SDMLB) bridges the Cartesian dichotomy between mind and body.

From a psychotherapeutic perspective, what is most interesting about these experiments is that they enable us to study the parameters of "reversible amnesia," which have been important criteria in understanding the phenomenology of psychoanalysis and therapeutic hypnosis (Rossi, 1986/1993, 1996). Most experiments in state-dependent memory and learning demonstrate that this "reversible amnesia" is only partial. That is, there is usually some memory and learning available even in the dissociated condition after the stress hormones return to normal levels) so most of the hypnotic literature documents that hypnotic amnesia is usually fragile and partial in character. A full amnesia that is completely reversible is relatively rare in state-dependent memory and learning experiments as well as in psychoanalysis and therapeutic hypnosis. In the historical literature of hypnosis and psychoanalysis this same fragile and partial character of reversible amnesia may have been responsible for many of the puzzling and paradoxical features of memory that remain the source of continuing controversy that challenges the validity of the various theories of depth psychology. Since the earliest days of psychoanalysis it has been noted that a sudden fright, shock, trauma and stress could evoke "hypnoidal states" that were somehow related to amnesia, dissociated and neurotic behavior. We now know they can activate the dynamics of mind-gene communication and state-dependent memory, learning and behavior (Rossi and Cheek, 1988).

**Psychoimmunology and the Mindbody Connection.** The most comprehensive demonstration of how psychosocial stress can modulate gene transcription was demonstrated by the research team of Ronald Glaser (1990, 1993) at Ohio State Medical School. His research traces the effects of psychological stress (experienced by medical students during academic examinations) on the transcription of the Interleukin-2 receptor gene and interleukin-2 production. These researchers documented the path of information transduction illustrated in Figure 2.3 from the (1) the limbic-hypothalamic-pituitary system's hormones (primary messengers) that trigger (2) cell receptors to initiate (3) a cascade of secondary messengers (cAMP) that mediate gene transcription which leads to messenger ribonucleic acid (the mRNA "blueprint" of the gene) production and the synthesis of new proteins that leads to the formation of (4) other hormonal messenger molecules that in turn flow back to the brain to modulate mind, memory, emotions and behavior in a state dependent manner.

*A major challenge for fundamental research in psychobiologically oriented psychotherapy is to document how a positive psychotherapeutic intervention designed to reduce psychosocial stress could lead to a facilitation of the transcription of the interleukin-2 receptor gene and its translation into the interleukin-2 protein that functions as a messenger molecule in the immune system. This research would become a new paradigm of mindbody healing.* It has profound significance for a general theory of mindbody communication and healing when we realize that other medical researchers (Rosenberg and Barry, 1992) have found that interleukin-2 is a messenger molecule of the immune system that signals white blood cells (cytotoxic T-cells) to attack pathogens and cancer cells. Thus, the traditional medical model of research represented by Rosenberg finds the same foundation of mindbody communication at the level of gene expression as the new field of psychoimmunology represented by Glaser. I propose that this will become the new criterion for evaluating all forms of mindbody healing in the future - biofeedback, body therapies, message, meditation, imagery, active imagination, hypnosis, prayer, ritual or whatever. Whatever the holistic method of mindbody healing, we can test whether it really facilitated mindbody communication and healing simply by taking a blood sample to determine whether a healing gene transcription actually took place with the very easy and reliable tests to determine whether immediate-early genes and mRNA is made so that new proteins and hormones could be synthesized for growth, healing and new phenomenological experiences of consciousness (Rossi, 1972/1985/2000).

**Applications.** - When the 90-120 ultradian cycle of mindbody communication of Figure 2.3 is unfolded over time we get graphs of the alternating ultradian rhythms of activity and rest on the genetic, endocrine and cognitive-behavioral levels as illustrated on the lower part of Figure 2.5. This coordination of the diverse systems of traditional psychobiology via their time parameters has been called the "Unification Hypothesis of Chronobiology" (Lloyd & Rossi, 1992). This new understanding of the chronobiology from the molecular-genetic to the cognitive-behavioral levels may be taken as a new data base for understanding the dynamics of mindbody communication and healing in psychotherapy.

The lower part of Figure Five summarizes the alternating 90-120 minute ultradian rhythms of the awake and sleep states that have their ultimate foundation in gene expression in a simplified manner. The ascending peaks of Rapid Eye

Movement (REM) sleep characteristic of nightly dreams every 90-120 minutes or so are illustrated along with the more variable ultradian rhythms of activity, adaptation and rest in the daytime. Figure 2.5 also illustrates how many hormonal messenger molecules of the endocrine system such as *growth hormone*, the activating and stress hormone *cortisol* and the sexual hormone *testosterone* have a typical circadian peak at different times of the 24 hour cycle. Because the non-linear chronobiological release of many of these hormones are recognized as having profound state dependent effects on memory, learning, emotions and behavior throughout the day, it is important to consider their relevance for psychotherapy.

An interesting example of the theoretical and practical implications of such chronobiological relationships between hormones and behavior is the so-called "human alarm clock" effect. It has been found that people who are able to awaken in the morning at a specific time experience a greater release of cortisol and adrenocorticotropin in their blood stream just before their alarm clock rings (Born et al., 1999). As illustrated in Figure 2.5, however, there is a normal peak in cortisol just before awakening in the morning. This implies that the *conscious intentionality* to awaken at a specific time in the morning is able to utilize a normally *involuntary circadian hormonal rhythm* to control the desired behavior of awakening at a certain time. We have long known that body processes can modulate conscious experience and behavior; the human alarm clock effect clearly documents the reverse - a conscious intentionality can modulate a normally involuntary circadian hormonal rhythm of messenger molecules throughout the body. ***It is precisely this reciprocal relationship between mind and body that is the deep psychobiological basis of information transduction, mind-body communication and healing in psychotherapy***.

The upper part of Figure 2.5 illustrates my conjecture that the natural unit of psychobiologically oriented psychotherapy may be a utilization of one natural 90-120 ultradian cycle of activity and rest. Research has documented how the ultradian peaks of cortisol secretion that lead to psychophysiological states of arousal every 90-120 minutes or so throughout the

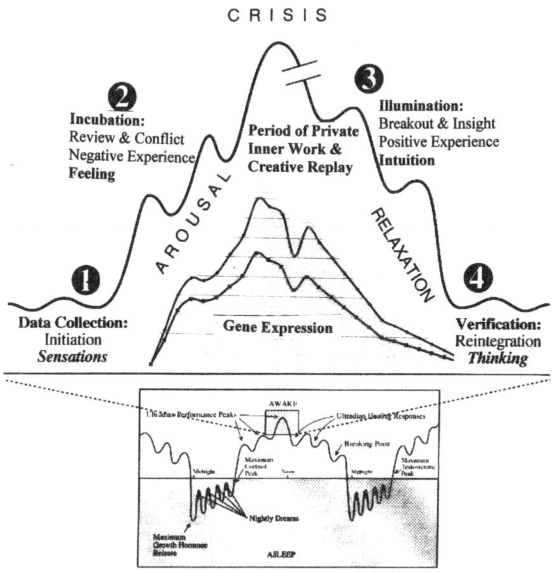


Fig. 2.5. - The four stages of the creative process in psychobiologically oriented psychotherapy. The lower diagram summarizes the alternating 90-120 minute ultradian rhythm of waking and sleep states of one day. The upper part of the diagram illustrates the four stage creative process of psychotherapy as the utilization of one of the natural 90-120 minute ultradian rhythms illustrated in the lower diagram.

day (that I label as "Ultradian Performance Peaks" in Figure 2.5) are typically followed after about 20 minutes by ultradian peaks of beta-endorphin that lead to rest and relaxation that I label as "Ultradian Healing Responses" (Rossi & Nimmons, 1991). It appears as if nature has built in a natural but flexible and highly adaptive ultradian rhythm of activity, rest and healing, the "work cycle of life" mentioned above, every 90-120 minutes.

What, exactly, is the "work" that is done in each 90-120 minute ultradian cycle? I propose that the essence of such work is the formation of new proteins for a creative response to changing environmental conditions, stress and healing. The basic implication of the deep psychobiological dynamics of new protein formation and healing in psychotherapy is that *what has been traditionally called "counseling" or "therapeutic sugges-*

*tion" may be, in essence, the accessing, entrainment and utilization of ultradian/circadian processes that respond to psychosocial cues.* Within this framework, the classical phenomena of hypnosis may be conceptualized as extreme manifestations and/or perseverations of time-dependent psychobiological processes that are responsive to psychosocial cues (Rossi, 1996). *What the biologist calls the "entrainment of ultradian and circadian rhythms on all levels from the cognitive-behavioral to the cellular-genetic by physical and psychosocial stimuli" is the deep psychobiological basis of what psychotherapists call "suggestion to facilitate creativity and mindbody healing."* This leads us to the formulation of new mathematical models of therapeutic hypnosis and psychotherapy (Rossi, 1972/1985/2000, 1996).

Table 2.1 outlines the four stages of the creative process as it is typically experienced every hour and a half or two in everyday life (Rossi and Nimmons, 1991). Throughout the day, particularly during the low phase of the ultradian rhythm, we all have a choice as to how we shall proceed with our natural psychobiological phases of consciousness, creativity, performance and healing. We can heed nature's call to take a healing break to experience what I call the "Ultradian Healing Response." This is often the best time to practice a natural form of self-hypnosis, meditation, deep self-reflection and holistic healing. If we persistently choose to ignore nature's call for rest and restoration at such times, however, we will fall into the "Ultradian Stress Syndrome" where we are prone to experience the genesis of psychosomatic symptoms and the typical dynamics of Freud's psychopathology of everyday life.

**Case example.** - Examples of how a deep psychobiological model of the four stage creative process may be utilized in practical psychotherapy have been described in detail (Rossi, 1972/1985/2000, 1985/1993,1996). In the case example presented here I propose a ***highly speculative outline*** of how the four stages of mind-body information transduction and healing illustrated in Figures 2.2, 2.3, and 2.5 may be experienced in the four stages of "the deep psychobiology of psychotherapy." It will probably require many years of genomic research to document the facts and fallacies of this view of mind-body communication and healing in a scientifically acceptable manner.

**Stage 1.** - *Initiation:* A woman in her thirties experiencing a period of emotional transition and stress suddenly begins to experience unusual and uncomfortable sensations of heat in her vagina. Medical examination indicates that she is having an outbreak of vaginal herpes for the first time in her life. She claims she has had no new sexual partners for over three years and her current partner has apparently been faithful. She is emotionally overwhelmed by this unexpected symptom that has the numinous effect of focusing her consciousness inward in an entirely natural manner. Recognizing this the therapist initiates her into *The Symptom Path to Enlightenment* (Rossi, 1996) by introducing her to symptom scaling, "*On a scale of one to ten where ten is the worst you have ever experienced that heat and five is average, just how strong is your sense of heat in your vagina right now?*" She replies that the heat is "*Seven right now,*" and crosses her legs with a facial grimace of distaste. The therapist asks her, "*Do you have the courage to really receive honestly just what you are feeling right now so you can fully experience what it leads to next?*" This seemingly simple but ***numinous*** question focuses her attention on her symptoms as mindbody signals that will eventually be transduceded into healing and emotional insights as we shall soon see.

**Stage 2.** - *Incubation and Arousal:* She responds with her impression that the symptom of herpes is the source of the heat she is feeling and it seems to be getting worst by the minute as she focuses on it with growing fascination. The therapist supports this development by slowly and quietly murmuring an incomplete sentence, "*I wonder if you can allow that to continue until...?*" Her eyes close as she apparently focuses inward. Her body tenses and she leans forward slightly over the next twenty minutes or so as she hesitantly becomes engaged in an "active imagination" wherein she whispers the following series of apparently spontaneous symptomatic transformations (that is, mindbody information transduction) and

**Table 2.1. - *The Ultradian Healing Response and the Ultradian Stress syndrome as they are experienced in everyday life.***

| THE ULTRADIAN HEALING RESPONSE | THE ULTRADIAN STRESS SYNDROME |
|---|---|
| 1. *Recognition Signals:* An acceptance of nature's call for your need to rest and recover your strength and well-being leads you into an experience of comfort and thankfulness. | 1. *Take-a-Break Signals:* A rejection of nature's call for your need to rest and recover your strength and well-being leads you into an experience of stress and fatigue. |
| 2. *Accessing A Deeper Breath:* A spontaneous deeper breath comes all by itself after a few moments of rest as a signal that you are slipping into a deeper state of relaxation and healing. Explore the deepening feelings of comfort that comes spontaneously. Wonder about the possibilities of mind-gene communication and healing with an attitude of "dispassionate compassion. | 2. *High On Your Hormones:* Continuing effort in the face of fatigue leads to the release of stress hormones that short-circuits the need for ultradian rest. Performance goes up briefly at the expense of hidden wear and tear so that you fall into further stress and a need for artificial stimulants (caffeine, nicotine, alcohol, cocaine, etc. |
| 3. *Mind-Body Healing:* Spontaneous fantasy, memory, feeling-toned complexes, active imagination, and numinous states of being are orchestrated for healing and life reframing. | 3. *Malfunction Junction:* Mistakes in performance, memory, and learning; emotional problems become manifest. You may become depressed or irritable and abusive to self and others. |
| 4. *Rejuvenation & Awakening:* A natural awakening with feelings of serenity, clarity and healing together with a sense of how you will enhance your performance and well-being in the world. | 4. *The Rebellious Body:* Classical psychosomatic symptoms now intrude so that you finally have to stop and rest. You are left with a nagging sense of failure, depression and illness. |

free associations with many pauses, "*Now the heat is shifting around a little to my butt on the left cheek.....now heat is moving through my body everywhere...its like a burning allergy.....my head hurts.....feels like an outbreak of psiorsis on my scalp.....I feel like I should confess it all to my mother like I did as a kid.....my right shoulder aches.....Why is my right side trembling?.....Why am I starting to cry?.....Why do I still try to get approval from my mother even when she never gave it but only punished me instead?.....I'm burning up with heat all over*!"

From a psychobiological point of view the heat she is experiencing indicates that she is going through a rather intense state of sympathetic system arousal as she experiences what has been traditionally called an "emotional catharsis or crisis." This is the state of "arousal" illustrated as stage 2 in Figures 2.2, 2.3, and 2.5 above. The hormonal stress messenger molecules of the limbic-hypothalamic-pituitary-adrenal system are being released into the blood stream where they travel to cells of the brain and body where they trigger receptors and initiate cascades of secondary messengers to signal the activation of immediate-early genes (IEGs). These IEGs turn on certain target genes to send their mRNA blueprints the cell's protein factories where stress and healing proteins are made. Many of the stress proteins and hormones produced at this time are the same as those that were released during previous real life trauma when they were responsible for encoding memories in a state-bound form. This new release of these stress hormones in the context of psychotherapy reactivates her original traumatic state just enough to reawaken the apparently lost state-bound memories with tears and the potential for a creative moment with a fresh perspective on her life.

**Stage 3.** - *Illumination and Insight:* For a few tense minutes she continues with, *"Burning! Burning! I know.....I know I have to leave* [her current boyfriend]. *I always knew it was only temporary, really, but now I really do have to leave.....He punishes me too, even when he doesn't know it.....my left knee is twitching uncontrollably.....Can't you make it stop?.....Oh, I'm tired of all this.....I will leave.....I'm getting sleepy.....I feel warm.....just warm now.....I really have to leave* [boyfriend]." Her body sags back and she remains silent for about three or four minutes as her face gradually becomes calm, smooth and relaxed.

A profound psychobiological transition is now taking place which is indicated with the two line break just past the crisis peak in the upper diagram of Figure 2.5. At this time no one really knows what is happening in this *Period of Private Creative Inner Work* at the moment of a creative cognitive-emotional breakthrough (as verbalized above when she realizes that she must leave her boyfriend). The author originally proposed *"The neuropeptide hypothesis of consciousness and catharsis"* to describe the dynamics of such creative moments. From the psychobiological perspective, the arousal and relaxation phases of inner work are mediated by the release of messenger molecules such as ACTH and B-endorphin.

The implication of more recent research suggests that immediate-early genes such as *c-fos and c-jun* are turned on in brain cells during emotional pain and arousal during the crisis of transition between stress and the release of relaxation hormones (Lydic, 1998). This arousal may lead to the synthesis of new proteins that will connect synapses bridging neural networks that will become the cellular-molecular basis of a new psychological insight and healing as illustrated in stage three of Figures 2.2, 2.3 and 2.5.

**Stage 4** - *Verification and Reintegration:* The therapist looks at the clock and with a mild sense of concern notices there is only ten minutes left to the session. He clears his throat and murmurs, "*Yes, and is that still going well*?" After a moment she shifts her feet, nods her head yes, adjusts her posture to a more normal sitting position, blinks a bit and finally opens her eyes as if awakening from a trance. The therapist then asks, "*And I wonder what number describes what your level of comfort is now*?" Somewhat surprised she acknowledges that her symptoms are now at a level of "one or two or maybe zero", that is, very close to complete comfort. She explains it is no longer a feeling of heat in her vagina, but rather a feeling of warmth, or is it a slight pressure, or a knowing somehow? From a psychobiological point of view she has moved past her state of arousal and catharsis in stage two and emotional insight in stage three to a state of comfort and parasympathetic relaxation in stage four that is so characteristic of what people typically feel when "a job is well done."

The therapist asks if she now knows what she has to do and she nods yes. She makes a few remarks about how she experiences a sense of relief in knowing that she now can make up her own mind with confidence about the important things in her life. She will leave her boyfriend and later she will tell others about it. By the next session a week later she reports that she has navigated the separation well. The herpes and burning sensations are apparently gone. At this time we can only speculate about how the shifts in the hormonal and immune system messenger molecules may have modulated the psychobiological state of her vaginal mucosa so that her symptoms and the virus go into remission. A year later she remains symptom free and all available evidence indicates that her life was effectively "turned around" in this session.

* * * *

This chapter outlines currently evolving views of how complex, self-organizing systems of communication across all levels from the molecular-genetic to the psychosocial could lead to a unified theory of mindbody communication and healing in psychotherapy. Research in the areas of the immediate-early genes, psychoimmunology and state-dependent memory is integrated with the chronobiology of ultradian rhythms as a new window into the psychobiology of creativity, stress and healing. A deep psychobiological perspective on psychotherapy that is consistent with much of the classical theory of psychoanalysis and psychosomatic medicine as well as the modern neuroscience of memory and learning at the cellular-genetic level is proposed. This new integration of theory and research leads to a psychobiological model of how we may utilize the creative process to facilitate mind-body healing in a manner that may seem visionary to some. Such a visionary ideal, however, may be what we need to develop truly new and inspired approaches to psychotherapy in the new millennium.

# III
# VISIONARY HYPOTHESES FOR A NEW MILLENNIUM

*This search for the deep psychobiological foundations of hypnosis begins with a review of some of the paradoxes of historical hypnosis and the impasse of current theory. It is proposed that further progress requires a deeper investigation of how psychosocial cues can modulate the mechanisms of healing at the CNS, autonomic, neuroendocrine and cellular-genomic levels. The dynamics of hypnotic communication and healing from the cognitive-behavior level to the cellular-genomic are outlined in four stages: (1) Information transduction between the experiences of consciousness and the limbic-hypothalamic-pituitary system; (2) The psychosomatic network of messenger molecules and their receptors; (3) The immediate early gene protein cascade; and (4) State dependent memory, learning and behavior. Neuroscience research is outlined for its contributions toward a mathematical model of how a psychobiological approach to the therapeutic applications of hypnosis and the placebo response could facilitate neurogenesis in the human hippocampus and healing at the cellular-genomic-protein level throughout the body. A series of ten hypotheses is proposed as a guide for theory and research in therapeutic hypnosis to facilitate healing and rehabilitation utilizing DNA chip technology in the new millennium.*

Neuroscience in this decade of the brain is setting the stage for an important updating of our understanding of the psychobiological foundations of hypnosis. We first review the problems and paradoxes of historical hypnosis as well as the impasse of current theories when they are limited to the cognitive-behavioral level. Ten hypotheses are then proposed to expand the domain of hypnosis to include the psychobiology of mind-body communication and healing on all levels from the cognitive-behavioral to the cellular-genomic. Implications are drawn from chronobiology and neurobiology to infer the types of research that are now needed to explore the psychobiological parameters of the therapeutic applications of hypnosis in the future.

**Historical Context**

***Activity-Passivity Paradox of Hypnosis***

In 1846 James Braid adopted the term "hypnotism" and is credited with defining its mechanism of action as "psychophysiology" with these words.

> "With the view of simplifying the study of reciprocal actions and reactions of mind and matter upon each other... the influence of concentrated attention, or dominant ideas, in modifying physical action, and these dynamic changes re-acting on the mind of the subject... I adopted the term *"hypnotism"* or nervous sleep (p. 369)... And finally as a generic term, comprising the *whole* of these phenomena which result from the reciprocal actions of mind and matter upon each other, I think no term more appropriate than *psychophysiology*." (Braid in Tinterow, 1970, p. 372).

It is surprising for some students to learn that it was Braid, the same physician who popularized the term "hypnotism," who apparently coined the word "psychophysiology" to describe how hypnosis achieved its therapeutic effects. In a later volume, *The Physiology of Fascination and the Critics Criticized* published in 1855, Braid notes his quandary about the role of fascination as well as voluntary and involuntary behavior in hypnosis.

> "The power possessed by serpents to fascinate birds has always been a source of interest and admiration...by what means is this remarkable result effected?...Is it a voluntary, or an involuntary process... After due consideration, I feel satisfied that the approach and surrender of itself by the bird, or other animal, is just another example of the monoideo-dynamic, or unconscious muscular action from a dominant idea possessing the mind" (Tinterow, 1970, 365).

Braid's concept of the *monoideo-dynamic* was an anticipation of the modern view that it is the *fixation of attention*, the patient's intense focus on the words and ideas of the clinician using hypnosis, that is the stimulus for the psychophysiological healing. Braid's questions about the role of *active, conscious, and voluntary* responsiveness versus more *passive, involuntary or unconscious* processes in hypnosis remains a central question in current theories. Until recently Braid's emphasis on the significance of fascination in hypnosis fell into the oblivion of folklore about the "evil eye." As we shall soon see, however, fascination during enriching life experiences may be fundamental in the focusing of attention, brain growth and the deep psychobiological dynamics of hypnosis.

The primitive state of psychophysiology during the first century of hypnosis meant that it could be conceptualized only as some sort of reflex or pathology. Bernheim (1886/1957, p. 138), the leader of the Nancy school in France, for example, described hypnosis as the "exaltation of the ideomotor reflex excitability, which effects the unconscious transformation of the thought into movement, unknown to the will ... The mechanism of suggestion in general, may then be summed up in the following formula: increase of the reflex ideo-motor, ideo-sensitivity, and ideo-excitability." The idea that hypnosis involved an *increase in "sensitivity" and "excitability"* is in striking contrast to the view of the Salpêtriêre school in Paris led by Charcot who maintained, to the contrary, that hypnosis was a pathological condition of *passivity* that progressed from *lethargy* and catalepsy to somnambulism.

This paradox in understanding the fundamental nature of hypnosis — *Is hypnosis heightened activity or passivity?* — continued into the next generation of leading researchers. Pavlov, for example, believed hypnosis was a state of *cerebral inhibition*, a kind of *"partial sleep"* while Clark Hull maintained the opposite view that hypnosis is a state of *arousal*.

> "We seem forced to the view that hypnosis is not sleep...Thus the extreme lethargic state is not hypnosis, but true sleep: only the alert stage is hypnotic. Lastly, evidence has been presented which indicates not only that condi-

> tioned reflexes may be set up during hypnosis, but that this may perhaps be accomplished with even greater ease than in the waking state. This probably disproves Pavlov's hypothesis that hypnosis is a state of partial sleep in the sense of a partial irradiation of inhibition." (Hull, 1933/1986, p. 221).

Milton H. Erickson, who was a student of Clark Hull, developed innovative approaches to hypnosis that could utilize either the passive, relaxed and sleep-like tendencies of his patients ("You don't even have to listen to my voice") or their more active, compulsive, "acting out" behavior such as pacing around the therapy office in an agitated manner. Erickson taught that the appropriate choice of hypnotic induction should be a function of the patient's mood, attitudes and behavior in the therapy session. Erickson described the great variety of approaches to hypnosis that made use of the full range of the patient's ongoing behavior as the "naturalistic" (Erickson, 1958/1980) or the "utilization" approach (Erickson, 1959, 1980) in his typically lengthy 90 to 120 minute therapeutic sessions. As we shall soon see 90-120 minutes corresponds to a basic rest activity cycle (Kleitman, 1969; Kleitman and Rossi, (1992) that is associated with many psychobiological processes on all levels from the cognitive-behavioral to the genetic that may be related to the therapeutic possibilities of hypnosis as well as the activity-passivity paradox.

One of the first efforts to understand the psychobiological basis of the activity-passivity paradox of hypnosis came from Erickson's colleague Bernard Gorton in his papers on the physiology of hypnosis (1957, 1958). A review of the existing literature suggested to Gorton that "vasomotor activity" and the autonomic nervous system (ANS) with its two main branches, the sympathetic system (arousal) and the parasympathetic system (relaxation) was the major avenue of the physiological effects of hypnosis. More recent research supports Gorton's view that there is "A positive correlation between hypnotic susceptibility and autonomic responsiveness during hypnosis..." (De Benedittis et al., 1994, p 140) and the nature of the "physiological responsiveness [arousal or relaxation] is dependent on the type of suggestions during hypnosis..." (Sturgis & Coe, 1990, p 205).

Even today, however, the use of hypnotic suggestion to modulate the active as well as the passive branches of the ANS is not well understood by researchers who report the "paradoxical" nature of their results. Weinstein and Au (1991), for example, reported that norepinephrine levels were significantly higher in a hypnotized group of patients undergoing angioplasty than in the control group. They report that this was "unexpected and seemed *paradoxical* (p 29) ... One would expect that if hypnosis does cause relaxation, then those patients who were hypnotized would have a lower arterial catecholamine level than their controls. This was not the case. Just the opposite occurred and is hard to explain" (p 35). This so-called paradox is hard to explain only if one assumes, as these and other authors do (Lazarus and Mayne, 1991), that hypnosis is essentially a state of relaxation. The importance of arousal in hypnosis was emphasized by Amigo (1994) who presented evidence for "self-regulation therapy" as a "cognitive-behavioral approach to hypnosis" that involves the "voluntary reproduction of the stimulant effects of epinephrine." Likewise Harris et al., (1992) found that "both branches of the autonomic nervous system may contribute to hypnotic susceptibility. This accumulating research on how hypnosis can utilize both branches of the ANS resolves the activity-passivity paradox of hypnosis and leads to an expanded model of the psychobiological domain of hypnosis.

## Expanding the Domain of Hypnosis: *The Deep Psychobiological Model*

The deep psychobiological model of hypnotic suggestion is consistent with a generation of research that firmly established, contrary to popular belief, that there is no transcendence of normal abilities in hypnosis (Wagstaff, 1986). Researchers openly acknowledge, however, that they have no adequate theory of the source and parameters of hypnotic performance. Naish (1986, p165-166), for example, has summarized this limitation of the cognitive-behavioral approach as follows: "As [hypnotic] susceptibility is normally assessed, a high scorer is one who *produces* the behavior, the *reason* for its production remains unknown...the claim was frequently made that cognitive processes are involved in the production of 'hypnotic' effects. However, the exact nature of these processes generally remained obscure." The psychobiological model of hypnosis proposes that what seems to be an extension of the normal parameters of mind-body performance skills via hypnosis is actually the optimization of the individual's normal range of abilities in response to the general process of adaptation to challenge and stress by the ANS and related systems.

The limitations of cognitive-behavioral level in conceptualizing suggestibility and hypnosis have been discussed by a number of researchers. Hilgard (1991, p. 45-46), for example, summarized the current dilemma with these words, "...hypnotic behavior cannot be defined simply as a response to suggestion... Although hypnotic-like behaviors are commonly responses to suggestion, the domain of suggestion includes responses that do not belong within hypnosis, and the phenomena of hypnosis covers more than specific responses to suggestion." Hilgard seems to be saying that the domain of hypnotic suggestion as currently defined on the cognitive-behavioral level is not adequate or complete. Eysenck (1991, p. 87) emphasizes the problem more pungently, "There is no single, unitary trait of suggestibility, no one uniform type of reaction to different kinds of suggestion in human subjects. There are several, or possibly many different suggestibilities that bear no relation to each other. These are uncorrelated and, in turn, correlate differentially with other cognitive and emotional variables. This finding is of considerable interest and importance... It does make books containing in their title the word 'suggestibility' of rather doubtful value!"

Researchers who began their careers by defining hypnosis as a cognitive-behavioral response to suggestion and then, after a lifetime of research, reversing themselves and concluding, in effect, (1) hypnosis cannot be operationally defined simply as responsiveness to suggestion, and (2) there is no unitary human trait of suggestibility, has led to a fundamental impasse in current theory. Balthazard & Woody (1992, p. 22), for example, have stated, "Although a fair amount of factor analytic work has been done with the hypnosis scales this work appears to be at an impasse methodologically, and it has failed to yield any consistent picture of the mechanisms that underlie performance on the hypnosis scales." Kirsch and Lynn (1995, p. 854) emphasize this impasse as follows. "*Is there a uniquely hypnotic state that serves as a background or gives rise to the altered subjective experiences produced by suggestion?* Having failed to find reliable markers of trance after 50 years of careful research, most researchers have concluded that this hypothesis has outlived its usefulness."

Paradox and impasse in science frequently indicate where theory and practice need to be revised and expanded in some fundamental way. The psychobiological model of hypnosis expands the domain of suggestion beyond the cognitive-behavioral level to include all systems of mind-body communication and healing at the molecular level that are responsive to psychosocial cues. It has been proposed that the major path of mind-body communication is the hormonal or *messenger molecule-cell receptor system* of the neuroendocrine, neuropeptide, autonomic and immune systems that mediate stress, emotions, memory, learning, personality, behavior and symptoms in a "Psychosomatic Network" (Pert et al. 1985, 1989; Rossi, 1972/1985/2000, 1986/1993, 1996) as illustrated in Figure One. This leads to our first hypothesis about the expanded psychobiological domain of hypnosis.

**Hypothesis 1:** ***The psychobiological domain of hypnosis consists of the utilization of the entire cybernetic loop of information transduction between the psychosocial environment, the central nervous system and the psychosomatic network of the autonomic, neuroendocrine and immune system at the organ, tissue and cellular-genomic-molecular levels.***

Figure 3.1 illustrates four major levels of the communication process of information transduction between mind, brain and

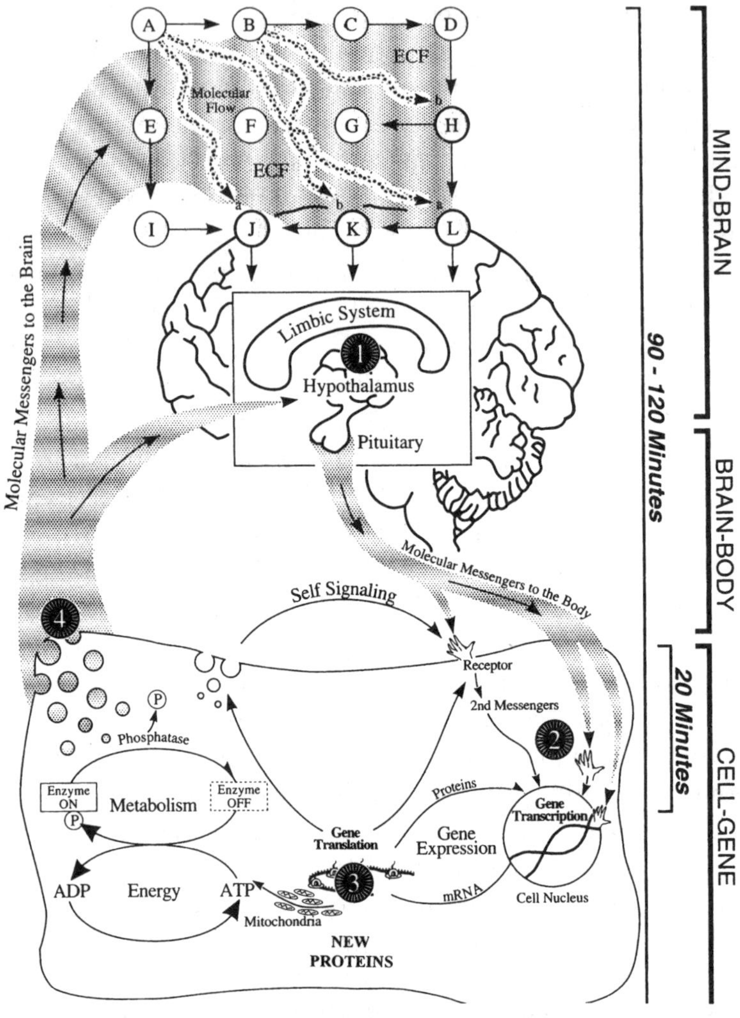


Fig. **3.1**. - The Four-Stage Mind-Body Communication loop illustrating the “Unification Hypothesis of Chronobiology” wherein the quasi-periodic rhythmical flow of messenger molecules of the neuroendocrine system (hormones associated with stress, relaxation, sexual hormones, immune system etc.) mediates communication between the mind-brain, brain-body and cellular genomic levels in ultradian time (less that 20 hours). (1) Information from the outside world encoded in the neurons of the brain is transformed within the limbic hypothalamic pituitary system into the hormones (messenger molecules) that travel through the blood stream to signal receptors on all cells of the brain and body. (2) The receptors on the surface of cells transmit the signal via 2nd messengers to the nucleus of the cell where immediate early genes signal other activity dependent genes to transcribe their code into messenger RNAs. (3) The messenger RNAs serve as blueprints for the synthesis of proteins that will function as (a) the ultimate healing structures - the materia of the body, (b) enzymes to facilitate energy dynamics and (c) receptors and messenger molecules for the informational dynamics of the cell. (4) Messenger molecules function as a type of "molecular memory" that can evoke psychological experiences of mind as state dependent memory, learning and behavior (SDMLB). SDMLB is expressed in the neural networks of the brain (illustrated as the rectangular array of letters A to L on the top) that receive molecular signaling through the Extra Cellular Fluid (ECF) of the brain that may be regarded as a field of complex, self-organizing communication processes that are the psychobiological basis of mind, meaning, memory, learning, psychosomatic medicine and healing. From Rossi, 2002.

body that are hypothesized as the psychobiological domain of hypnosis. Let us review what is currently known about the psychobiological mechanisms of hypnosis at each level and the types of research that are now needed for further progress.

**Level 1: *The CNS and the Mind-Brain.***

Papez (1937) traced the anatomical pathways by which emotional experience of the brain was transduced into the physiological responses of the body in a circuit of brain structures that are now generally recognized as the limbic-hypothalamic-pituitary system. The Scharrers (1940) documented how secretory cells within the hypothalamus could mediate molecular information transduction between brain and body. Cells within the hypothalamus transduce the electrochemical impulses of neurons of the cerebral cortex that encode the phenomenological experience of "mind" into the hormonal or "primary messenger molecules" of the endocrine system. These primary messenger molecules of the neuroendocrine system flow through the blood stream to signal receptors on cells of the brain and body to modulate their responsiveness in a cybernetic loop of information transduction illustrated in Figure One. The current consensus is that corticotrophin releasing factor (CRF) is the central psychobiological coordinator of the generalized arousal and stress response (Felker and Hubbard, 1998). CRF signals the anterior pituitary to induce proopiomelanocortin (POMC) that is cleaved into about a dozen messenger molecules including adrenocorticotropic hormone (ACTH) and beta-endorphin that have varying effects upon the psychobiology of arousal, stress, memory, learning and relaxation (Brush & Levine, 1989).

Recent experiments document how such psychobiological mechanisms operate in the transition between waking and sleeping in a manner that may be of relevance for the therapeutic applications of hypnosis. Born et al. (1999), for example, investigated what we may call the "human alarm clock effect" which enables some people to awaken at a specific time in the morning without using an alarm clock. Awakening from a night of sleep is related to a daily (circadian) rhythm in the release of the pituitary and adrenal hormones. This circadian rhythm is in turn made up of a series of ultradian rhythms wherein there is a peak in the release of ACTH and cortisol every 90-120 minutes through out the day and night. Normally this release of ACTH and cortisol increases during the later stages of sleep and reaches a daily peak just before the time of awakening in the morning. Born et al. (1999) demonstrated that the conscious intentionality or anticipation of awakening at a specific time in the morning could shift the ultradian peak in the release of ACTH to that specific time. That is, a conscious intentionality or expectancy of awakening at a specific time could pervade sleep and modulate the expression of a normally autonomous or involuntary rhythm of hormone regulation across the limbic-hypothalamic-pituitary system at locus one in Figure 3.1. This is a clear example of information transduction between the phenomenological experiences of consciousness and the flow of information through the body via hormones (also called "messenger molecules" or "information substances") (Rossi, 1986/1993, 2002).

It would be of great interest to determine whether hypnosis achieves its mind-body effects by utilizing the same type of psychobiological mechanisms at the hormonal level when a posthypnotic suggestion is administered to awaken from a night of sleep at a specific time.

This type of research would integrate hypnosis with the vast database and methodology of modern neuroscience (Barabasz, 2002; Rainville et al., 1997, 1999). Such research could, for example, provide experimental data to assess a mathematical model of the chronobiological mechanisms of hypnosis as proposed later in Figure 3.9.

Such research motivates us to take a closer look at the typical dynamics of some of the other major systems of mind-body communication that may be operative in the psychobiological domain of hypnosis. Figures 3.2 through 3.7 illustrate half a dozen interrelated life processes from the cellular-genomic level to the psychosocial including self-hypnosis. A superficial glance suggests these systems are all very different. A more careful examination, however, indicates that much of their apparent difference is due to the fact that (1) they were sampled with different numbers of data points on (2) different time scales that (3) are not strictly regular in their periodicity but manifest, rather, what is now called "quasi-periodic." Quasi-periodic behavior is found when two or more rhythms with different periods interact to produce oscillations that never repeat themselves exactly. Many physical systems such as the two-coupled pendulum, most biological processes, and brain dynamics are quasi-periodic (Rossi, 1996).

*Quasi-periodicity is the signature of the nonlinear dynamics of communication in the highly adaptive processes of life on all levels from the psychosocial to the cellular-genomic* (Lloyd and Rossi, 1992; Newtson, 1994; Vallacher and Nowak, 1994). What all psychobiological systems have in common is that they may be modeled by the quasi-periodic dynamics of circadian (about 24 hours) and ultradian rhythms (less than 20 hours). This leads to conjectures about the non-linear dynamics of future mathematical models of the entire range of psychobiological states that are now loosely called "therapeutic hypnosis."

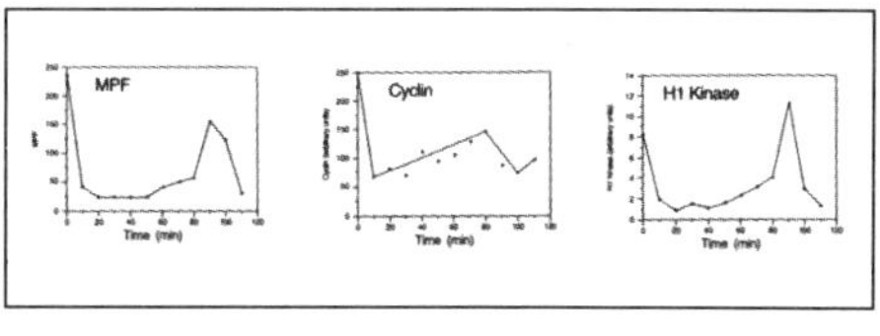


Fig. 3.2 The Cellular Genomic Level. In this series of graphs it can be seen that Kleitman's (1969; Kleitman & Rossi, 1992) Basic Rest Activity Cycle - typically a 90-120 minute quasi-periodic ultradian

rhythm - is fundamental in cell growth and replication. The approximately 20 minute peak in Maturation Promoting Factor (MPF), the protein Cyclin and the enzyme H1 kinase act in concert to signal the final stage of genetic replication and cell division (mitosis) (From Murray et al., 1989). Some researchers propose that this may be the basic ultradian pacemaker that sets all other levels such as the metabolic, neuroendocrinological, cognitive-behavioral and the socio-cultural as illustrated below.

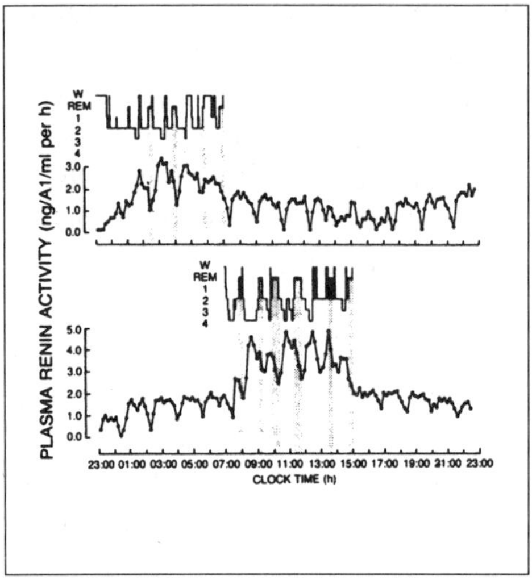


Fig. 3.3 The Endocrine-Behavior Level. An example of the interaction between the cognitive-behavioral and hormonal levels. The typical 90-120 minute ultradian rhythms in the pulsate expression of the endocrine renin-angiotensin-aldosterone system have their higher amplitude peaks shifted to the right by an 8-hour delay of the sleep-wake cycle in the lower graph. The small inset on the upper part of each graph illustrates the relationship between plasma renin and the stages of waking consciousness , Rapid Eye Movement Sleep (REM Dreaming state) and the four major levels of sleep depth (From Brandenberger, 1992).

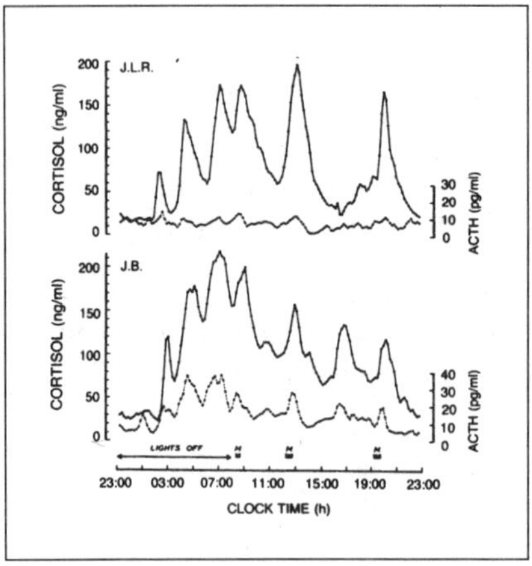


Fig. 3.4a Neuroendocrine Level - ACTH & Cortisol. Hormonal profiles of the ultradian rhythms in ACTH and Cortisol in two subjects when blood samples were taken at 10 minute intervals over a 24 hour period (From Brandenberger, 1992). While there are obvious differences, both individuals manifest 90-120 minute pulsate rhythms with varying amplitudes. The lower profile, which is perhaps the more typical, illustrates how these two hormonal messenger molecules that mediate states of arousal tend to their highest peaks in the early morning hours and gradually dwindle in the afternoon and evening when our energy levels are lower.

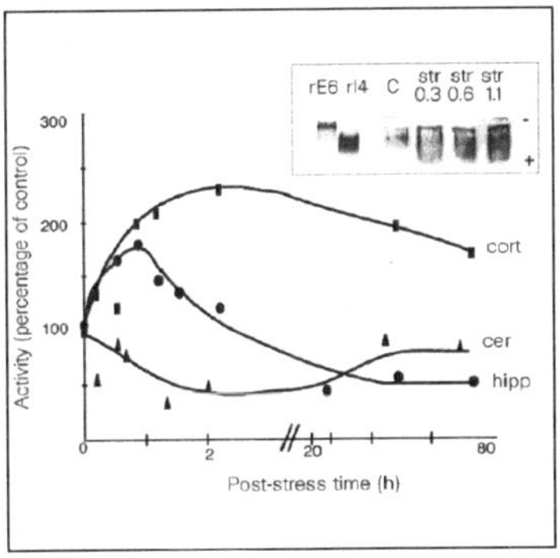


Fig. 3.4b The ultradian and circadian time parameters of an acute stressor (forced swimming in mice). Notice the initial peak of stress induced acetylcholinesterase activity during the typical time frame of Kleitman's 90-120 minute Basic Rest Activity Cycle (BRAC) in the hippocampus (hipp) that is followed by a peak in the cortex (cort) when the symptoms of PTSD become manifest. This figure is reproduced with permission from Kaufer et al., 1998.

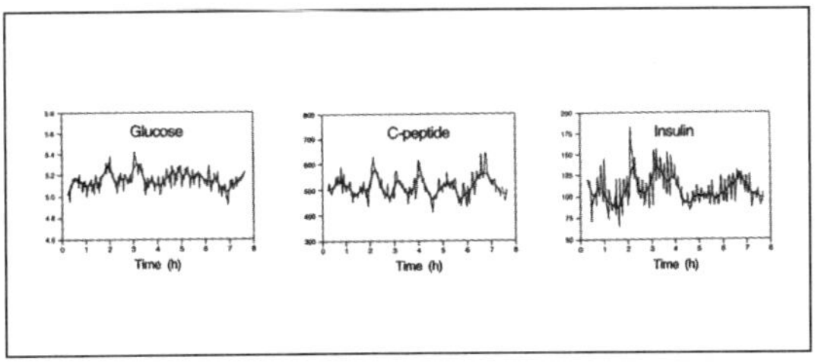


Fig. 3.5 The Endocrine-Energy-Metabolic Level. These three graphs illustrate how the blood Glucose (left), C-peptide (middle) and insulin (right) obtained at two minute intervals during an 8-hour fasting period have rapid 10-15 minute oscillations that are shown superimposed on the approximately 90-120 ultradian rhythms which were obtained by a best-fit curve with a regression algorithm (From Sturis et al., 1992).

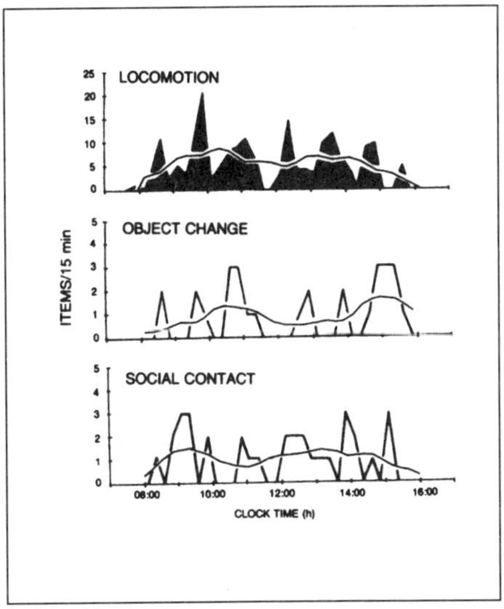


Fig. 3.6 The Behavioral-Social-Cultural (Psychosocial) Level. Time series illustrating the ultradian rhythms of the locomotion (about 80 minutes), object change (about 65 minutes) and social contacts (about 103 minutes) for the Indian child "Ram" in a natural setting in a small village in India. Similar ultradian rhythms are found in a child of !Ko Bushmen and social synchronization in a village community of Colombian Indians as well as free play in highly urbanized children in Germany. (From Meier-Koll, 1992).

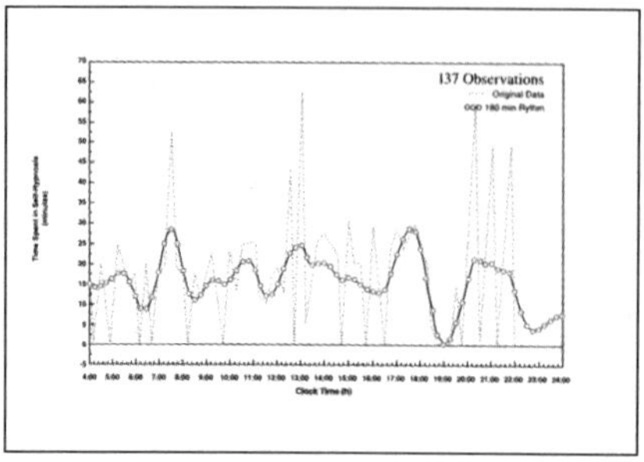


Fig. 3.7a The Cognitive-Behavioral-Hypnosis Level. An illustration of the quasi-periodic 180 minute ultradian rhythms of self hypnosis that were recorded by 9 subjects in their diaries in a pilot study that now requires independent confirmation (From Rossi, 1992).

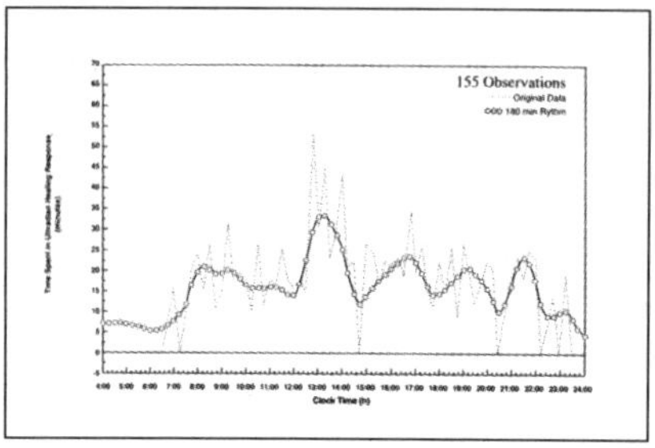


Fig. 3.7b The Basic Rest Activity Cycle (BRAC) found in the time series of 7 subjects who were instructed to simply enjoy an "ultradian healing response" whenever they felt a need to throughout the day in a pilot study that now requires independent confirmation (From Rossi, 1992).

**Hypothesis 2:** ***The clinical applications and experimental investigations of hypnosis utilize a wide variety of psychobiological states ranging from "High to Low Phase Hypnosis." The entire range of the classical phenomena of hypnosis and their therapeutic applications may be mathematically modeled as nonlinear, quasi-periodic Yerkes-Dodson Functions of arousal, creative work, and relaxation.***

The nonlinear relationship between performance and arousal illustrated on the right of Figure Eight, called the Yerkes-Dodson Function (1908), and is one the earliest and most well established laws in psychology. All life processes go through an initial phase of arousal, they reach a peak where performance is optimized, and then relax back to a basal level. Guastello (1995) has updated the significance of the Yerkes-Dodson function in psychophysics and sensory-perceptual systems as well as the nonlinear dynamics of creativity, work and social organizations. We hypothesize that this intensively researched Yerkes-Dodson function models the arousal (high phase) and relaxation (low phase) aspects of hypnosis as well. An outline of the parameters of this mathematical model adapted from chronobiology is presented in Figure 3.9. Figure 3.10 illustrates how this quasi-periodic model could integrate many of the apparently paradoxical phenomena and conflicting theories of high and low phase hypnosis in a complementary manner (Rossi, 1996, 2002).

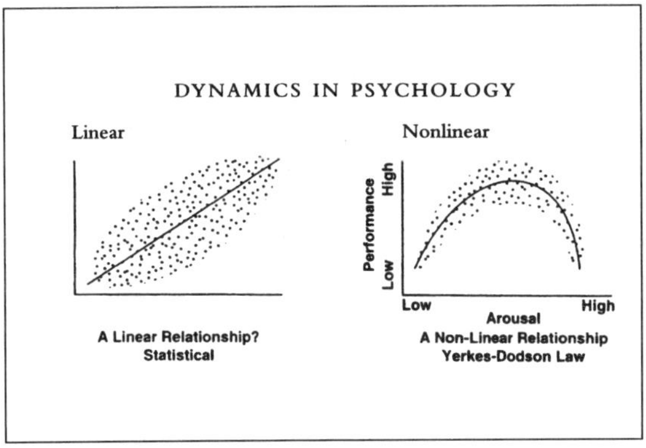


Fig. 3.8 Linear and nonlinear dynamics in psychobiology. The typical linear approach is illustrated on the left with the straight line cutting through a cloud of data points is the "best linear fit," of the data. Traditionally each of the data points is said to be a combination of a measurable psychological factor and experimental error or "noise." The nonlinear dynamics systems approach, however, recognizes that many apparently random deviations of so-called "noise" actually may be the signature of quasi-periodicity and deterministic chaos in the natural and highly adaptable ultradian rhythms of mind-body communication and healing that are responsive (easily entrained by) psychosocial cues of stress, trauma, arousal, and relaxation. This responsiveness, adaptation, or entrainment of biological processes by psychosocial cues (communications) is the essence of psychosomatic medicine, mind-body healing, and so-called "alternative, complementary or holistic medicine."

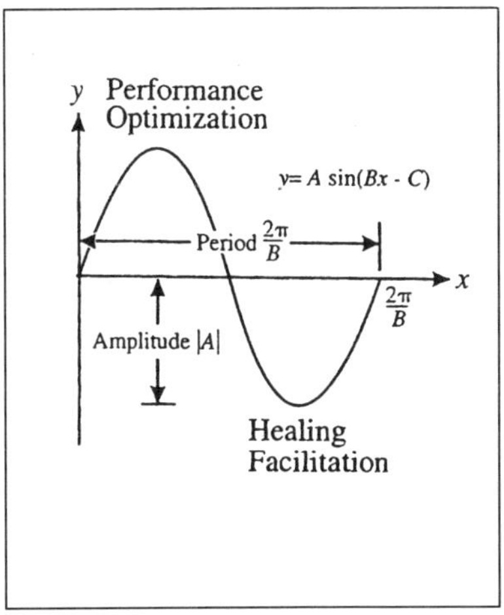


Fig. 3.9 Towards a quasi-periodic mathematical model of hypnosis. The parameters of a possible mathematical model of how the therapeutic applications of hypnosis may entrain and utilize the quasi-periodic psychobiological dynamics of adaptation to stress, arousal, and relaxation are as follows.

*Amplitude* refers to the absolute value of the height or depth of a quasi-periodic rhythm of behavior: it measures how far rhythm deviates from its mean level. The amplitude may correspond to the degree to which hypnosis may optimize performance variables that require psychobiological arousal (sympathetic system) or healing (parasympathetic system).

*Period* is the time required for one complete cycle of adaptation on all levels from the cognitive-behavioral to the cellular-genomic. The period is a quasi-periodic parameter of psychobiological rhythms that may be contracted, stretched or "phase locked" in the therapeutic applications of hypnosis. The frequency is the reciprocal of the period.

*Phase* identifies parts of natural quasi-periodic psychobiological rhythms of adaptation that may be accessed, entrained and utilized by hypnosis. The crest or peak phase is associated with the activation (sympathetic system) dynamics of "high phase hypnosis," while the trough phase of relaxation (parasympathetic system) may be entrained by "low phase hypnosis." The number C/B, called the phase shift, is a measure of the degree to which certain portions of chronobiological behavior can be entrained and modulated with hypnosis.

*Entrainment or synchronization* refers to the interaction of psychobiological rhythms (x and y below) with psychosocial cues such as hypnosis (z below). Many therapeutic applications of hypnosis may be conceptualized as the entrainment of portions of the Basic Rest-Activity Cycle (BRAC) that are utilized for enhancing performance or facilitating healing.

$$k^2\ddot{x} + k\mu_x(-1 + x^2)\dot{x} + \omega_x^2 x + F_{yx}k\dot{y} = 0 \qquad (k = \pi/12)$$

$$k^2\ddot{y} + k\mu_y(-1 + y^2)\dot{y} + \omega_y^2 y + F_{xy}k\dot{x} = F_{zy}z$$

$$z = F_{zy}\cos(k\omega_z t + \theta_z)$$

A mathematical model adapted from the field of chronobiology (Kronauer, 1984) is used to illustrate how the hypnotic entrainment of quasi-periodic psychobiological rhythms may operate. In the equations below Fzy is the "influence" or entrainment coefficient; w x and w y are the natural frequencies of quasi-periodic psychobiological rhythms x and y; k may be a constant associated with hypnotic susceptibility; that is, the entrainability of a particular psychobiological process by highly focused psychosocial cues.

Note that this is not yet a complete mathematical model of therapeutic hypnosis! It is merely the initial conceptual stage where we are assuming that therapeutic hypnosis actually engages the natural ultradian psychosomatic network on all levels from mind to gene expression and neurogenesis as proposed by the new discipline that I call "Psychosocial Genomics: How psychosocial processes can modulate immediate early, activity dependent, and behavior state related gene expression." (Rossi, 2002). A series of iterative steps integrating theory and experiment are now required to assess, modify, and retest the value of this possible math model of therapeutic hypnosis (Borrelli and Coleman, 1998)

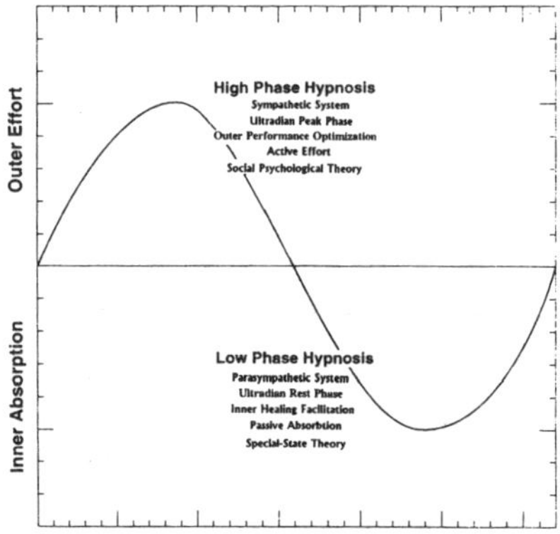


Fig. 3.10 The continuum of therapeutic hypnosis to be assessed by DNA chip technology ranges from the quasi-periodic ultradian peaks of performance of (1) high phase hypnosis with its active focus on problem solving as described by social-psychological theorists to (2) the apparently passive periods of deep inner absorption and ultradian healing with low phase hypnosis emphasized by special state theorists. A complete cycle of hypnotic work usually consists of an initial high phase of sympathetic system arousal as the subject becomes engaged in problem solving. With the resolution of a problem the subject spontaneously slips into low phase hypnosis wherein parasympathetic system dominance is experienced as relaxation and healing (Rossi, 1996, 1972/1985/2000, 2002).

What is the type of data that could be used to assess this mathematical model of high and low phase hypnosis? While the therapeutic applications of hypnosis have traditionally focused on relaxation or low phase hypnosis, a pair of research reports by Hautkappe & Bongartz (1992) and Unterweger, Lamas & Bongartz (1992) supports the idea that hypnosis involves a significant *"work function"* that operates differently in high and low hypnotic susceptibility subjects. Hautkappe & Bongartz (1992) found that heart rate variability was a useful physiological parameter for discriminating high and low hypnotic susceptibility; high susceptible hypnotic subjects have less heart rate variability (p. 82-83). This implies, "High susceptible subjects do not have to work as hard on passing a suggestion as do low susceptibles" (Unterweger et al., 1992, p. 87). High hypnotic susceptibility is apparently associated with a more efficient psychobiological use of information and/or energy. From a historical perspective this may be interpreted as consistent with Braid's conception of hypnosis as facilitating "monoideism" (Tinterow, 371).

In Ericksonian terms high susceptible subjects have higher

"response attentiveness" or selective focus so their mind-body system does not require an indiscriminate massive arousal to do certain tasks (Erickson and Rossi, 1979). Erickson pioneered the use of psychological shocks and creative moments (Rossi, 1973) to focus attention in what we would now call "high phase" hypnosis. More recent clinical and experimental research by Barabasz and Barabasz (1996) has documented how such "alert hypnosis" can facilitate the neurotherapy (neural biofeedback) of children with attention deficit hyperactivity disorder (ADHD) by enhancing their skills in shifting from a predominantly theta to beta modes of functioning as measured by their eeg. Further research is now needed to explore the relationships between high and low susceptible hypnotic subjects with the use of high and low phase hypnosis. One wonders, for example, whether high susceptible hypnotic subjects are more efficient in turning on and focusing the activating (sympathetic branch or high phase hypnosis) as well as the relaxation (parasympathetic branch or low phase hypnosis) of their neuroendocrine system. What are the relative merits of (1) The Stanford and Harvard Hypnotic Susceptibility Scales that may induce low phase hypnosis with their emphasis on sleep; (2) The Barber Scale the and Spiegle Hypnotic Induction Profile that may induce high phase hypnosis with their emphasis on focused attention, versus (3) Rossi's (1986) Indirect Trance Assessment Scale (ITAS) that seeks to eliminate a bias toward either high or low phase hypnosis in the therapeutic applications of hypnosis?

At the limbic-hypothalamic-pituitary level any powerful physical or psychosocial stimulus leads within minutes to the release of CRF which in turn initiates an ACTH-Cortisol-Beta-endorphin cascade that coordinates a variety of adaptive psychobiological processes on the organ, tissue and cellular-genomic levels. One of the most prominent of these is the 90-120 Basic Rest Activity Cycle (BRAC) associated with the quasi-periodicity of hormonal systems of mind-body regulation during waking (Lloyd and Rossi, 1992, 1993), sleep and dreaming (Kleitman, 1969; Kleitman and Rossi, 1992; Rossi 1972/1986/2000). Available evidence indicates that the highly adaptable quasi-periodic 90-120 minute ultradian cycle consists of an initial phase of psychobiological arousal, work and performance mediated by the ACTH-cortisol phase of the BRAC. The relaxation phase of the BRAC is then mediated by beta-endorphin that signals a molecular cascade at the cellular-genomic-protein level to facilitate restoration, rehabilitation and healing. Iranmanesh et al. (1989, p. 1019), for example report that "...cortisol was considered to lead B-endorphin by 20 or 30 minutes. We conclude that B-endorphin is released physiologically in a pulsate manner with circadian and ultradian rhythmicity and a close temporal coupling to cortisol." Such research is consistent with the hypothesis that arousal or high phase hypnosis is associated with the cortisol peak of the BRAC while relaxation or low phase hypnosis is associated with the subsequent B-endorphin response. Research is now required to continuously monitor the release of ACTH, cortisol, B-endorphin and related messenger molecules during the quasi-periodic phases of hypnosis as described previously (Rossi, 1988,1990 a & b, 1996; Rossi and Cheek, 1988). This leads to a pair of hypotheses about relationships between the quasi-periodic parameters of adaptation, stress, hypnosis and healing.

**Hypothesis 3:** ***The communication systems between mind and body (the messenger molecule receptor dynamics of the CNS, ANS, neuropeptide and immune systems of the psychosomatic network) are coordinated by and serve the adaptive gene expression-protein synthesis cycle at the cellular level. Many of these interrelated systems of mind-body communication and healing are manifest in the observable behavior of the quasi-periodic Basic Rest Activity Cycle (BRAC) that may be entrained and utilized in the clinical applications of hypnosis.***

The *proof of the principle* that we can now trace pathways of communication between mind and body to the genomic level is evident in emerging virtual models of systems biology where researchers are creating interactive computer simulations of the molecular dynamics of the entire cell (Davidson et al., 2002; Kitano, 2002). (http://www.sbml.org// http://www.sbml.org// http://www.genome.ad.jp//).

Noble (2002), for example, describes the current status of "Modeling the heart - from genes to cells to the whole organ" and behavior that has implications for the deep psychobiology of therapeutic hypnosis.

> "Successful physiological analysis requires an understanding of the functional interactions between the key components of cells, organs, and systems, as well as how these interactions change in disease states. This information resides neither in the genome nor even in individual proteins that genes code for. It lies at the level of protein interactions within the context of subcellular, cellular, tissue, organ, and system structures. There is therefore no alternative to copying nature and computing these interactions to determine the logic of healthy and diseased states. The rapid growth in biological data bases; models of the cells, tissues, and organs; and the development of powerful computing hardware and algorithms have made it possible to explore functionality in a quantitative manner all the way to the level of genes to physiological function of whole organs and regulatory systems.
>
> The amount of biological data generated over the past decade by new technologies has completely overwhelmed our ability to understand it. Genomics has provided us with a massive "parts catalogue" for the human body; proteomics seeks to define these individual "parts" and the structures they form in detail. But there is as yet no "user's guide" describing how these parts are put together to allow these interactions that sustain life or cause disease. In many cases, the cellular, organ, and system functions of genes and proteins are unknown, although clues often come from similarity in the gene sequences. . .
>
> An important strength of models based on reconstructing the functional properties of proteins is that it is possible for the models to reach down to the genomic level, for example, in reconstructing the effects of particular mutations when these are characterized by changes in protein function. An example of this approach is the use of multistate (Markov) models of the sodium channel in which models of the wild-type ["normal" as typically found in nature] and a mutant sodium channel were formulated and validated. . . This causes major prolongation of repolarization and the development of arrhythmogenic early afterpolarizations at slow pacing rates, a behavior that is consistent with the clinical presentation of bradycardia-related arrhythmogenic episodes during sleep or relaxation in LQT3 patients." (Pp. 1678-1679)

These relationships between systems of molecular interactions at all levels from the rhythms of the gene expression- protein cycle (figure 1) to mind and behavior such as "sleep and relaxation" point to new ways of validating our deep psychobiological model therapeutic hypnosis. The current challenge is to include the cognitive-behavioral approaches of therapeutic hypnosis in the emerging virtual models of the cell that are being developed in systems biology (Davidson et al., 2002). These virtual computer models will become our new tool for bridging the so-called "Cartesian gap between mind and body." Meanwhile we can review some basic research on hypnosis that is consistent with the experimental associations found between "arrhythogenic episodes during sleep or relaxation," described above, as but one example of the more general finding of "chronic desynchronization" found in psychosomatic problems that may be ameliorated by therapeutic hypnosis.

**Hypothesis 4:** ***Psychobiological stress engendered by the chronic desynchronization of circadian and ultradian dynamics such as the quasi-periodic 90-120 minute Basic Rest Activity Cycle by traumatic and/or excessive work loads is a major etiology for psychosomatic problems that may be ameliorated by the therapeutic applications of hypnosis.***

Well replicated work on the normal psychobiological dynamics of the major systems of mind-body regulation and healing is evolving out of time parameter research on many levels ranging from the molecular biology of the cell cycle to the neuroendocrinal and the behavioral levels as illustrated in Figures Two through Seven (Lloyd and Rossi, 1992, 1993). It now appears that many of the highly adaptive processes of psychobiological regulation that manifest a natural quasi-periodic circadian/ultradian variability are modifiable by psychosocial cues and hypnosis (Rossi, 1982, 1986/1993, 1996). This implies that what has been traditionally called "hypnotic suggestion" may be, from a chronobiological perspective, the accessing and utilization of the quasi-periodic variability of ultradian and circadian processes on all levels from the cognitive-behavioral to the molecular that respond to psychosocial cues. Within this framework, many of the classical phenomena of hypnosis may be conceptualized as extreme manifestations and/or preservations of quasi-periodic psychobiological processes that are responsive to psychosocial cues. That is, what the clinician using hypnosis calls "therapeutic suggestion" is what the chronobiologist calls "the entrainment of biological processes by psychosocial cues."

This matching of the domains of chronobiological and hypnotic phenomena leads to important falsifiability tests about the relevance of quasi-periodic parameters in hypnotic work. In brief, if one could find a single chronobiological phenomenon (aspects of memory, emotions, learning, behavior and self-regulation, for example) that can be entrained by psychosocial cues that is not responsive to hypnotic suggestion then one would have disproved the relevance of quasi-periodic parameters for the psychobiological model of hypnosis.

The reverse is also true. If one could find a single psychobiological process modifiable by hypnotic suggestion that does not have a natural quasi-periodic ultradian/circadian rhythmicity then one would have disproven the relevance of chronobiological parameters for the psychobiological model of hypnosis. Rossi (1996) has discussed an unintended but highly informative example of this falsifiablity test with the event-related P300 brain wave potential. Initially the P300 wave was found to be modifiable by hypnosis (Barabasz and Lonsdale, 1983; Spiegel and Barabasz, 1988).

Later in unrelated research the P300 wave was found to have a 90-120 minute ultradian periodicity (Escera et al., 1992). This is exactly what our matching of the domains of chronobiological and hypnotic phenomena that are responsive to psychosocial cues would predict.

Research consistent with the significance of quasi-periodic parameters in hypnosis has been reported by a number of investigators in the past decade. Aldrich and Bernstein (1987) who found that "time of day" was a statistically significant factor in hypnotic susceptibility initially assessed it. They reported a bi-modal distribution of scores on The Harvard Group Scale of Hypnotic Susceptibility in college students with a sharp major peak at 12 noon and a secondary, broader plateau around 5 to 6 p.m. Further research found a very prominent circadian rhythm with a peak between noon and 1 p.m. in the self-hypnosis as well as an ultradian periodicity of about 90 to 180 minutes throughout the day that approximates Kleitman's 90-120 minute Basic Rest Activity Cycle (BRAC) (Rossi, 1992). The pilot data of Figure Seven suggests that self-hypnosis usually lasts about 20 minutes.

It is interesting and probably not coincidental that much research assessing the therapeutic value of the various modalities of alternative medicine such as acupuncture, biofeedback, imagery, meditation, music, therapeutic touch, etc. also use a core 20 minute therapeutic period (Green and Green, 1987). While this core therapeutic period may be condensed or extended depending on practical exigencies, it is rarely extended beyond the Kleitman's typical 90-129 Basic Rest Activity Cycle. Similar quasi-periodic parameters were found associated with hypnosis (Brown, 1991 a & b; Lippincott, 1992,1993; Osowiec, 1992; Rossi, 1982,1992; Sommer, 1993; Wallace, 1993) and imagery (Wallace and Kokoszka, 1995).

Two studies (Mann and Sanders, 1995; Saito and Kano, 1992) were correct in emphasizing that a narrow interpretation of Rossi's original chronobiological hypothesis - that hypnotic susceptibility was a function of strictly periodic or statistically uniform ultradian rhythms (Rossi, 1982) - would be incorrect.

The experimental results of these two studies, however, are entirely consistent with the current hypothesis about the quasi-periodic or nonlinear dynamics of hypnotic susceptibility. Further research utilizing more efficient mathematical algorithms for detecting the synchronization of psychobiological rhythms (Schäfer et al., 1998) are now required to examine how hypnosis can assess and entrain the natural quasi-periodic dynamics of mind-body communication and healing.

**Level 2: *Stress, Immediate-Early Genes, and Psychoimmunology.***

The most revealing demonstrations of how psychosocial

stress can modulate the actual mechanisms of gene expression in the immune system are a series of papers by Glaser et al. (1990, 1993). Recently this research team (Kiecolt-Glaser, 2001) and others (Gruzelier et al., 2001) have demonstrated how hypnosis could successfully modulate cellular immune dysregulation. Their research traces the effects of psychological stress (experienced by medical students during examinations) in down regulating the transcription of the interleukin-2 receptor gene and interleukin-2 mRNA production. Since interleukin-2 is a messenger molecule of the immune system, its down regulation by psychological stress is the first demonstration of how the immune system's optimal functioning can be impaired at the cellular-genomic level by psychosocial cues. Glaser's research gains even more profound significance for a general theory of mind-body communication and healing when we realize that other independent medical researchers (Rosenberg and Barry, 1992) found that interleukin-2 is a messenger molecule of the immune system that "tells" white blood cells to attack pathogens and cancer cells. This means that traditional medical research represented by Rosenberg and mind-body medicine represented by Glaser have found the same bottom line of mind-gene communication and healing in psychoimmunology.

This molecular-genomic essence of communication in psychoimmunology is now know to be mediated by immediate early genes (IEGs) that typically turn on within a minute of the onset of psychological arousal and stress and operate for about and hour and a half or so (Schlingensiepen et al., 1995). Further evidence of how the immune system can be impaired at the cellular-genomic level was provided when Glaser et al. (1993) found that academic stress led to the down regulation of the two IEGs (also called proto-oncogenes) *c-myc* and *c-myb* in peripheral blood leukocytes. The IEG *c-myc* is part of an informational loop at the cellular level that activates oncogenes that are involved with breast cancer as well as stomach and lung cancers and leukemia. Such research detailing how psychosocial cues can modulate immediate early gene expression provides at least a clue about the mechanisms of hypnosis that could optimize the functioning of the immune system to facilitate the remission of cancer as is occasionally reported in a well documented manner (Crasilneck, 1997).

The most obvious research frontier in this area of psychoimmunology is to document *the reverse of the Glaser protocol outlined above*. We need to design experimental protocols to assess whether an hypnotic intervention designed to reduce psychosocial stress could lead to a facilitation of the expression of the Interleukin-2 receptor gene and mRNA.

in 2002. In the very same issue of *The American Journal of Clinical Hypnosis* in which this paper appeared there was an abstract of a experimental study by Castes et al., (1999) that took a step in this direction. Castes et al. (1999) reported that a 6-month program of relaxation, guided imagery, and self-esteem workshops with a group of asthmatic children led to significantly reduced number of illness episodes and the use of bronchodilator medication compared with a control group. The experimental group also showed an increase in the gene expression of the T-cell receptor for interleukin-2 (precisely as predicted above as a test of the mind-gene pathway of healing that would be *the reverse of the Glaser protocol*) as well as a significant increase in natural killers cells and other immune system factors associated with psychosocial stress.

Recent advances in genomic technology (Brown, 1999) now make further research on the positive utilization of psychosocial genomics easily within the reach of clinicians and researchers in hypnosis. Gene expression microarrays present up to 10,000 genes (including the interleukin-2 gene) on a single "DNA chip" that can be used to assess the patterns of gene expression associated with varying experimental conditions. A prototype of such research assessing interleukin-2 gene expression as well as the ultradian time dynamics of early, intermediate and late activated genes in human peripheral blood mononuclear cells (including the major cells of the immune system such as T and B-lymphocytes, monocytes, natural killer cells, and dendritic cells) is available on the internet (Incyte, 1999). Most research using DNA chip technology at the present time has been carried out to assess purely biological variables at the molecular level in medical and pharmaceutical studies. We propose that in the near future, however, this DNA chip technology will be adapted as the ultimate test of the efficacy of all psychosocial variables in facilitating mind-body healing and therapeutic hypnosis at the level of gene expression and new protein formation.

**Hypothesis 5:** ***The final common path to healing by the traditional model of physical medicine as well as many therapeutic applications of hypnosis involves the expression of genes coding for the formation of proteins that are the bottom line of healing on the material, energetic and informational levels.***

The source of this hypothesis is in the history of our understanding of the role of proteins in life processes and healing. Originally the albuminoids, called "proteins" by G.J. Mudler in 1838, were described as the structural or *material* substance of life. Proteins were later recognized as playing an essentially *energetic* role as enzymes catalyzing the metabolism of the cell. Currently proteins are understood to function also as *informational and computational* processors of life at the molecular level (Brey, 1995). The nonlinear dynamics of proteins are described as the informational, autopoietic or self-organizational systems of life at the cellular-genomic-protein level (Kauffman, 1995).

*The hypothesis of gene-protein dynamics as the final common path to healing integrates traditional physical medicine with the mind-body models of "alternative or holistic medicine." The facilitation of gene-protein dynamics becomes an important criterion for evaluating all forms of therapeutic communication and healing - biofeedback, body work, emotional catharsis, EMDR, imagery, active imagination, hypnosis, meditation, prayer, ritual, yoga or whatever - with a common yardstick.* Whatever the therapeutic method, we can test whether it has really facilitated healing with relatively simple DNA chip assays to determine whether the appropriate immediate early genes and their target genes are expressed in the form of mRNAs that serve as "blueprints" for the synthesis of proteins. A dramatic example of a psychosocial intervention that can modulate gene expression and growth protein formation is provided by Schanberg (1995) who found that premature and isolated human babies had significant gains in weight development and sympatho-adrenal maturation when they were administered maternal touch. He reports in an animal model (preweanling rat pups) that the absence of nurturing touch suppresses ODC [ornithine decarboxylase] gene transcription by interfering with a cells ability to transduce the hormone receptor activated signal. This is takes place when the lack of adequate maternal behavior (lack of physical contact and touch) leads to a down regulation of specific IEGs for synthesizing the protein for this growth regulating enzyme.

Touch and "therapeutic passes," of course, have a long and honorable place in the history of hypnosis (Edmonston, 1986). It is estimated that 40,000 nurses and caregivers in the United States have eagerly embraced "therapeutic touch" with an erroneous theory attributing its value to "energy fields" (Rosa et al., 1998) in a striking parallelism with the false attribution of healing to magnetism in Mesmerism 200 years ago. Can hypnosis now reclaim its psychophysiological heritage in Braid by investigating the psychobiology of hypnosis at the genomic-protein level? The current research literature implies that both touch and verbal suggestion can initiate immediate early gene protein cascades to facilitate growth and healing. As we shall soon see, further research in this area with the new DNA chip technology could become a new methodology for differentiating between the relative merits of the many approaches to hypnosis and their therapeutic applications.

**Level 3: *Proteins, Time, Trauma, Stress and Healing.***

The time required to make new proteins in adaptive responses to stress and trauma provides an important window into the informational and healing dynamics of mind-body communication at level three in Figure One. Todorov's research (1990) on how cells maintain their stability in response to environmental demands reveals three major phases of gene transcription and translation in response to physical trauma and stress. Immediate-early genes are turned on first to initiate the formation of M (Metabolism) proteins *within a hour* to produce proteins required for energy dynamics (ATP in Figure One). The next R (Ribosomal) block of proteins are turned on to facilitate a heightened state of mRNA translation processes of healing *within a few hours*. The third stage produces the N (Nuclear) proteins such as DNA polmerases and histones required for deeper healing in a slower process requiring *12 hours to a day or more*. The previously cited model of gene expression (Incyte, 1999) is a more recent example of these ultradian time dynamics in the expression of early-activated gene expression (30 minutes to an hour), intermediate-activated genes (one to two hours), and late-activated genes (two to four hours).

A new research frontier for the psychobiological model of hypnosis is to demonstrate how these typical ultradian time stages *physical* stress at the genomic-protein level are found in response to sever *psychological* trauma and stress as well. This response to trauma and stress is but one illustration of the quasi-periodic dynamics for a complete ultradian healing cycle noted in Figures One through Seven. Figure 4a compares the typical ultradian rhythms of ACTH and cortisol which mediate the stress response at the neuroendocrine level in two human subjects during Kleitman's Basic Rest Activity Cycle (90 - 120 minutes) in the normal routines of everyday life. Figure 4b illustrates the ultradian and circadian time parameters of an acute stressor (forced swimming in mice). Notice the initial peak of activity in the hippocampus followed by a later peak of activity in the cortex when symptoms of PTSD typically become manifest as described by Kaufer et al. (1998).

> "Acute traumatic stress may lead to post-traumatic stress disorder (PTSD), which is characterized by delayed neuropsychiatric symptoms including depression, irritability, and impaired cognitive performance. Curiously, inhibitors of the acetylcholine-hydrolysing enzyme acetylcholinesterase may induce psychopathologies that are reminiscent of PTSD. It is unknown how a single stressful event mediates long-term neuronal plasticity. Moreover, no

> mechanism has been proposed to explain the convergent neuropsychological outcomes of stress and of acetylcholinesterase inhibition. However, acute stress elicits a transient increase in the amounts released of the neurotransmitter acetylcholine and a phase of enhanced neuronal excitability. Inhibitors of acetylcholinesterase also promote enhanced electrical brain activity, presumably by increasing the survival of acetylcholine at the synapse. Here we report that there is similar bidirectional modulation of genes that regulate acetylcholine availability after stress and blockade of acetylcholinesterase. These calcium-dependent changes in gene expression coincide with phases of rapid enhancement and delayed depression of neuronal excitability. Both of these phases are mediated by muscarinic acetylcholine receptors. Our results suggest a model in which robust cholinergic stimulation triggers rapid induction of the gene encoding the transcription factor c-Fos. This protein then mediates selective regulatory effects on the long-lasting activities of genes involved in acetylcholine metabolism. . . .*The molecular mechanisms translating a traumatic life experience into long-term neuropsychological sequelae are expected to involve complex changes in gene regulation.*" (p. 373, italics added.)

These ultradian time parameters for gene expression in response to acute stress reported by Kaufer et al. are similar to those reported in the utilization of therapeutic hypnosis to ameliorate trauma and stress (Rossi, 2002). Dabney Ewin (1986), a former president of The American Society of Clinical Hypnosis, has documented, for example, how hypnotic suggestions for cooling administered within two hours (a typical ultradian time BRAC) of a severe burn can reduce inflammation and facilitate healing to a much greater degree than when hypnotic suggestion is used more than two hours after the burn. Research is now needed to assess the hypothesis that the mechanism of this therapeutic application of hypnosis may be in modulating gene expression and the formation of "stress proteins" whose over-production after two hours complicates the healing process in burn patients (e.g., bradykinin-like substances that mediate inflammation, Pardue et al., 1989).

**Level 4:** ***State Dependent Memory, Learning and Behavior (SDMLB).***

Level four of Figure 3.1 illustrates how messenger molecules that have their origin in the processing of larger protein "mother-molecules" such as POMC may be stored within cells as a kind of molecular memory. Messenger molecules from the peripheral cells of the body, such as epinephrine and norepinephrine from the adrenals, are released into the blood stream. There they can complete the SDMLB loop of information transduction from all parts of the body to the brain's neural networks illustrated by the block of letters A through L at the top of Figure One. It has been theorized that messenger molecules can diffuse as much as 15 mm through the extracellular fluid (ECF) to any site in the cerebral cortex (Schmitt, 1984) to modulate memory, emotions and behavior at the cellular-genomic level within neurons of the brain (Routtenberg and Meberg, 1998).

**Hypothesis 6:** ***The state dependent pathways of mind-body communication and healing (SDMLB) encoded by the messenger molecule receptor systems of the CNS, ANS, neuropeptide and immune systems of the psychosomatic network are two way streets. Just as purely biological approaches to the messenger molecules receptor system can modulate cognitive-emotional experience, so can the accessing and focusing of cognitive-emotional experience with hypnosis modulate the psychosomatic network to facilitate healing.***

Neurobiological research indicates that most forms of learning (Pavlovian, Skinnerian, imprinting, sensitization, etc.) utilize messenger molecules of arousal and stress (such as ACTH, epinephrine etc.) in the construction and reconstruction of memory (Izquierdo et al., 1988). Insofar as these classical forms of learning use messenger molecules, they *ipso facto* have a state dependent (SDMLB) component (McGaugh, 1989). SDMLB can be turned on or off by purely biological approaches (injecting a messenger molecule such as ACTH or epinephrine) as well as psychological-behavioral approaches (positive or negative reinforcement with food, electric shock etc.). When memory and new learning is encoded under conditions of high emotional arousal, shock, surprise, stress or trauma, it tends to become state dependent (SDMLB) to that psychobiological condition. This state dependent memory becomes dissociated or apparently "lost" after the person apparently recovers when the stress hormones or drugs are metabolized out of the system. Reactivating stress in another context, however, has a tendency to reestablish the original SDMLB encoding condition to reactivate the cognitions, emotions and behaviors associated with it with varying degrees of memory.

It is important to recognize how SDMLB completes the information transduction loop of mind-body communication in Figure One so that the psychosomatic network that may be accessed and modulated by hypnosis becomes a two way street. One theoretical objection to the idea that the psychosomatic network is a two way street is that the brain has long been regarded as a "privileged organ" that has a "blood-brain barrier" that normally protects the brain from toxic substances that may be circulating throughout the rest of the body. This would imply that many messenger molecules from the body could be blocked from entering the brain by the blood-brain barrier. Recent research, however, has demonstrated that during highly stressful emotional conditions the blood-brain barrier is lowered so that many messenger molecules and other substances can enter the brain (Soreq and Friedman, 1997).

The two way street hypothesis proposes a major mechanism of how the molecules of the body can modulate mental experience as well as how mental experience such as hypnotic suggestion can modulate the expression of genes and other molecules of the body. *SDMLB is the quintessential psychobiological concept that bridges the so-called Cartesian dichotomy between mind and body.* What is most significant about research in SDMLB is that it enables us to study the parameters of "reversible amnesia," which is the fundamental psychobiological phenomenon that theories of hypnosis and psychoanalysis have always tried to explain (Rossi, 1996). Most experiments in SDMLB demonstrate that "reversible amnesia" is only partial (that is, there is usually some memory/learning available even in the dissociated condition after the messenger molecules return to normal levels). Likewise much hypnotic literature documents that hypnotic amnesia is usually fragile and partial in character. Since the earliest days of psychoanalysis it has been noted that a sudden fright, shock, trauma and stress could evoke "hypnoidal states" that were somehow related to amnesia, dissociated and neurotic behavior. A full amnesia that is completely reversible, however, is relatively rare in SDMLB research as well as in psychoanalysis and therapeutic hypnosis (Rossi and Cheek, 1988). In the historical literature of hypnosis and psychoanalysis this same fragile and partial character of reversible amnesia has been responsible for many of the paradoxes of dissociation and memory. SDMLB is the first experimental model that can account for the paradoxes of dissociation in a manner consistent with current neurobiological research on the effects of stress on mind, memory and healing (Rossi and Ryan, 1986).

**Hypothesis 7:** ***The clinical application of hypnosis, psychoanalysis and the therapeutic arts in general engages the patient's state dependent mental experience (SDMLB) that is encoded and continually modulated by the molecular genomic dynamics of the CNS, ANS,*** **neuropeptide and immune systems of** ***the psychosomatic network. This is the psychobiological foundation of the classical concepts of depth psychology such*** **as** ***dissociation, reversible amnesia, repression, and emotional complexes.***

These hypothesized relationships between the molecular genomic dynamics of the psychosomatic network and SDMLB suggest a new research frontier for the psychobiological investigation of classical psychoanalytic concepts of *repression, dissociation, and emotional complexes* that is in keeping with the dissociated control theory of hypnosis (Spiegel, 1998; Woody and Farvolden, 1998) as well as the cortical and subcortical pathways of hypnosis and the placebo response (Price, 1998;

Rainville et al., 1997). A new paradigm for such research has been provided by Cahill et al. (1994) who compared the effects of the beta-adrenergic receptor antagonist propranolol hydrochloride on the long-term memory for an emotionally arousing versus an emotionally neutral short story. Their results were consistent with the hypothesis that the messenger molecules of beta-adrenergic (fight or flight) dynamics of SDMLB mediate the enhanced memory associated with the emotionally arousing metaphors of a short story. Such research could become a more effective psychobiological approach for investigating the relative merits of thc direct and indirect, metaphorical approaches to hypnosis (Erickson et al., 1976)

**Frontiers in the Psychobiology of Hypnosis: Behavioral State-Related Gene Expression and Activity-Dependent Gene Expression in Psychosocial Genomics.**

Many lines of research suggest that immediate-early genes (IEGs) are the newly discovered mediators between nature and nurture at the cellular-genomic level (Merchant, 1996; Rossi, 1996, 1997; Tölle et al., 1995). Immediate-early genes act as transducers allowing signals (sometimes called primary response genes or third messengers) from the external environment to regulate the adaptive transcription of target gene expression at the cellular level. Immediate-early genes can initiate a series of molecular-genomic transformations that can transduce relatively brief signals from the environment into enduring changes in the physical structure of the developing nervous system and the formation of new memory and learning throughout life (Morimoto and Jacob, 1998; Tölle et al., 1995). To coin a metaphor, IEGs seem to function as a "steering committee" mediating between stimuli from the outside world and the inner conditions of the cells in processes of creative adaptation at the genomic-protein level. Novel, arousing or stressful environmental stimuli, for example, can induce the expression of immediate early genes such as c-fos and c-jun and the proteins they code for within neurons of the brain. In response to conditions prevailing within the neurons, the fos and jun proteins associate to form the transcription factor activator protein-1 (AP-1) DNA-binding complex.

As Bentivoglio and Grassi-Zucconi (1999, p. 236), who have done considerable laboratory research on the role of IEG expression in sleep and wakefulness describe it, "The AP-1 recognition sequence is found in a variety of promoter regions of target genes and modulates their transcription, thus mediating neuronal responses to many different extracellular signals." As we have seen above, many of these "extracellular signals" have their ultimate origin from psychobiologically arousing signals from the outside world that activate the expression of immediate early genes. We now need research to document how such extracellular signals that activate IEGs may include the heightened expectancy provided by the psychobiologically arousing and arresting psychosocial cues of hypnosis. This orchestration of target gene expression by outer world psychosocial signals is translated into the formation of proteins that carry on the adaptive life dynamics of *energy, information*, and *structure* at the molecular level - what is commonly called "stress," "work" and "healing" by the psychotherapist. "Behavioral state-related gene expression" (Bentivoglio and Grassi-Zucconi, 1999, p. 245) via IEGs such as c-fos and c-jun, among others, is associated with a wide range of normal behavioral states of consciousness, sleep and dreaming as well as heightened states of psychological arousal, novelty, physical stress and trauma.

While more than 100 IEGs have been reported, many of their functions in activating the target genes associated with behavioral states in health and illness remain unknown. The complex range of interrelated biological and psychological functions that immediate-early genes are already known to serve, however, recommends a central role for IEGs and their target genes in the deep psychobiology of hypnosis that can now be explored with the new DNA chip technology. Hypnosis has been characterised as a continuum of mind-body states (DeBenedittis et al., 1994; Rossi, 1996). Mass screenings of large samples of subjects of varying degrees of hypnotic susceptibility with the new DNA chip technology would be an ultimate way of precisely specifying what we mean by defining hypnosis as an altered state. We would expect that varying patterns of genes expression would be associated with the continuum of phenomenological states we call hypnosis.

Similar mass screenings of medical patients with a variety of stress related dysfunctions would reveal what patterns of gene expression are associated with the various psychosomatic problems. It would then be a straight forward task to match the patterns of gene expression in hypnosis with those found in stress related medical conditions to determine which patients are most likely to benefit from hypnosis. It should not escape our notice that this would provide a new scientific criterion for assessing the therapeutic efficacy of hypnosis and related approaches to mind-body healing. Whatever the approach to medical, psychological or "alternative" healing the crucial question to answer remains the same: What patterns of gene expression are facilitated to optimize what pathway of mind-body healing? This question is of central importance in rehabilitation medicine.

***Rehabilitation via Activity-Dependent Gene Expression*.**

Facilitating gene expression, neurogenesis, and stem cell differentiation and maturation into newly functioning tissues via activity-dependent cognitive-behavioral approaches is a basic mechanism of healing that makes rehabilitation possible. Patients with severe trauma resulting in loss or paralysis of sensory-motor functions due to physical injury, cardio-vascular accidents, stroke, etc., for example, can recover their abilities via occupational and physical therapy that works primarily by *activating their behavior*. Until recently the molecular-genomic mechanism of rehabilitative medicine by behavioral activation has not been understood. *Our new neuroscience hypothesis is that behavior activation initiates activity-dependent gene expression. This activity-dependent gene expression initiates healing by the generation of proteins that facilitate neurogenesis and stem cell differentiation into new tissues.* We do not yet know the full range and limitations of this new neuroscience cognitive-behavioral approach to healing and rehabilitation. Intense research efforts are now underway to discover the optimal conditions whereby cognitive-behavioral stimulation - and presumably therapeutic hypnosis - can evoke *activity-dependent gene expression, neurogenesis and physical healing in rehabilitative medicine* (Johansson, 2002).

**Hypothesis 8:** ***The therapeutic applications of hypnosis facilitate the psychosocial genomics of encoding new experience, memory, learning and behavior at the neural-genomic-protein level in the quasi-periodic ultradian/circadian time parameters of long term potentiation.***

Most arousing environmental stimuli that have been studied can induce IEGs within minutes in neurons of the CNS, their concentrations typically peak within fifteen to twenty minutes and their effects usually last for an hour or two. These are the same time parameters of the psychobiological model of memory and learning called "Long Term potentiation" (LTP). It is now known that LTP takes place in many regions of the brain associated with stress and emotional learning (McKernan and Shinnick-Gallagher, 1997). LTP operates in the same quasi-periodic ultradian time frame of about 90-120 minutes (Bailey et al. 1996) that is typical of many dynamical processes of mind-body healing and hypnosis discussed above. We hypothesize that changes in the orchestration of patterns of gene transcription and new protein formation initiated by the therapeutic applications of hypnosis could lead to lasting changes in the central nervous system by converting short term memory to long lasting learning and behavior (Tully, 1996).

IEGs are now used as markers or indicators of changes in neuronal activity in depression and schizophrenia. Anti-psychotic drugs are currently being designed to modulate the effects of IEGs on pathways leading to the production and utilization of neurotransmitters such as dopamine, serotonin and noradrenaline that are implicated in the "dopamine hypothesis" of schizophrenia (Merchant, 1996). Most drugs dealing with pain as well as related addictive drugs such as cocaine, amphetamine and the opiates are also mediated by immediate-early genes. A few brief quotes from Ziegelgaensgerger (1998) at the Max Planck Institute of Psychiatry, Munich , Germany outlines many of the molecular pathways by which the psychological experience of pain evokes activity dependent gene expression as follows.

> "The earliest short-term responses following nociceptor [PAIN] activation are reflected in rapid changes of neuronal discharge activity in a variety of pharmacologically and anatomically distinct systems in the central nervous system. *In these systems long-term changes most commonly require alterations in gene expression. The activity dependent modulation of gene expression is a characteristic feature of highly integrated systems such as the pain matrix*...Each of the levels of integration of nociceptive information probably receives and is the origin of modulatory mechanisms conveyed by afferent segmental or descending pathways. Besides "classical" neurotransmitters, biologically active molecules such as peptide hormones, neurosteroids, adenosine, trophic factors or cytokines released synaptically or non-synaptically from terminals, neighboring neurons, glia cells or components of the immune system or from the circulation participate in the integration of somatosensory information...A major facilitator effect of the central nervous system responding to noxious stimuli involves the interaction between L-glutamate and substance P, a neuropeptide long thought to have a role in pain perception. *Immediate early genes (IEGs) are thought to participate as third messengers in the late phase of the stimulus-transcription cascade. They code for transcription factors and alter gene expression* and translation into the corresponding protein products such as enzymes, receptors or neurotransmitters. The amount of several IEG coded proteins, produced by central neurons, is proportional to the degree of synaptic excitation following somatic and visceral acute noxious stimulation and is reduced by morphine application before the stimulation. Protein phosphorylation of ligandgated channels appears as *a major mechanism in the regulation of neuronal plasticity." (Italics added here)*

The implication of Ziegelgaensgerger's summary of the many levels involved in pain experience from mind to molecule is that immediate early genes are central in mediating the psychobiology of pain and its associated emotions such as depression and behavioral addictions. There are as yet no studies of the effects of hypnosis on IEGs but recent reviews on the relation of c-fos and nerve growth factor-induced A (NGFI-A) in the wake-sleep cycle (Bentivoglio and Grassi-Zucconi, 1999) implies they may be related to the quasi-periodic parameters of hypnosis. It has been found, for example, "that the expression of c-fos during waking is strictly dependent on the level of activity of the noradrenergic system...high levels of c-fos during forced and spontaneous waking and...low levels during sleep" (Cirelli et al., 1998, pp. 46). While most research has been done with animals, it is tempting to hypothesize that stimulation of the noradrenergic system and IEG expression may be the molecular-genomic basis of Braid's (1855/1970) "psychophysiology of fascination" as a pathway to healing in the clinical applications of hypnosis. This new field of mind-gene communication that we might call "psychosocial genomics" – how psychosocial cues can signal mind-body responsiveness at the level of gene transcription and the formation of new proteins in health and optimal performance as well as illness – has very broad implications for a new understanding of the phenomenology of creative life experience (Rossi, 1972/1985/2000, 2002).

**Hypothesis 9**: ***Enriching life experiences that evoke psychobiological arousal with positive fascination and focused attention during creative moments of art, music, dance, drama, humor, spirituality, numinosity, awe, joy, hope, expectation, and social rituals can evoke immediate early gene protein cascades to optimize neurogenesis, mind-body communication, and healing via the differentiation of stem cells into new, healthy tissues by the therapeutic applications of hypnosis.***

The entire history of holistic approaches to healing including ancient and modern spiritual rituals of exorcism, shamanism, fire walking and the still "mysterious methods" of acupuncture, body work, therapeutic touch, and biofeedback that evoke a positive experience of wonder and expectation are the experiential data base for this hypothesis about the role of psychological arousal and fascination in the clinical applications of hypnosis (Rossi and Nimmons, 1991). We speculate that psychobiological healing during ecstatic religious experiences of the "*numinosum,*" consisting of a combined sense of *fascination*, the *mysterious* and the *tremendous* (Otto, 1923/50), has much in common with traditional and modern rituals of healing associated with the self-help groups, twelve step programs and the so-called "miracle cures" in clinical demonstrations of hypnosis (Barber, 1990). We hypothesize that just as negative states of emotional arousal can evoke the CNS, ANS, and neuropeptide systems of the psychosomatic network to initiate an IEG cascade leading to the synthesis of stress proteins and illness, so can positive psychological experiences initiate cascades of healing at the IEG protein level. This implies that the experience of positive fascination, novelty, mystery, surprise and insight experienced in hypnosis (Rossi, 1973) could access and facilitate an IEG cascade leading to the synthesis of healing proteins (Rossi, 1972/1986/2000).

In mature mice and primates the experience enriched environments and novelty initiate IEG cascades leading to the formation of new proteins and 15% more granule cell neurons along with increased synapses and dendrites in the dentate gyrus of the hippocampus that encode memory and learning (Kempermann et al., 1997; Gould et al., 1998). When combined with voluntary exercise such as running the number of newborn cells is doubled (Van Praag et al., 1999). Neurogenesis (the growth of new neurons) is now documented as taking place in the adult human hippocampus (Eriksson et al., 1998; Gould et al., 1999) as well. The significance of such experimental findings is that it provides an urgent rational for our new psychobiological paradigm to investigate the therapeutic applications of hypnosis in facilitating neurogenesis in the hippocampus of the human brain as well as mind-body healing at the cellular-genomic-protein level throughout the body. This implies a more general hypothesis about the role of IEGs in the positive experiences of arousal in high phase hypnosis and the placebo response.

Current research suggests that stem cells normally reside quietly in most tissues of the mature organism. When there is a traumatic injury or stress that damages healthy cells, stem cells are able to receive molecular signals to differentiate and produce new health cells to replace those that have been injured and died. Stem cells thus function as "mother natures menders" (Vogel, 2000). Stress, trauma, injury, and diseases of many types leave a trail of molecular signals that activate the gene expression-protein synthesis cycle in the stem cells still residing in malfunctioning tissues. Nishimura et al. (2002) have summarized their research on the differentiation of stem cells into hair follicles as a clear example of how stem cells respond to molecular signals in their immediate environment (their niche).

> Stem cells—which have the capacity to self-renew and generate differentiated progeny—are thought to be maintained in a specific environment known as a niche. The localization of the niche, however, remains largely obscure for most stem-cell systems. Melanocytes (pigment cells) in hair follicles proliferate and differentiate closely coupled to the hair regeneration cycle. Here we report that stem cells of the melanocyte lineage can be identified, using *Dct-lacZ* transgenic mice, in the lower permanent portion of mouse hair follicles throughout the hair cycle. It is only the population in this region that fulfils the criteria for stem cells, being immature, slow cycling, self-maintaining and fully competent in regenerating progeny on activation at early anagen (the growing phase of hair follicles). Induction of the re-pigmentation process in *K14-steel factor* transgenic mice demonstrates that a portion of amplifying stem-cell progeny can migrate out from the niche and retain sufficient self-renewing capability to function as stem cells after repopulation into vacant niches. Our data indicate that the niche has a dominant role in the fate determination of melanocyte stem-cell progeny.

A more recent example of the use of stem cells is to facilitate recovery in a mouse model of multiple sclerosis (Steinman, 2003; Pluchino et al., 2003). It was found that injected adult neural precursor cells ("neurospheres" or adult stem cells) into the blood stream or spinal fluid can find their way to neurons damaged by multiple sclerosis and repair damage to the brain and spinal cord.

*I now hypothesize that the molecular messengers generated by stress, injury, and disease can activate immediate early genes within stem cells, so that they then signal the target genes required to synthesize the proteins that will transform (differentiate) the stem cells into mature, well-functioning tissues.* These new tissue cells can then replace injured, aging, and dysfunctional cells that die by a process of apoptosis—so-called "cell suicide" that takes place due to senescence, stress, injury, genetic mutations, etc. (McLaren, 2000; Temple, 2001). Research is now needed to document the pathways by which molecular messengers associated with injury and stress and their modulation by therapeutic hypnosis could facilitate the differentiation of stem cells leading to healing during positive psychological experiences.

**Hypothesis 10: *The Novelty Immediate Early Gene Protein Placebo Effect. Placebos entrain pathways of mind-body communication between positive psychosocial expectation and the state dependent dynamics of the CNS, ANS, and neuropeptide systems of the psychosomatic network that triggers immediate early gene cascades leading to the synthesis of proteins at the structural, energetic and informational levels of healing in stem cells of the brain and body.***

Placebo research is consistent in finding about 30% of subjects report themselves as experiencing a therapeutic benefit (Harrington, 1997; Quitkin et al., 1996). With psychiatric drugs such as the selective serotonin uptake inhibitors there is a relapse rate of only 45% after one year. This means that 55% of the subjects reported a significant placebo response even after one year (Moller and Volz, 1996). Is the placebo response merely wishful thinking or does this research reflect a 30% to 55% psychobiological healing effect that is measurable on the cellular-genomic-protein level? Since it is now known that psychobiological arousal, such as that provided by an enriched environment, can generate an IEG response, we propose that a "*novelty immediate early gene protein placebo effect*" could generate an IEG cascade leading to the synthesis of proteins for a genuinely healing placebo response at the cellular-genomic level within stem cells (Rossi, 1997, 2002).

We hypothesize that a heightened, positive sense of fascination, emotional arousal and expectation associated with a novel, brightly colored sugar pill can be just as effective as a new, mysterious therapeutic ritual introduced by a healer coming from a far away country. It is known that increased corticotrophin releasing factor (CRF mRNA) appears in the paraventricular nucleus of the hypothalamus to initiate an adaptive arousal response within minutes of experiencing an acute physical or psychological stressor (Brush and Levine, 1989; Lloyd and Rossi, 1992). We propose that a placebo can initiate these psychobiological dynamics to entrain the quasi-periodic parameters of a healing response at the cellular-genomic-protein level within stem cells. This accounts for the fact that the placebo response appears within minutes but may also disappear when its novelty effect wears out - just as is the case when the mind-body CRF-POMC-ACTH-Cortisol-Gene-Protein cascade is metabolized out of the system when novelty or a stressor is removed. Research with the new DNA chip technology is now needed to explore the parameters of how enriched life experiences and the novelty immediate early gene protein placebo effect could be facilitated with the therapeutic applications of hypnosis.

A recent brain imaging study by Petrovic et al. (2002) documents how placebos engage the same brain circuits as pain killing drugs.

> "It has been suggested that placebo analgesia involves both higher order cognitive networks and endogenous opiate systems. The rostral anterior cingulated cortex (ACC) and brainstem are implicated in opioid analgesia, suggesting a similar role for these structures in placebo. In this [PET] study, we confirm that both opioid and placebo analgesia are associated with increased activity in the rostral ACC. We also observe a covariation between activity in rostral ACC and brainstem during both opioid and placebo analgesia, but not during the pain only condition. These findings indicate a related neural mechanism in placebo and opioid analgesia." (p. 923)

Since we now know that pain and analgesia both engage gene expression (Rossi, 2002), this study is a model for future psychogenomic research as implied in Holden's (2002) commentary on the Petrovic study.

> "Both the genuine analgesic and the placebo led to increased blood flow in areas of the brain known to be rich in opioid receptors: the brainstem and the rostral anterior cingulated cortex (ACC), which exchanges information with a network of brain regions, including the orbitofrontal cortex, a relatively sophisticated part of the brain know to process emotions. Furthermore, those people who responded most to the placebo – according to their ratings on a scale of 0 to 100 of how much it reduced pain – also showed more rostral ACC activation from the drug. *This... provides new fodder for the hypothesis that 'high placebo responders have a more efficient opioid system.* " (p. 947, italics added).

The use of a 0 to 100 scale in the Petrovic study is identical to the use of "symptom scaling" to facilitate and assess the analgesic and healing effects of therapeutic hypnosis (Rossi, 200 2). The Petrovic study contradicts the implications of a previous generation of researchers that failed to relate hypnosis to the placebo and opioid analgesia (Hilgard and Hilgard, 1983). More recent research by Rainville et al. (1997, 1999), however, is consistent with Petrovic in documenting how brain areas that mediate pain include the human anterior cingulate, which is implicated in the experience of hypnosis. These studies indicating the same brain localization of the analgesic effects of molecules, placebo and therapeutic hypnosis are consistent with the emerging science of psychosocial genomics that could unify modern molecular medicine with alternative and complementary medicine into a unified theory of mind-body healing via therapeutic hypnosis, the placebo and psychotherapy in general.

* * * *

We have reviewed how many of the problems and paradoxes of historical hypnosis and the impasse of current theory on the cognitive-behavioral level could be resolved by a deeper understanding of the natural psychobiological time parameters of mind-body communication and healing on the cellular-genomic-protein level. We propose an expansion of the psychological domain of therapeutic hypnosis to include the utilization of the entire cybernetic loop of information transduction between the psychosocial environment, the central nervous system and the psychosomatic networks of the autonomic, neuroendocrine and immune systems. Four levels of the general process of mind-body communication via messenger molecules and their receptors have been integrated into a broad picture of how psychosocial experiences such as hypnosis and the placebo response could initiate immediate early gene protein cascades to facilitate healing, memory, learning, creativity and behavior. Ten hypotheses integrating the traditional model of physical medicine with the psychobiological dynamics of therapeutic hypnosis are outlined as a guide for research using the new DNA chip technology to trace the pathways of mind-body communication on all levels from the cognitive-emotional to the cellular-genomic. The implication of such research is that the therapeutic applications of hypnosis in the new millennium may include the facilitation of neurogenesis in the human brain as well as mind-body healing and rehabilitation at the cellular-genomic-protein level throughout the body.

# IV
# BRAIN PLASTICITY, GENE EXPRESSION AND HEALING IN TERAPEUTIC HYPNOSIS

The historical lineage of therapeutic hypnosis in James Braid's "psychophysiology", Pierre Janet's "physiological modification", and Milton Erickson's "neuro-psycho-physiology" is extended to include current neuroscience research on brain plasticity (growth) that engages activity-dependent gene expression, neurogenesis, and stem cells in memory, learning, behavior change, and healing. Three conditions that optimize gene expression and neurogenesis—novelty, environmental enrichment, and exercise—could integrate fundamentals of the theory, research, and practice of therapeutic hypnosis. Continuing research on immediate-early, activity-dependent, behavior state-related, and clock gene expression could enhance our understanding of how relaxation, sleep, dreaming, consciousness, arousal, stress and trauma are modulated by therapeutic hypnosis. It is speculated that therapeutic and post-hypnotic suggestion could be focused more precisely with the time parameters of gene expression and neurogenesis that range from minutes and hours for synthesizing new synapses to weeks and months for the generation and maturation of new, functioning neurons and their synaptic connections in the adult brain.

## The Historical Continuum of Therapeutic Hypnosis and Neuroscience

Research in current neuroscience indicates, contrary to 100 years of dogma, that the human brain is capable of generating new brain cells throughout the life cycle (Gross, 2000). *Novelty, environmental enrichment, and exercise* can activate gene expression leading to the differentiation of new neurons from neural stem cells in the adult mammalian brain during salient life experiences that are associated with the generation and reconstruction of memory, learning, and behavior (Eriksson et al., 1998; Gage, 2000a, 2000b; Van Praag et al., 2002). Kempermann and Gage (1999) have summarized the implications of such neuroscience research:

> Contrary to dogma, the human brain does produce new brain cells in adulthood... With continued diligence, scientists able to trace the molecular cascades that lead from a specific stimulus, be it an environmental cue, or some internal event, to particular alterations in genetic activity and, in turn, to rises or falls in neurogenesis.
> Then they will have much more of the information needed to induce neuronal regeneration at will. Such a therapeutic approach could involve administration of key regulatory molecules or other pharmacological agents, delivery of gene therapy to supply helpful molecules, transplantation of stem cells, *modulations of environmental or cognitive stimuli,* alterations in physical activity, or some combination of these factors (p. 53, italics added).

With the words "modulations of environmental or cognitive stimuli" Kempermann and Gage are intuiting a new neuroscience update of what many would regard as the historical goal of therapeutic hypnosis. James Braid (1855/1970), for example, described the "Physiology of Fascination", an early concept of therapeutic hypnosis, in this way:

> With the view of simplifying the study of reciprocal actions and reactions of mind and matter upon each other... the [hypnotic] condition arose from influences existing within the patient's own body, viz., the influence of concentrated attention, or dominant ideas, in modifying physical action, and these dynamic changes re-acting on the mind of the subject. I adopted the term "hypnotism" or nervous sleep for this process... And finally as a generic term, comprising the whole of these phenomena which result from the reciprocal actions of mind and matter upon each other, I think no term more appropriate than "psychophysiology" (pp. 369-372).

Because of the limitations of psychophysiological techniques of Braid's time, however, his concept of therapeutic hypnosis remained a theoretical hope rather than a verifiable method to guide clinical practice. If we are willing to intuit an association between what Braid called "The Physiology of Fascination" and what now is called *Novelty* by current neuroscientists, however, we can recognize a deep psychobiological lineage between the historical foundations of hypnosis (Tinterow, 1970) and current neuroscience research on gene expression and neurogenesis in memory, learning, and healing.

This psychobiological lineage was reviewed and expanded by Pierre Janet's (1919/1925) historical and clinical study of psychological healing:

> The most important phenomena that have been adduced to substantiate the alleged power of hypnotic suggestion are the *physiological modifications which suggestion is said to have caused.* Normally the power we can exercise over our visceral organs and our physiological functions is greatly restricted, and is usually indirect. Hypnotic suggestion, we are told, can influence these organs markedly and directly. A precise verification of these phenomena, and the discovery of the exact mechanisms of their causation, would do a great deal to demonstrate the power of suggestion (Pp. 308-309, italics added).

Janet also noted:

> This historical summary, though cursory and incomplete, is an attempt to display the remarkable evolution of the methods of treatment by gymnastics and education. At the onset, the endeavor was to treat by these means various elementary and circumscribed disorders affecting well-known motor-functions. Then the claims of the gymnastics were enlarged, and they were applied as remedial measures to functions of a more lofty and complicated character. The aim was to practice the gymnastics of emotion and attention just as at an earlier stage there had been practiced the gymnastics of walking and speech... The treatments we have just been discussing were considered under the caption "Education and Re-education" (Pp. 732).

If we are willing to sense and feel an essential similarity between Janet's physiologically oriented concept of "education and reeducation" and "gymnastics" with what neuroscience currently calls "environmental enrichment" and "exercise" in modulating gene expression and stimulating neurogenesis, we may be approaching Janet's scientific ideal about how "the discovery of the exact mechanisms of their causation, would do a great deal to demonstrate the power of suggestion" (Janet 19197 1925. p. 309).

Milton Erickson (1948/1980) formulated a more recent expression of this historical lineage when he described the "neuro-psycho-physiological" dynamics of "hypnotic psychotherapy," in this way:

> Direct suggestion is based primarily, if unwittingly, upon the assumption that whatever develops in hypnosis derives from the suggestions given. It implies that the therapist has the miraculous power of effecting therapeutic changes in the patient, and disregards the fact that *therapy results from an inner re-synthesis of the patient's behavior achieved by the patient himself.* It is true that direct suggestion can effect an alteration in the patient's behavior and result in a symptomatic cure, at least temporarily. However, such a "cure" is simply a response to the suggestion and does not entail that re-association and reorganization of ideas, understandings, and memories so essential for an actual cure. *It is this experience of re-associating and reorganizing his own experiential life that eventuates in a cure, not the manifestation of responsive behavior, which can, at best, satisfy only the observer...*
> *By such indirect suggestion the patient is enabled to go through those difficult inner processes of disorganization, reorganization, reassociating, and projecting of inner real experience to meet the requirements of the suggestion and thus the induced anesthesia becomes a part of his experiential life instead of a simple, superficial response...* [M.H.E. illustrates this with a clinical case.] Accordingly, the suggestion was offered that she might develop a generalized anesthesia in terms of her own experiences when her body was without sensory meaning to her. *This suggestion was intentionally vague since the patient, knowing the purpose of the hypnosis, was enabled by the vagueness of the suggestion to make her own selection of those items of personal experience that would best enable her to act upon the suggestion.* She

> responded by reviewing mentally the absence of any memories of physical stimuli during *physiological sleep,* and by reviewing her dreams of walking effortlessly and without sensation through closed doors and walls and floating pleasantly through the air as a disembodied spirit looking happily down upon her sleeping, unfeeling body. By means of this review she was able to initiate a process of reorganization of her experiential life. As a result she was able to develop a remarkably effective anesthesia, which fully met the needs of the subsequent delivery. *Not until sometime later did the therapist learn by what train of thought she had initiated the neuro-psycho-physiological process by which she achieved anesthesia* (pp. 38-39, italics added).

This quote is an example of the essential dynamics of what Erickson called the *naturalistic and utilization approach* to the *neuro-psycho-physiological* dynamics of therapeutic hypnosis where the ultimate locus of control remains within the patient rather than the therapist (Erickson, 1958/1980, 1959/1980). Erickson encouraged his early colleagues (Gorton, 1949, 1957, 1958) to pursue this neuro-psycho-physiological approach that is currently developing a variety of successful experimental paradigms that are documenting the clinical efficacy of hypnosis (Barabasz, 2002; Gruzelier, Smith, Nagy, & Henderson, 2001; Rainville, Duncan, Price, Carrier, & Bushnell, 1997, 1999). Current neuroscience, however, is extending the domain of psychophysiology far beyond its traditional boundaries, to include what is now called "gene expression" and "neurogenesis" in the construction and re-construction of memory throughout a lifetime.

## Fundamentals of Gene Expression and Neurogenesis in Neuroscience

That human experiencing on a psychological level is so intimately associated with gene expression on a deep biological level is one of the most surprising findings of current research in genomic neuroscience (Chin & Moldin, 2001; Pfaff, Berrettini, Joh, & Maxson, 2000). Until now, most research in the Human Genome Project (Collins, 2000) has been conducted on the purely biological and biochemical levels to uncover genes and their molecular structure. As this initial structural phase of the Human Genome Project (Collins, 2000) approaches a conclusion with the identification of a currently estimated 35,000 human genes, however, attention is now shifting to the second or *annotation phase* that explores the functions of genes. *Functional genomics* explores how networks of genes are turned on and off in response to signals from all parts of the body as well as the outer environment. *The popular but over-simplistic idea of genetic determinism is incomplete: It needs to be amended to include the complementary concept that human experience and behavior can modulate gene expression* (Moore, 2001; Ridley, 1999, 2001). While there is as yet no direct experimental documentation of how psychotherapy can modulate gene expression, Stahl (2000) summarizes the implications of current research:

> But can behavior modify genes? Learning as well as experiences from the environment can give rise to changes in neural connections. In this way, human experiences, education, and *even psychotherapy may change the expression of genes* that alter the distribution and strength of specific synaptic connections. *Thus genes modify behavior and behavior modifies genes.* Psychotherapy may even induce neurotropic factors to preserve critical cells and innervate new therapeutic targets of alter emotions and behaviors (p. 37, italics added).

In a series of pioneering papers, Eric Kandel (1983, 1989, 1998, 2000), the Nobel Laureate for Physiology in 2000, documented how the experience *of activity-dependent gene expression* could account for many of the classical forms of memory and learning as well as clinical phenomena of psychopathology such as chronic anxiety depression, neurosis, and schizophrenia. Kandel's (1998) theoretical and experimental work is one of the clearest and most optimistic expressions of the possibilities of using the associations between gene expression, neurogenesis, and human experiencing as a new psychobiological foundation of psychosomatic medicine and psychotherapy. Kandel (1998) makes an important distinction between (1) classical Mendelian genetics that is concerned with the transmission of hereditary traits from parents to children from one generation to another, and (2) the developmental and adaptive "regulation of gene expression by social factors" (p. 460). The latter is fundamental for understanding the possible role of gene expression, neurogenesis, and psychosocial genomic healing in therapeutic hypnosis (Rossi, 2002a, 2002b).

Evidence possibly suggestive of hypnotizability as a hereditary trait that is transmitted from parents to children by classical Mendelian genetics was noted by Hilgard and Hilgard( 1983):

> A careful study showed that this possibility cannot be ruled out, because there is a low but positive correlation between parent and child, and because identical twins were indeed more alike in their hypnotic responsiveness than either fraternal pairs or non-twin siblings. A study of this kind cannot be fully decisive because identical twins, who look alike are always of the same sex, share a more common environment than fraternal twins. However, the results leave open the possibility of a hereditary component (p. 10).

In addition to this still open question of a (1) Mendelian hereditary component of hypnosis, I propose an extension of Kandel's concept of (2) the everyday modulation of gene expression by social factors, counseling, and psychotherapy to include therapeutic hypnosis. I speculate that therapeutic hypnosis involves "social influences" that can modulate gene expression during salient life experiences but "they are not incorporated in the sperm and egg and therefore are not transmitted genetically" (Kandel, 1998, p. 460) from parent to as is the case with in classical Mendelian genetics.

While current terminology is still fluid, for the purposes of this paper, I will distinguish between (1) the term *genetics* that refers to the Mendelian transmission of hereditary traits across generations that is the subject of *behavioral genetics* and (2) the term *genomics* that refers to the developmental and adaptive modulation of gene expression by many environmental stimuli, stressors, and psychosocial factors *within the individual's daily life experience* (Ridley, 1999, 2001).

Because this paper is primarily concerned with the possibility of the *psychosocial* modulation of gene expression by therapeutic hypnosis, I will focus primarily on what I call *psychosocial genomics—the emerging science of how sensory, psychological, social, and cultural signals and stressors can modulate gene expression and vice versa in the deep psychobiology of* health and illness (Rossi, 1990, 1994,1999,2000b, 2000c, 2001a, 2001b, 2002a 2002b).

Five areas of research exploring this new perspective of psychosocial genomics are briefly outlined here together with my current speculations about their possible association with therapeutic hypnosis (Rossi, 2002a, 2002b). This initial effort to define the domain of psychosocial genomics as the deep psychobiological foundation of therapeutic hypnosis as well as complementary and alternative medicine is not complete in any sense, however.

## 1. Immediate Early Gene Expression, Psychosocial Cues, and Adaptation

A special class of genes—called *immediate early genes*—can respond to psychosocial cues and significant life events in an adaptive manner within minutes (Cirelli, Pompeiano, & Tanoni, 1996,1998). Bentivoglio and Grassi-Zucconi (1999) have summarized their research on the dynamics of immediate early genes (lEGs) and psychological experience as follows:

> The study of [lEGs] indicates that sleep and wake, as well as synchronized and desynchronized sleep [REM state dreaming], are characterized by different genomic expressions, the level of lEGs being high during wake and low during sleep. Such fluctuation of gene expression is not ubiquitous but occurs in certain cell populations in the brain... IEG induction may reveal the activation of neural networks in different behavioral states. Although stimulating, these findings leave unanswered a number of questions. Do the areas in which lEGs oscillate during sleep and wake subserve specific roles in the regulation of these physiological states and in a general "resetting" of behavioral states? Is gene induction a clue to understanding the alternation of

> sleep and wake, and REM and non-REM sleep?... Could behavioral state-related IEG induction underlie, at least in part, learning mechanisms? The oscillation of IEGs affects the expression of target genes, and thus brings about other questions: May the transcriptional cascade explain the biological need and the significance of sleep? Does this explain the molecular and cellular correlates of arousal, alertness, and, more in general, of consciousness? We do not have final answers to these questions (p. 249).

I now speculate that these profound questions raised by biological researchers regarding the psychosocial genomics of consciousness, dreaming, sleep, and the "resetting of behavioral states" have significant research implications for the state concept of hypnosis (Barabasz, 2002). Immediate early genes have been described as the newly discovered mediators between nature and nurture: they receive signals from the environment to activate the genes that code for the formation of proteins, which then carry out the adaptive functions of the cell in health and illness (Rossi, 1996, 2000c, 2002a, 2002b). From this deep psychobiological perspective it seems reasonable to explore how immediate early genes are likely to be key players in the induction, deepening, and therapeutic phases of hypnosis.

## 2. Clock Genes, Mind-Body Rhythms, and Psychoneuroimmunology

Psychosocial genomics finds a major experimental database in the so-called "clock genes" that coordinate the chronobiology of life at the cellular level (Dunlap, 1999). In a recent review Reppert and Weaver (2002) have discussed the relevance of the circadian timing of clock genes for understanding human behavior:

> Defining the molecular basis of circadian timing in mammals has profound implications. In terms of fundamental brain mechanisms, the circadian system is among the most tractable models for providing a complete understanding of the cellular and molecular events connecting genes to

> behavior. Through dissection of the genetic basis of circadian behavior may help to decipher the connections for more complex behaviors. Understanding the molecular clock could increase our knowledge of how gene mutations of the molecular clock could contribute to psychopathology (for example, major depression and seasonal affective disorder). Similarly, such understanding should lead to new strategies for pharmacological manipulation of the human clocks to improve the treatment of jet lag and aliments affecting shift workers, and the clock-related sleep and psychiatric disorders (pp. 935-936).

Clock genes set the time cycles of many common everyday psychobiological states such as waking, sleeping and dreaming. The time parameters of the gene expression-protein synthesis cycle set the pace and rhythm of many patterns of mind-body communication, health, and healing.

The fundamental life processes of metabolism, homeostasis, growth, energy, information flow, behavior, memory, and healing in health and illness are invariably associated with the rhythms of gene expression and the synthesis of their "cognate" proteins over circadian (24 hours) and ultradian time frames (less than 20 hours [Kleitman & Rossi, 1992; Lloyd & Rossi, 1992,1993; Rossi, 2002a])— typically Kleitman's 90-120 minute basic rest-activity cycle.

The fact that Erickson's hypnotherapeutic sessions typically lasted an hour and a half or two is anecdotal evidence for a possible association between the typical time frames for the gene expression—protein synthesis cycle and the effectiveness of his therapeutic approaches (Rossi, 1986,200 la, 2002a, 2002b).

There is a wide range of time frames for the activation and expression of many clock genes associated with the central nervous system, the endocrine system and the immune system. *Early activated genes,* for example, reach a peak of expression in about an hour, while *intermediate activated genes* peak in about two hours. Early activated genes are of special interest because they mediate some of relationships between mind and body in *psychoneuroimmunology.*

Psychosocial stress, for example, can turn off the early-activated interleukin-2 gene so that the immune system cannot communicate well and we are left more vulnerable to all sorts of opportunistic infections (Glaser et al., 1990; Glaser, Lafuse, Bonneau, Atlinson, & Kiecolt-Glaser, 1993; Kiecolt-Glaser, Marucha, Atkinson, & Glaser, 2001).

Positive psychosocial experiences, on the other hand, can turn on the interleukin-2 gene within an hour or two to facilitate molecular communication, healing, and health (Castes et al., 1999).

*Late activated genes,* by contrast, require as much as four hours to reach their peak levels of expression. The turning on and off of cascades of gene expression begins within a few minutes of receiving important psychobiological signals and may continue for hours, days, weeks—or even a lifetime (Incyte, 1999; Kaufer Friedman, Seidman, & Soreq, 1998; Rossi, 1999, 2002a, 2002b).

Research is now needed to explore the significance of these ultradian time parameters of gene expression for the efficacy of therapeutic hypnosis in psychoneuroimmunology (Gruzelier et al., 2001; Kiecolt-Glaser et al., 2001, Rossi, 2002a).

## 3. Behavioral State-Related Gene Expression in the Sleep-Awake Continuum

Bentivoglio and Grassi-Zucconi (1999) introduced the concept of "behavior state-related gene expression" to describe their research on immediate early genes in the molecular dynamics of waking, sleeping, and dreaming.

> *In relation to behavioral state-related gene expression, it is interesting to note that Fos [the protein coded for by the immediate early gene c-fos] is induced during wake in thalamic "non-specific" structures, in which it increases after a sustained wake, thus paralleling changes in the physiological activity of these neurons...* In the brainstem, Fos-immunostained neurons were found after a period of wake in the dorsal raphe, locus coeruleus and mesopontine nuclei, central gray, and superior and inferior colliculi. Such activation of the tectal neurons could be related to their role in *sensory processing and in the integration of multi-sensory and sensorimotor functioning. On the other hand,*

> *the central gray is involved in emotional behavior-related sensory processing. The role of state-dependent behavior of the monoaminergic brainstem structures, which contain neurons in which Fos is induced during wake is extensively discussed...* (pp. 245-246, italics added).

Different states of behavior and consciousness—waking, sleeping, dreaming, emotions, motivation, and stress—are all associated with different patterns of behavioral state-related gene expression. From this perspective behavioral state-related gene expression is a fundamental link between psychology and biology. This leads me to speculate that behavioral state-relat-

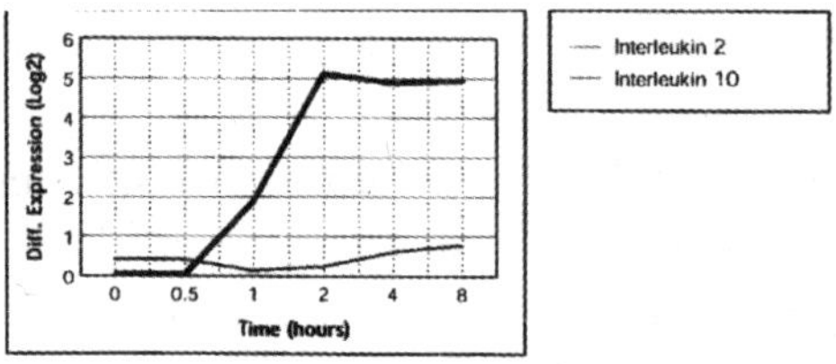


**Figure 4.1.** - A comparison of the ultradian time frames for expression of the interleukin-2 and interleukin-10 genes. Notice the how the increase in expression of the interleukin-2 gene, which is responsive to psychosocial stress in humans, takes place in the ultradian time frame of about an hour or two typical of psychotherapy. Expression of the interleukin-10 gene, which apparently is not responsive to psychosocial stress, by contrast does not increase significantly. This implies, but does not prove directly, that significant modifications of the psychoneuroimmune system could take place in the typical time frames of psychotherapy, therapeutic hypnosis, and many forms of the healing arts. (From Rossi, 2002a. Original graphs are from rats with permission from Incyte Corporation, 1999)

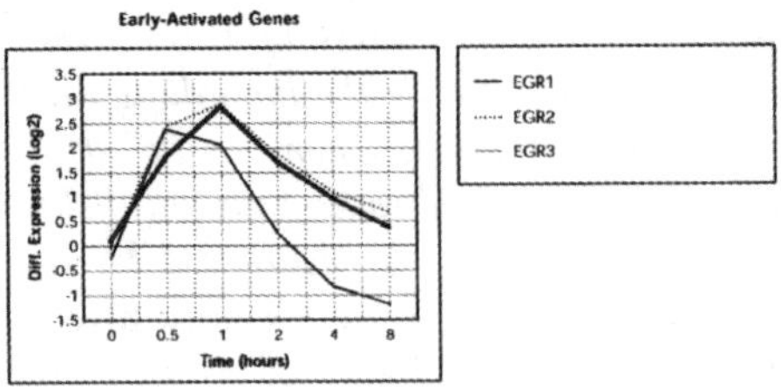


**Figures 4.2.** - Early-activated genes (epidermal growth responsel, 2 & 3) are expressed within Kleitman's typical ultradian cycle of 90-120 minutes in response to extra-cellular signaling (Reprinted with permission from Incyte Corporation, 1999).

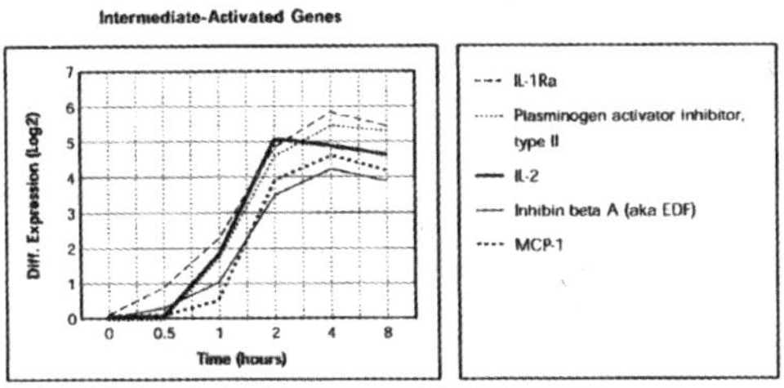


**Figure 4.3.** - Intermediate-activated gene expression that peaks within 2 to 4 hours. Some of these genes are associated with psychoneuroimmunology (Reprinted with permission from Incyte Corporation, 1999).

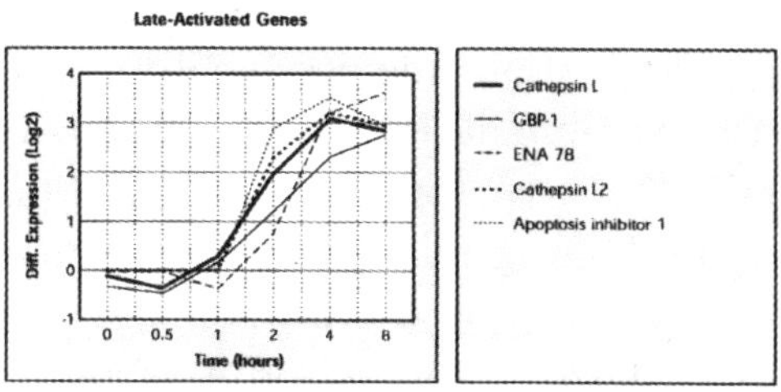


**Figure 4.4.** - Late-activated genes illustrate the longer ultradian timeframes (4 to 8 hours) under intense investigation that have not been related to psychoneuroimmunology at the present time (Sillman, 1999). (Reprinted with permission from Incyte Corporation, 1999).

ed gene expression may be of essence in exploring the deep psychobiology of consciousness in therapeutic hypnosis. The general hypothesis is that behavioral state-related gene expression can be modulated with psychosocial cues and cultural rituals such as therapeutic hypnosis to facilitate health, performance, and healing (Rossi, 2000c, 2002a). Research paradigms utilizing the new DNA microarray technology have been proposed to assess the degree to which behavioral state-related gene expression is actually engaged in the wide range of states from relaxation to arousal and stress that are modulated by therapeutic hypnosis (Rossi, 1999,2000b, 2002a, 2002b).

## 4. Stress Related Gene Expression, Psychosomatics, and Healing

Genes whose expression is associated with stress are a category of psychosocial genomics that may become of special interest to therapeutic hypnosis.

The relationship between Posttraumatic Stress Disorder and cholinergic gene expression, for example, has been documented experimentally by Kaufer et al. (1998) and Meshorer et al. (2002) as reviewed by Rossi (2002b). I have hypothesized that stress induced gene expression is the basic but generally unrecognized psychogenomic source of psychosomatic problems that are addressed by therapeutic hypnosis, alternative, and complementary medicine (Rossi, 2000c, 2002a). Figure 4.5a, for example, illustrates how the stress of maternal separation and a lack of touch results in a sharp decline in expression of the ODC gene (that codes for the synthesis of the Ornithine Decarboxylase enzyme) within 2 hours. This leads to a suppression of growth hormone and DNA synthesis in most organ tissues of the body. Figure 4.5b, by contrast, illustrates how simple touch can restore ODC gene expression in the brain and heart within the same ultradian time frame.

I speculate that this association between touch, ODC gene expression, growth hormone, and DNA synthesis is an example of the unrecognized psychosocial genomic mechanisms evoked by a wide variety of approaches to healing evident in the "passes" of historical hypnosis (Edmonston, 1986), "therapeutic touch" (Fields, 1995), and the many "holistic" body therapies found in diverse cultures (Rossi, 2000c, 2002a).

## 5. Activity-Dependent Gene Expression, Neurogenesis, and Healing

Kandel et al. (1983,1989,1998,2000) found in memory and learning experiments that repeated salient stimulation can turn on gene expression to produce messenger RNA (mRNA) that codes for the production of new proteins in neurons. Since gene expression and protein synthesis is dependent on activities of the organism, this is called "activity-dependent gene expression." As reviewed previously (Rossi, 2002a, 2002b), short term memory (1 to 20 minutes) involves rapid neurotransmission from one neuron to another. Long term memory, by contrast, requires the engagement of activity-dependent gene expression over a period of 90-120 minutes. Long term memory and learning engages a molecular communication system in a circular loop of information transduction from the environment to the nucleus of the neuron where the genes are located (Kandel et al. 2000; Stahl, 2000). This activation of gene transcription that is the basis of neurogenesis is called the *"genomic action potential"* (Clayton, 2000). As is typical of the activity-dependent gene expression and protein synthesis cycle through out the brain and body, it requires an ultradian periodicity of about 90-120 minutes (Rossi, 2002a).

Activity-dependent gene expression generates the proteins and growth factors that signal *stem cells,* normally residing in the brain, to differentiate into newly functioning neurons with new connections between them (Gage, 2000a, 2000b; Galli et al., 2000). The direct isolation of human central nervous system stem cells (Uchida et al., 2000), together with their capacity for self-renewal and migration to the loci of pathology in adult brain (Aboody et al., 2000), suggests that they are a natural source of healing that would be of central interest in a psychobiologically oriented theory of therapeutic hypnosis. The search for research paradigms in this area in the future may derive from current neuroscience experiments relating psychological experiences to activity-dependent gene expression and neurogenesis.

## Three Psychosocial Factors that Optimize Gene Expression, Neurogenesis, Memory, and Healing

Actiyity or behavior is the linchpin that connects psychological experience to the three factors that optimize brain plasticity gene expression, neurogenesis, and healing:

1) *novelty* (Eriksson et al., 1998)

2) *environmental enrichment* (Kempermann, Kuhn, & Gage, 1997; Van Praag, Kempermann, & Gage, 2000), and

3) *physical exercise* (Van Praag et al., 1999, 2002).

*Activity-dependent gene expression facilitates the generation of functional neurons in the hippocampus of the brain, which encodes new memory, learning, and behavior* (Gould et al., 1999a, 1999b; Van Praag et al., 2002). Most significantly, Gould, Tanapat, Reeves and Shors (1999B) document how neurogenesis takes place in three neocortical association areas (prefrontal, inferior temporal, and posterior parietal cortex) that are involved in behavioral plasticity (that is, behavioral change) and associated transformations of psychological experience. Gould et al. (1999a) conclude, "these new neurons, which are continually added in adulthood, may play a role in the functions of association neocortex" (p. 548).

Although other reputable researchers have confirmed neurogenesis in the hippocampus and olfactory bulb, however, they have not been able to substantiate Gould et al.'s claim of neurogenesis in the normal adult primate neocortex (Kornack & Rakic, 2001). The current controversy about the validity and extent of neurogenesis in the adult primate association neocortex requires resolution before the most optimistic implications of such neurogenesis could be accepted as a new neuroscience foundation for the theory and practice of psychotherapy, therapeutic hypnosis, and the therapeutic arts in general (Nowakowski & Hayes, 2000; Gould & Gross, 2000).

Van Praag et al. (2002) have presented the most recent evidence in this controversy about the functional significance of neurogenesis in the adult hippocampus.

> There is extensive evidence indicating that new neurons are generated in the dentate gyrus of the adult mammalian hippocampus, a region of the brain that is important for learning and memory... We report that newly generated cells in the adult mouse hippocampus have neuronal morphology and can display passive membrane properties, action potentials and functional synaptic inputs similar to those found in mature dentate granule cells. Our findings demonstrate that newly generated cells mature into functional neurons in the adult mammalian brain... *we have identified one-month-old neurons with functional properties similar to those of mature dentate granule cells in the adult hippocampus. We also show that newly generated neurons are initially smaller and reach a more mature*

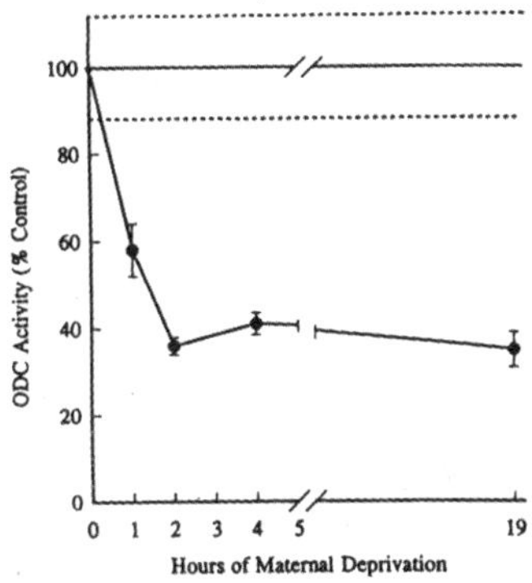


**Figure 4.5a:** Maternal separation and deprivation of physical touch results in decreased ODC gene expression in the 10-day-old rat brain within 10 to 15 minutes. Within 2 hours ODC gene expression is less than 40% of normal. Reprinted with permission from Schanberg, S. (1995). Genetic bases for touch effects. In T. Field (Ed.), *Touch in early development*. New York: Lawrence Erlbaum & Associates.

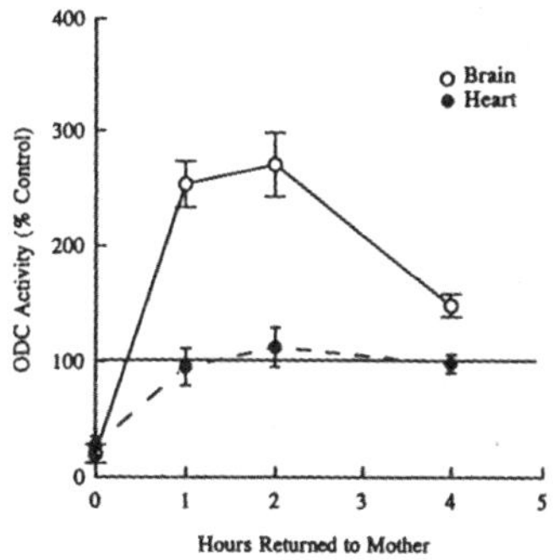


**Figure 4.5b:** The recovery of ODC gene expression in the brain and heart after a 2-hour deprivation and the return to the mother. Notice the statistically significant overcompensation that approaches 300% in 2 hours and remains elevated after 4 hours. This "therapeutic effect" of maternal touch is an example of psychosocial genomics—how sensory, psychological, social, and cultural signals and stressors can modulate expression in the deep psychobiology of health and illness. Reprinted with permission from Schanberg, S. (1995). Genetic bases for touch effects. In T. Field (Ed.), *Touch in early development*. New York: Lawrence Erlbaum & Associates.

*morphology after four months* (p. 1030, italics added).

To understand how such research on the time parameters of neurogenesis could have important implications for the theory, research, and practice of therapeutic hypnosis, psychotherapy, and counseling in general (Gross, 2000; Kandel, 1998), we need to develop new experimental paradigms to explore the relationships between activity, psychobiological arousal, gene expression, neurogenesis, memory, learning, stress, and healing.

Figure 4.6. - Proposed stages in the genesis of new synapses with long term potentiation (LTP). Within 10 minutes of appropriate stimulation there is a series of molecular signals that activate AMPA receptors and an increase in their single channel conductance. Within 30 minutes the size of the post-synaptic density increases and divides in the receiving neuron. Within 60 minutes some synapses divide to generate multi-spine synapses where 2 or more post-synaptic spines connect with the presynaptic bouton. Concomitant retrograde communication from the post to the pre-synaptic bouton increases so that the total number of functioning synapses proliferate over time (90-120 minutes to hours). This implies that the activity-dependent generation of new synapses for the construction and re-consturction of the neural networks of memory, learning and behavior may be possible within the typical time frames of many approaches to therapeutic hypnosis and psychotherapy in general. Reprinted with permission from Luscher, Nicoll, Malenka, & Muller (2000), Synaptic plasticity and dynamic modulation of the postsynaptic mebrane. *Nature Neuroscience* 3, pp. 545-567.

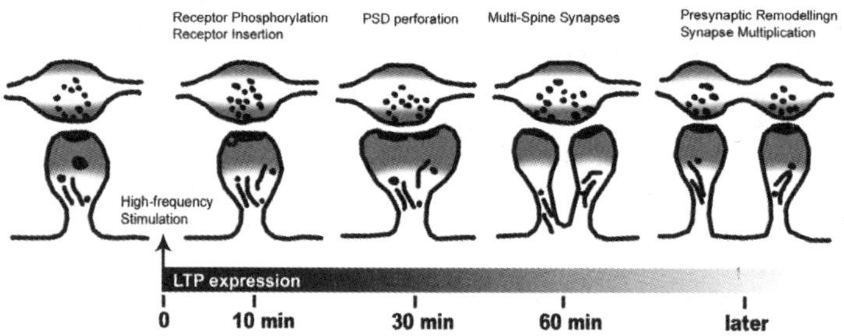


## Neurogenesis, Long-term Potentiation and Psychobiological Arousal

Neurogenesis has been related to studies of memory and learning at the neurobiological level via "long term potentiation" (LTP). Johnston and Maio-Sin Wu (1995) define LTP "as an enduring, activity-dependent increase in synaptic efficiency... *which persists on the order of an hour or more... initiating gene expression*" (p.459, italics added).

A visual model of the dynamic process by which LTP generates the formation and growth of new synapses to encode activity-dependent memory, learning, and behavior has been proposed recently by Liischer et al., (2000). As illustrated in

Figure 4.6, LTP generates an increase in the number of synapses connecting one neuron with another within an hour. Within the first 10 minutes of LTP induction, there is an activation of the receptors that are involved in synaptic communication via neurotransmitters. Within the first 30 minutes, the size of the synaptic spine increases leading to an increase in the size of the "post-synaptic density" that will become the receiving site of neurotransmission. Within an hour, some synapses divide in two as illustrated in Figure 4.6. This leads, in turn, to further growth in pre-synaptic multiplication and remodeling to create new neural networks for encoding memory, learning, and behavior change. I propose this as the neurobiological essence

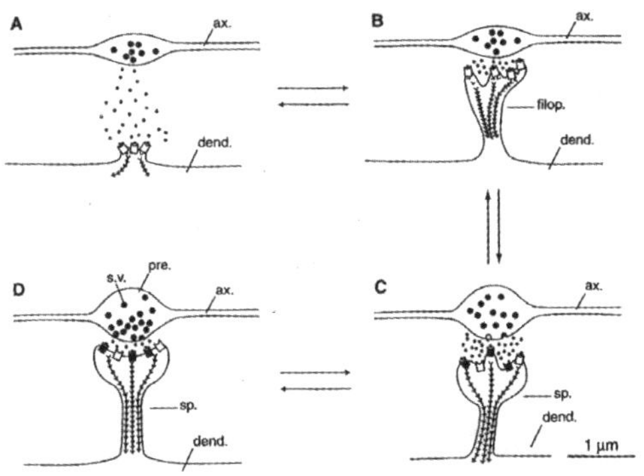


Figure 4.7. - A hypothetical illustration of the dynamics of glutamate receptors in the reversibility of synaptic development in the construction and reconstruction of neural networks. (A) Stimulation of NMDA receptors (white boxes in the receiving dendrite [dend.]) generates branching actin strings at the newly developing synaptic sites leading to (B) the outgrowth of motile filopodia (filop.) making connection with the pre-synaptic axon (ax.). The acquisition of AMPA receptors (black boxes) and the formation of synaptic contacts is accompanied by reduced motility. (D) Activation of AMPA receptors further suppresses motility and stabilizes spine (sp.) morphology. Neurotransmitter molecules of glutamate are illustrated as black spots inside the pre-synaptic vesicles (s.v.). The reversible arrows between stages A, B, C and D mean that these connections form new structures and neural networks that remain reversible and plastic (changeable) in time with changing patterns of activity-dependent stimulation. Reprinted with permission from Matus, A. (2000). Actin-based plasticity in dendritic spines. *Science*, 290,754-758.

of therapeutic hypnosis, psychotherapy, and many of the alternative and complementary paths to healing (Rossi, 2000c, 2002a). It is somewhat startling to realize how such experimental research could serve as a neurobiological model for therapeutic hypnosis within the typical time frame of a clinical session of about an hour. Does this imply that a shorter period is not sufficient for the psychogenomic dynamics of therapeutic hypnosis? Not really. Anecdotal reports of Milton Erickson's use of surprise and "psychological shock" to evoke psychobiological arousal in therapeutic hypnosis and post-hypnotic suggestion (Erickson, 1964/1980; Rossi, 1973, 2002a, 2002b), for example, indicates that a brief but memorable experience of a few seconds or minutes can change one's perspective and behavior for a lifetime. A variety of clinical research paradigms utilizing the new DNA microarray technology for assessing the efficacy of therapeutic hypnosis in the typical time frames of gene expression and neurogenesis have been outlined (Rossi, 1999,2000a)

## Stages and Reversibility of Activity-Dependent Synaptogenesis in Hypnosis?

A closer look at the details of the actual growth process at the synaptic level by Matus (2000) fills in some of the details of the mechanisms proposed by Liischer et al (2000). Matus begins by noting that during brain development, neurogenesis proceeds in a number of distinct stages. The first stage, which takes place during embryogenesis, involves the initial creation of neural networks under the guidance of neurotropic messenger molecules. The second stage takes place after birth, during a relatively short critical period when synaptic connections are refined by activity-dependent sensory and motor experience. The third and fourth stages, which are active throughout the lifespan, engage activity-dependent memory and learning as well as the reversible synthesis and re-synthesis of synapses as illustrated in Figure 4.7.

Figure 4.7 illustrates reversible formation of synapses at the level of dendritic growth in the shifting dynamics of neural networks as proposed by Matus. Stages A and B in Figure 4.7. illustrate the initial steps of dendritic outgrowth, which appears to be stimulated by the activation of NMDA receptors that are known to be involved in memory and learning. Stages C and D of Figure 4.3 illustrates the experience or activity-dependent gene expression leading to the generation of proteins in the formation and placement of AMPA receptors that stabilizes the growth of synaptic connections between neurons.

A Current research suggests that the four transition states (A, B, C, & D in Figure 4.7) in this sequence of development are reversible depending on the strength and maintenance level of activity-dependent stimulation at the pre-synaptic terminal. *It is of greatest importance to recognize psychological implications of the reversible arrows between the four phases ofneurogenesis at the level of dendritic synaptic development and stabilization. This reversibility illustrates a fundamental psychogenomic mechanism for*

(1) *the creation and maintenance of memory and learning*

(2) *the molecular dynamics of forgetting, amnesia and dissociation as well as*

(3) *the change, transformation and re-synthesis of memory and behavior associated with further activity-dependent gene expression and neurogenesis.*

Matus (2000) summarizes the interaction between the psychological level of memory and learning and the molecular level of neurogenesis at the dendritic level in this way:

> Such a scheme may help explain experience-dependent [that is, activity-dependent] shaping of neuronal circuits because it would make sense to require a mechanism that depends on strong stimulation, like that of NMDA receptor activation, for the initiation of new connections but to then support them with a mechanism sensitive to low rates of neuronal activity, like that of AMPA receptor activation. Learning responses could thus be maintained during periods when they are not actually being used. Similarly, should synaptic stimuli fall below a certain threshold indicative of disuse, it would be appropriate for synaptic connections to be broken *and re-formed in new configurations* (p. 757, italics added).

At the present time we can only speculate about the implications of such "experience-dependent shaping of neuronal circuits" for facilitating the deep psychobiology of therapeutic hypnosis. When "synaptic connections [are] broken" we wonder, for example, if this can be the psychobiological basis for ordinary forgetting as well as a number of classical phenomena associated with hypnosis such as amnesia and dissociation (Janet, 1919/1925). When "synaptic connections [are] re-formed in new configurations" we wonder if this could be the psychobiological foundation for the re-association, reorganization, and re-synthesis of psychological experience during therapeutic hypnosis proposed by Milton Erickson (1948/1980). A variety of new research paradigms have been outlined (Rossi, 2000b, 2000c, 2002a; Rossi & Cheek, 1988) to assess the fundamental psychobiological dynamics of experience or activity-dependent gene expression and neurogenesis could account, at least in part, for the efficacy of therapeutic suggestion pioneered by James Braid, Pierre Janet, and Milton Erickson.

### Brain Plasticity, Synaptogenesis, and Hormones in the Female Cycle

Perhaps the most vivid reality of brain plasticity and synaptogenesis on human consciousness, mood, and behavior is experienced by woman during their monthly menstrual cycle. Figure 4.8 illustrates how estrogen has a dramatic trophic (growth) effect on the neurons of the hippocampus of adult female experimental animals within days and across a single menstrual (estrus) cycle (Stahl, 2000). During the early phase of the estrus cycle estrogen levels rise and induce the formation of dendrites and synaptogenesis. This makes evolutionary sense since the process of ovulation and possible pregnancy would lead to an increase in survival if the mother's brain actually grows by producing more synaptic connections and an associated increase in brain functioning that could help her young prosper. Alas, the new dendrites and synapses disappear again in the later part of the cycle with the rise in progesterone. This naturally has important implications for the mood, behavioral, and mentation changes related to PMS.

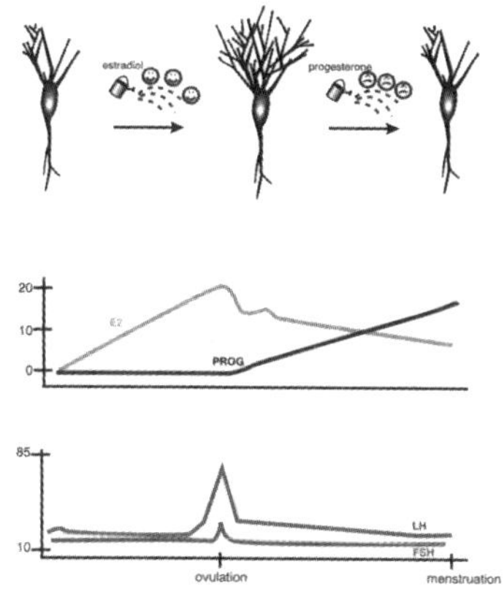


Figure 4.8 The increase in the growth of dendrites and synaptogenesis in the first half of the menstrual cycle under the impact of estrogen (estradiol) and their loss by down regulation in the second half of the cycle with the dominance of progesterone. (With permission, Stahl, 2000)

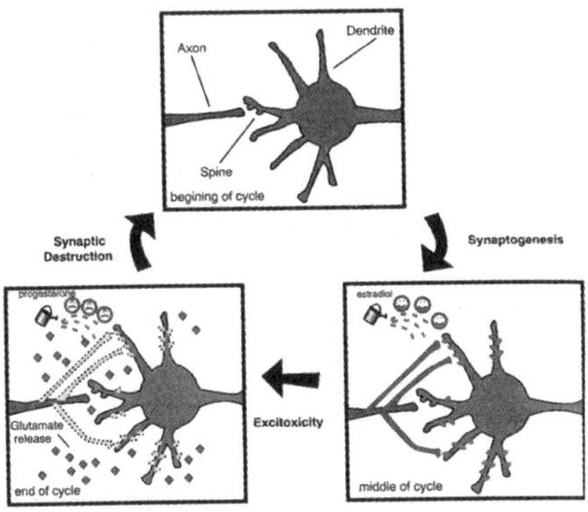


Figure 4.9 A dramatic illustration of the cycle of the synaptogenesis and synapse destruction. It is hypothesized that estrogen leads to synaptogenesis during an initial low level of glutamate activation. With the accumulation of greater glutamate and progesterone levels, however, the heightened excitotoxciity results in the destruction of the same dendritic spines and synapses. (With permission, Stahl, 2000)

This is a rich field for research. What, for example, are the possible effects of such estrogen-dependent synaptogenesis on experience of therapeutic hypnosis and psychotherapy? Is it possible that, creativity, insight, learning, memory, and the training of new skills in the first half of the estrus cycle could proceed more easily under the impact of rising estrogen levels? If the training continued with optimum intensity in the second half of the cycle could that lead to the preservation of some of these newly acquired synaptic connections? What are the implications of this cyclic brain growth for relationships and family life? We obviously have more questions than answers about the dramatic reality of brain plasticity on the human condition as illustrated in figures 4.8 and 4.9.

## Implications of Gene Expression. Brain Platicity and healing for Therapeutic Hypnosis Today

There is no research that directly validates the hypothesized associations between brain plasticity gene expression, neurogenesis, and healing with the practical techniques of therapeutic hypnosis that are in use today. New experimental paradigms combining recent studies on the psychobiological localization of hypnotic experience in the brain and body (Barabasz, 2002; Gruzelier et al., 2001; Kiecolt-Glaser et al., 2001; Rainville et al., 1997, 1999) with the most successful research designs for identifying the locus of gene expression and neurogenesis (Van Praag et al., 2002) are needed. Until such research is carried out, any attempt to generalize the proposed associations between the gene expression, neurogenesis, healing, and psychotherapy (Gross, 2000; Kandel, 1998) to the psychobiological domain of therapeutic hypnosis remains speculative. Given this caveat, however, researchers, clinicians and students may choose to explore the implications of current neuroscience research for understanding and facilitating the efficacy of therapeutic hypnosis in the following ways.

- Exploring the new theory, experimental techniques, and research base of current neuroscience has a potential to vastly expand the scientific status and domain of therapeutic hypnosis. Such research could integrate interdisciplinary research among psychology, biology, and medicine with the "psychophysiological" foundation of therapeutic hypnosis pioneered by James Braid, Pierre Janet, and Milton Erickson.
- The fundamentals of gene expression, neurogenesis, and healing by the differentiation and maturation of stem cells may be the common denominator that integrates modern molecular medicine with alternative and complementary medicine. The ambiguous borderlines between therapeutic hypnosis and the cognitive-behavioral therapies, as well as the holistic and spiritually oriented therapeutic arts of many cultures throughout

human history, could be clarified with a common yardstick that measures precisely what each approach contributes to facilitating healing via gene expression, neurogenesis, and stem cell healing.

- The time parameters of gene expression and neurogenesis in the synthesis and re-synthesis memory, learning and behavior ranging from seconds to minutes, hours, days, weeks and months can sharpen the clinician's awareness of the potential value of the many time-related techniques of therapeutic hypnosis. These include the momentary value of "psychological shock" and surprise to optimize psychobiological arousal and its evocation of immediate-early, behavioral state-related, activity-dependent, and clock gene expression—as well as the traditional use of relaxation for the mitigation of stress-related gene expression in facilitating healing within minutes to hours.

- Research on the proposed associations between gene expression, neurogenesis, healing, and therapeutic hypnosis could enhance our understanding of how to optimize the duration of suggestions to facilitate healing with the varying states of highly focused attention, relaxation, dreaming, and sleep. Therapeutic hypnosis and post-hypnotic suggestion could be focused more precisely with the natural time parameters of gene expression and neurogenesis, for example, that can range from minutes and hours to facilitate the differentiation of new synapses, to 4 weeks for the generation of young functioning neurons, to 4 months for their maturation. *The implication is that if therapeutic suggestion is to endure over time (more than the typical short term memory of 15 or 20 minutes), it must engage gene expression to facilitate long term potentiation, synaptogenesis, neurogenesis, and healing.* Such research could lead to new criteria for the resolution of the controversies about the duration of the effectiveness of therapeutic hypnosis in facilitating optimum performance, habit change, symptom amelioration, and healing.

- Therapeutic hypnosis has traditionally facilitated the motivation (fascination, arousal, interest) for *"activity"* and the actual means (techniques, behavioral prescriptions, approaches) for *"doing"* to help people help themselves. Indeed, facilitating activity or behavior could be described as the essential purpose of "suggestion." Therapeutic suggestions are offered in the hope of helping people *do something* to facilitate their learning, problem solving, and healing. Current neuroscience research now traces the deep psychobiology of cognitive-behavioral activity to "activity-dependent gene expression," which can facilitate "activity-dependent neurogenesis and healing." This implies that *novelty, environmental enrichment, and exercise—the* three factors that optimize gene expression, neurogenesis, and healing—could become new guidelines for crafting therapeutic suggestions with hypnosis as well criteria for evaluating their effectiveness.

# V
# THE BIOINFORMATICS OF PSYCHOSOCIAL GENOMICS: EXPECTANCY AND SURPRISE IN THERAPEUTIC HYPNOSIS

This conceptual review explores some speculative associations between the neuroscience of expectancy and surprise during stress and therapeutic hypnosis. Current neuroscience is exploring how novel interactions between the organism and the environment initiate cascades of gene expression, protein synthesis, neurogenesis, and healing that operate via Darwinian principles of natural variation and selection on all levels from the molecular-genomic to the subjective states of consciousness. From a neuroscience perspective, the novel and surprising experiences of consciousness appear to have as important a role as expectancy in memory, learning and behavior change in the psychobiology of therapeutic hypnosis. This paper explores how we may integrate the psychosocial genomics of expectancy and surprise in therapeutic hypnosis as a complex system of creative adaptation on all levels of human experience from mind to gene expression.

## The Conceptual Review in Science

The "Conceptual Review" is a new form of scientific exposition introduced recently by Blagosklonny and Pardee (2002) as follows:

> The conceptual review should take its place as an essential component of scientific research. At this opportune time, conceptual biology is being born. Millions of easily retrievable facts are being accumulated in databases, from a variety of sources in seemingly unrelated fields, and from thousands of journals. New knowledge can be generated by

> "reviewing" these accumulated results in a concept-driven manner, linking them into testable chains and networks... Connecting separate facts into new concepts is analogous to combining the 26 letters of the alphabet into languages. One can generate enormous diversity without inventing new letters. These concepts (words), in turn, constitute pieces of more complex concepts (sentences, paragraphs, chapters, books). We call this process "conceptual" research, to distinguish it from automated data-mining and from conventional theoretical biology... Can a review provide new knowledge? A review can constitute a comprehensive summary of the data in the field—this type of writing educates but does not directly generate new knowledge. But a "conceptual" review, on the other hand, can generate knowledge by revealing "cryptic" data and testing hypotheses by published experiments... Conceptual research can encompass many fields without limitation. In comparison with labor-based research, conceptual research is more cost-effective; indeed, verification of a hypothesis using existing data does not limit research to scientists in well-resourced fields or countries. Hypothesis-driven, experimental research will continue to be a cornerstone of biology, but it should strike up a partnership with the essential components of theoretical and conceptual research (pp. 373-374).

While the conceptual review was developed in the context of biology, it is adopted here as a new approach to integrating important research developments in current neuroscience with therapeutic hypnosis. First, however, some significant caveats and limitations of this conceptual review must be emphasized. Most of the research from neuroscience that is reviewed here has been done with model organisms ranging from snails, flies, and rats to sub-human primates. The justification for this is that the fundamental mechanisms of biology and neuroscience are "conserved". This biological conservation means that at the basic levels of molecules, cellular and neural mechanisms—the fundamental processes of life—are similar across all organisms of the phylogenic scale. Some readers may well question whether this so-called "conservation in biology" really can be adapted to the cognitive-behavioral level of therapeutic hypnosis. While this remains a controversial question, for the purposes of this conceptual review we shall accept some of the assumptions and data from a number of prominent neuroscientists such as Kandel (1983,1989,1998,2000), Kaufer, Friedman, Seidman, and Soreq (1998), and Meshorer et al. (2002) whose research apparently bridges the levels between the molecular-genomic, hormonal and cognitive-behavioral. Kandel (1998) maintains, for example:

> *Insofar as psychotherapy or counseling is effective and produces long-term changes in behavior, it presumably does so through learning, by producing changes in gene expression that alter the strength of synaptic connections and structural changes that alter the anatomical pattern of interconnections between nerve cells of the brain...* Stated simply, *the regulation of gene expression by social factors makes all bodily functions, including all functions of the brain, susceptible to social influences* (p. 460, italics added).

The evidence and implications for the *"regulation of gene expression by social factors"* has been described as the new discipline of "psychosocial genomics" that is proposed as a psychobiological approach to therapeutic hypnosis (Rossi, 1999, 2000a, 2000b, 2000c, 2002).

The second major limitation of this conceptual review is semantic. It is not always clear that terms such as "stress" and "adaptation" mean the same thing when applied by neuroscientists to genomic and cellular mechanisms as they do to therapists who use these terms on a human experiential, cognitive-behavioral level. Can "oxidative stress" that is deleterious to molecular, genomic and neural mechanisms (by attaching oxygen molecules to proteins, for example, to render them nonfunctional) during traumatic experiences really be equated to "cognitive and emotional stress" humans experience in difficult situations of everyday life? We will approach this controversial question by reviewing the types of research that illustrate how acute and posttraumatic stress is related to gene expression in both animals and humans. Kaufer et al. (1998), for example, document how "acute stress facilitates long-lasting changes in cholinergic gene expression" by forced swimming in rats. Meshorer et al. (2002) discuss at least 3 known molecular mechanisms whereby stress could modulate gene expression humans experiencing Posttraumatic Stress Disorder (PTSD).

The third limitation of the conceptual review is that it is not a comprehensive survey of the scientific literature in any sense. Rather, it is a highly selective and focused interpretation of existing research that generates new hypotheses that need to be assessed by further research and clinical practice. In this paper I explore how the psychological experiences of expectancy and surprise may function as complementary mechanisms on the levels of synaptic neurotransmission and gene expression respectively. I then speculate how these psychobiological mechanisms may be evident in the use of expectancy and surprise in a verbatim transcript of a hypnotic induction by Milton H. Erickson. These speculations point to the type of research that is now needed to determine whether the hypothesized associations between the biological level (gene expression and synaptic neurotransmission) and the psychological level (surprise and expectancy) in therapeutic hypnosis are merely metaphorical or indicative of deep psychobiological mechanisms of mind-body communication and possible healing.

## The Bioinformatics of Psychosocial Genomics

To understand the meaning and implications of psychosocial genomics we need to understand a few basic concepts of bioinformatics as they are currently used in molecular biology. It is important to realize that genes exist within long strands of DNA which are made up of several billion base pairs (in humans) of the 4 nucleotides (Adenine, Cytosine, Guanine, & Thymine) that make up the molecular code of life. The *(1) sequence of these nucleotides* that make up DNA is a code that determines the order in which amino acids will be strung together to build the *(2) structure of proteins,* which in turn determine the *(3) physiological functions* of the organism. When the splicing of genes is changed by stress it means that the (1) *sequence of nucleotides* is changed so that (2) the *structure of proteins* is changed in a manner that leads to (3) corresponding changes in *physiological functions* of the organism for good (a little stress can facilitate healing) or for ill (too much chronic stress results in dysfunction).

This relationship between *sequence, structure, and function* is a basic concept of molecular biology that is now called, "bioinformatics" (Claverie & Notredame, 2003). To this standard textbook concept of *bioinformatics* I now add a 4th link so that *human experience* on the cognitive-emotional-behavioral level completes the bioinformatic cycle of information transduction between mind and matter as illustrated in figure one. It is of essence for mind-body medicine and psychotherapy to realize that the bioinformatic levels of *sequence, structure, functbn, and experience* are all related in figure one. *This means that if you enter the cycle at any level with any of the various approaches to mind-body medicine, therapeutic hypnosis, and psychotherapy, it is possible to modulate and possibly heal dysfunctions on any other level.*

The second fundamental concept of bioinformatics is to understand a few surprising facts about genes and how they are expressed (that is, turned on and off). The essential regions of genes are *sequences of nucleotides* called, "exons" that code for the *structure of proteins.* These exons, however, are frequently interrupted by *non-coding* regions of "junk DNA" (that is, nucleotide sequences that originally were erroneously thought to be non-functional) that are called "introns" (Gibbs, 2003). Before genes can be expressed (transcribed) these non-coding introns must be cut out and eliminated so that the *coding exons can be spliced together to make up a useful piece of messenger ribonucleic acid (mRNA).* When the exons are spliced together they are shipped out the cell nucleus to ribosomes (the protein factories of the cell) where the *sequence of nucleotides* in the mRNA is translated into a string of aminoacids that fold into the *structure of proteins* that determines their *physiological function.*

Physiological functions, of course, are related to psychophysiology as reviewed in the previous chapters. The research of Meshorer et al. (2002), for example, recently documented how the splicing of exons (the protein coding parts of genes) is modulated, at least in part, by the molecular messengers of trauma and stress (the hypothalamic-pituitary-ad renal hormones of Selye) that are experienced on the psychological

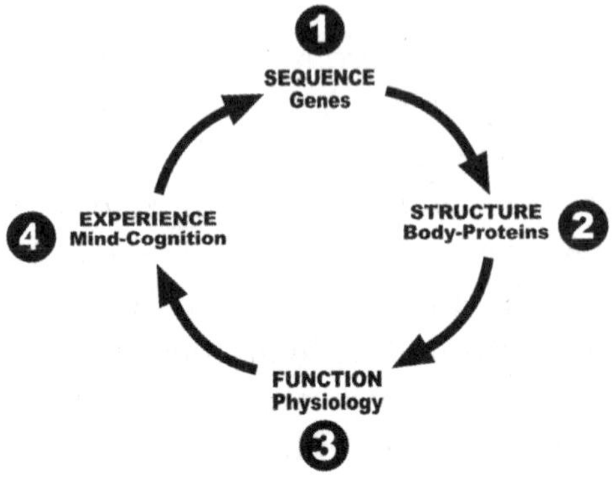


**Fig. 5.1.** - The adaptive bioinformatic cycle of sequence, structure, function, and experience that integrates mind-body communication and healing with therapeutic hypnosis and psychotherapy. This could more properly be called, "The Creative Psycho-Bio-Informatic Cycle" since I added the level of experience to the standard text-book bioinformatic progression of sequence, structure, and function.

level that *humans experience* as PTSD. This is just one example of a *psychosocial genomic* mechanism that completes the cycle of information transduction between the *sequence and structure of matter* with the *physiological functions of the body* and the *experiences of the mind* (Rossi, 2002 a). As noted above, *healing can be initiated by entering the bioinformatic cycle of figure 1 at any of the four levels of sequence, structure, function, or experience.* Modern molecular medicine usually chooses entry at the purely physical and molecular-biological levels of *sequence, structure,* and *physiological functions.* The alternative and complementary approaches to mind-body medicine (including the cognitive-behavioral approaches of psychotherapy, meditation, therapeutic hypnosis etc., and some spiritual practices) by contrast, tend to focus on the psychobiological levels of *function,* especially *psychological experience,* in facilitating the psycho-bioinformatic cycle of mind-body communication and healing in figure one. This is particularly evident in the experiences of expectancy and surprise in whatever context they may arise.

### Expectancy and Surprise in Motivation, Memory, and Learning

A recent paper by Waelti, Dickensen, and Schultz (2001) reports new findings on the psychobiological mechanisms of expectancy and surprise in motivation, memory, and learning that may have important implications for current theories of therapeutic hypnosis. They introduce their research contrasting the relative merit of expectancy versus surprise in memory, learning, and behavior by noting that current theories assume that predictive learning occurs whenever a stimulus is paired with a reward or punishment. Their research, however, demonstrated that temporal contiguity between a stimulus and a reinforcer is not enough for learning as follows:

> *...a discrepancy between the reinforcer that is predicted by a stimulus and the actual reinforcer is also required.* This discrepancy can be characterized as a "prediction error". *Presentations of surprising or unexpected reinforcers generate positive prediction errors, and thereby support learning... Expected reinforcers do not generate prediction errors and therefore fail to support further learning even when the stimulus is consistently paired with the reinforcer...* neuronal messages encoding prediction errors can act as explicit teaching signals for modifying synaptic connections that underlie associative learning, (p. 43, italics added)

This is a radical finding, indeed! Classical theories of memory, learning, and behavior change were based on the continuity of stimulus, response, and reward as described by Pavlovian theory, behavior theory, and especially Hebb's (1949) activity-dependent theory of how connections are formed between neurons (Rossi, 2002). There are no surprises in these classical theories of learning; there is no essential role for creative experiences of fascination and mystery that were associated with early theories of healing in the history of hypnosis (Braid, 1855/1970). Waelti et al. (2001) formulate the molecular dynamics of a sense of surprise as teaching signals for associative learning in this way.

> *During initial learning, when rewards occur unpredictably, dopamine neurons are activated by rewards. They gradually lose the response as the reward becomes increasingly predicted... Dopamine neurons also respond to novel, attention generating, and motivating stimuli, indicating that attentional mechanisms also contribute* (p. 43, italics added).

Students of the creative arts, humanities, and culture have

long felt bored with the classical psychological theories of human memory and motivation. Zeki (1999,2001), for example, has described how all forms of artistic creativity are expressions of the dynamics of Darwinian variation and the selection of subjective states of consciousness and experience that have neurobiological roots. There was very little in classical theories of learning, however, that could account for the sense of surprise, fascination, and wonder that is so motivating in human experience. There was no satisfying understanding of the human sense of adventure, quest, spiritual striving, and the experience of the *numinosum* (Otto, 1926/1950)—the sense *of fascination, mystery* and the *tremendous*— long associated with hypnosis. The research of Waelti and colleagues, however, opens the window to a new world-view of the relationships between the most interesting and motivating experiences of consciousness and the molecular dynamics of memory and learning that are described as the "*novelty-numinosum-neurogenesis effect*" (Rossi, 2002). In this sense, it not so much what is expected and easily predictable in human affairs that is motivating, but the exact reverse. That which is surprising, unknown, and unpredicted garners our attention and sets us forth on the human quests for problem solving and creative adventure in the novelty-numinosum-neurogenesis dynamics of mind-body communication and healing explored by the new discipline of psychosocial genomics. In what follows we will focus on recent research on the psychosocial genomics of both expectancy and surprise that require investigation in future research on the deep psychobiology of therapeutic hypnosis.

## The Psychosocial Genomics of Expectancy and Surprise

Psychosocial genomics has been described as the emerging science of how psychological, social, and cultural signals modulate gene expression and vice versa in health, trauma, stress, and illness, as well as the therapeutic arts (Rossi, 1999, 2000a, 2000b, 2000c, 2002). Here we build a conceptual bridge between the molecular level of psychosocial genomics and its associated experiences of expectancy and surprise at the psychological level. The "response expectancy theory" of hypnosis is based on evidence of how psychological experiences of positive and negative expectancy can modulate the "brain state" and its "physiological substrate" such as "blood pressure, pulse, etc." (Kirsch, 2000; Kirsch & Lynn, 1997). Psychosocial genomics extends the reach of response expectancy theory to the molecular dynamics of brain states and their physiological substrates at the levels of gene expression (gene transcription and translation). Gene expression, in response to signals ranging from the psychosocial to the molecular (e.g., hormones of the neuro-endocrine, autonomic and immune systems etc.), generates the proteins (the molecular machines of life) for adaptive homeostasis, neurogenesis and optimal performance as well as healing via therapeutic hypnosis (Rossi, 1990,1994,1996,2000a, 2000b, 2000c, 2001,2002).

## Expectancy and the Synaptic Dynamics of Priming

A decade of research has linked psychosocial and expectancy theories of hypnosis to the phenomena of positive and negative priming on the psychological level (Barnier, Bryant, & Riscoe, 2001; David, King, & Borkhardt, 2001; David & Brown, 2002; Dorfman & Kihlstrom, 1994; Zeig, 1990,1999). In the hypnotic context, Kihlstrom (1998) has defined priming as

> ...the facilitation of perceptual-cognitive processing of a stimulus (known as the target) by presentation of a prior stimulus (known as the prime). In *repetition priming,* prime and target are identical (e.g., water-water); in *semantic priming,* prime and target are related in terms of meaning (e.g., ocean-water) (p. 467).

What, however, are the neuro-molecular dynamics of the brain state and physiological substrate of priming that underlies its potential efficacy in therapeutic hypnosis?

In a recent series of papers on the molecular neurobiology of "priming plasticity" researchers have summarized the roots of psychological experience and expectancy in this way:

> *Nerve cells communicate by using chemical messengers, which are released from neurons after a "priming" step. It seems that priming may be the key to controlling the strength of chemical transmission.* The roots of cognition, behavior, learning and memory are embedded in the brain's intricate network of nerve cells and their specialized points of contact, the synapses. Synapses can convert electrical impulses into chemical signals and back again, as well as modulate the strength of the transmitted signals. *This ability to modify the strength of transmission—known as synaptic plasticity—is thought to be the cellular basis of the brains ability to compute, learn and remember. A goal of many neurobiologists is to understand the molecular basis of synaptic plasticity* (Dobrunz & Garner, 2002, pp. 277-278, italics added).

In a series of papers Schoch et al. (2002) and Castillo, Schoch, Schmitz, Sudhof, & Malenka (2002) have determined that a synaptic protein called RIM, among others, is involved in a key regulatory step of synaptic plasticity facilitated by priming the synaptic vesicles between neurons to release their neurotransmitters. Changes in synaptic strength initiated by priming can range between milliseconds to minutes for short-term plasticity and they can persist for hours, days or months (long-term plasticity). This research indicates that regulation of the priming step of the neurotransmitter release has important consequences for memory, learning, problem solving, and behavior change at the synaptic level.

Further research is now needed to determine how neurotransmitter release by priming is regulated during the psychosocial experiences of everyday life in general, and therapeutic hypnosis in particular. It has been hypothesized that the similarities in synaptic strength due to priming (milliseconds to minutes for short-term plasticity or hours, days or months for long-term plasticity) could be an answer to the question of why hypnotic suggestion has a similar range of therapeutic efficacy (Rossi, 1996, 2002). Current research (Castillo et al., 2002; Dobrunz & Garner, 2002; Schoch et al., 2002) supports the view that psychosocial expectancy facilitates synaptic neurotransmission at the synaptic level of "vesicle priming" at loci one and two in Figure 5.1.

As we learn more about the molecular-genomic dynamics of

priming in facilitating gene expression and protein synthesis in short-and long-term synaptic plasticity in memory and learning as illustrated in Figure 5.1, it could provide the psychobiological foundation for developing more reliable methods of utilizing psychosocial expectancy for facilitating the duration of therapeutic suggestion.

### Surprise and the Molecular-Genomic Dynamics of Stress and Healing

In view of the above evidence of the association between expectancy, vesicle priming at the synapse and neurotransmission, how are we to make sense of the bold assessment offered above by Waelti et al. (2001): *"Presentations of surprising or unexpected reinforcers... support learning,* whereas... *Expected reinforcers... fail to support further learning even when the stimulus is consistently paired with the reinforcer"* (p. 43). Careful study of Figure 1 suggests that the apparent paradox can be resolved by realizing that different researchers are referring to different levels or loci in the molecular neurobiology of memory, learning and behavior. Waelti et al. (2001) emphasize how new memory and learning require "surprising or unexpected reinforcers" to evoke the fundamental process of gene expression and protein synthesis that is the organic substrate of psychological experience (memory, learning etc.) at locus 3 in Figure 5.1. Research on how expectancy facilitates synaptic neurotransmission at the level of "vesicle priming" summarized by Dobrunz & Garner (2002), on the other hand, describe events of memory, learning and behavior at loci 1 and 2 in Figure 5.1. This leads me to propose the hypothesis that the complementary processes of expectancy and surprise are both components of a "complex adaptive system" in human experience (Rossi, 1996, 2002). It usually makes little sense to try to determine which comes first in the circular dynamics of non-linear, complex processes such as therapeutic hypnosis. In an early effort to outline the micro-dynamics of trance and suggestion for didactic purposes, however, Erickson & Rossi (1976) initially proposed that it was the experience of "shock and surprise" that could initiate an "attitude of expectancy," which, in turn led to a "hypnotic response." We will illustrate these dynamics later in a verbatim transcript illustrating Erickson's neuro-psycho-physiological approach to facilitating hypnotic experience.

Current research has identified at least three different mechanisms whereby environmental factors such as trauma, stress and surprise can interact with the genomic level at locus 3 in Figure 5.1:

1. Modulation of gene expression;
2. Alternative splicing of genes to produce different proteins, and
3. Translocation of neuritic mRNA during long-term neuronal hypersensitivity.

Since these three psycho-genomic mechanisms are fundamental in integrating mind-gene communication (information transduction) during health, stress and healing (Rossi, 1990,1994,2002) we will summarize each in turn.

1. *Surprise, stress and healing modulate gene expression.* Kaufer et al. (1998) have presented evidence of how acute stress facilitates long-lasting changes in cholinergic gene expression in a mouse model on two levels: (1) On the behavioral level of forced swimming stress, and (2) on the enzymatic-genomic level of gene expression in excised brain sections as follows:

> ...mice subjected to either forced swimming stress or inhibitors of the acetylcholine-hydrolysing enzyme acetylcholinesterase (AChE) exhibit dramatic increases in levels of messenger RNA encoding the early immediate transcription factor c-Fos in the brain. *In vitro,* sagittal corticohippocampal brain slices exposed to AChE inhibitors showed enhanced neuronal excitability and similar increases in cortical c-fos gene expression within 10 min... As modulated gene expression occurs in vitro, independently of the pituitary-adrenocortical axis, local mechanisms in the central nervous system are apparently enough to mediate this response (pp. 373-4).

Citing a variety of supporting experimental evidence they generalize their findings to the human experience of PTSD as follows:

> Acute traumatic stress may lead to posttraumatic stress disorder (PTSD), which is characterized by delayed neuropsychiatric symptoms including depression, irritability, and impaired cognitive performance... Here we report that there is a bi-directional modulation of genes that regulate acetylcholine availability after stress and a blockade of acetylcholinesterase... Our results suggest a model in which robust cholinergic stimulation triggers rapid induction of the gene encoding the transcription factor c-Fos. This protein then mediates selective regulatory effects on the long-lasting activities of genes involved in acetylcholine metabolism. The molecular mechanisms translating a traumatic life experience into long-term neuropsychological sequelae are expected to involve complex changes in gene regulation (p. 373).

Figure 5.2 is an illustration of the interesting ultradian (less than 20 hours) and circadian (about a day) time parameters of the long-term changes in acetylcholinesterase activity following stress.

Insofar as therapeutic hypnosis can be used to facilitate recovery from PTSD within the time parameters of Figure 5.2, we may hypothesize that in appropriate circumstances hypnosis modulates the psychosocial genomics of gene expression.

2. *Alternative gene expression via stress-induced gene splicing.* Meshorer et al. (2002) summarize their research on the molecular-genomic mechanisms of stress with a mouse model of forced swimming similar to that of Kaufer et al. (1998). They generalize their results to PTSD as follows.

> Traumatic stress is often followed by long-term pathological changes. In humans, extreme cases of such changes are clinically recognized as posttraumatic stress disorder (PTSD). Although the immediate response to acute stressful insults has been extensively studied, the molecular mechanisms leading to the long-term neuronal hypersensitivity that is characteristic of PTSD are yet unknown. Stimulus-induced changes in alternative splicing have recently emerged as a major mechanism of neuronal adaptation to stress, contributing to the versatility and complexity of the expression patterns of the human genome (p. 508, italics added).

**Fig. 5.2.** - The psychosocial genomics and time parameters of short- and long-term memory, learning and behavior change. It is hypothesized that psychosocial expectancy operates via the level of neurotransmission at the synaptic vesicles between neurons. Surprise, stress and trauma are hypothesized to operate at the levels of gene transcription in the nucleus of cells and gene translation (making proteins) on the ribosomes in the cytoplasm of the cell (from Rossi, 2002).

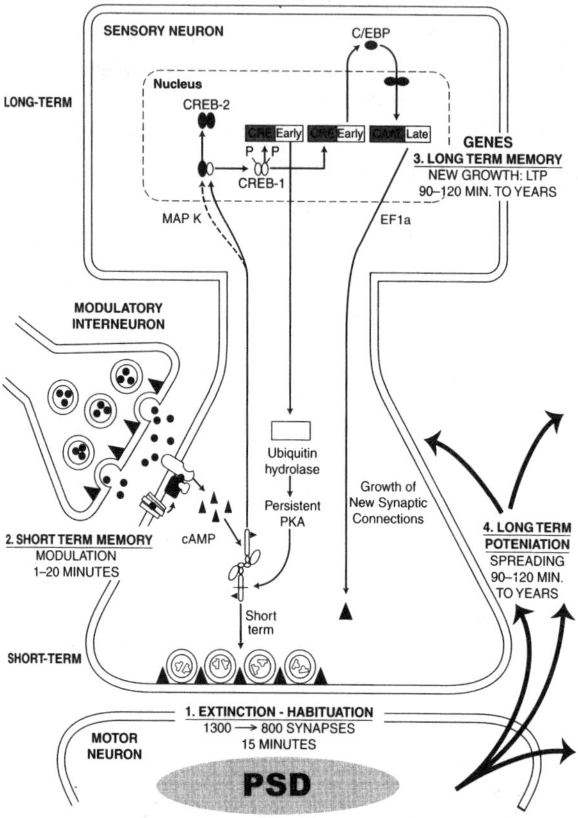


The growth, decay, resynthesis, and remodeling of new synaptic connections in the brain is a basic mechanism for information processing and storage. The post-synaptic density (PSD) contains recptors for neurotransmitters and scaffolding proteins that organize signal transduction. "Much to our surprise, the turnover rate of total PSD protein occurred on the order of only a few hours, suggesting that the entire complement of synaptic proteins in mature neural circuits are replaced multiple times a day!" (Ehlers, 2003, p. 800).

Some background in the molecular dynamics of gene expression is required to understand the implications of this research for our evolving conceptions of the psychosocial genomics of mind-body communication in health, stress and healing. Even today textbooks of molecular biology (Lodish et al., 2000) as well as behaviorally oriented accounts of gene expression (Moore, 2001; Ridley, 1999) give varying definitions of what genes actually are and how they operate. The original "dogma of molecular biology" implied that stimuli from the environment (on all levels from the molecular messengers [hormones] to the psychosocial environment) signaled genes to "turn on" (gene transcription) to make messenger RNA (mRNA), which is a molecular code that is transported out of the nucleus of the cell to the ribosomes in the cytoplasm of the cell where they are transcribed into proteins.

In their original form on the strands of DNA within the nucleus, however, the coding regions for proteins are mixed in non-coding regions, which are called "cistrons". To make meaningful mRNA the non-coding regions of cistrons (called "introns") must be separated from the coding regions (called "exons"). The meaningful exons are then spliced together to make the mRNA that codes for a protein. What is surprising is that the exons from a single cistron can be spliced together in different ways to code for different proteins. If we define a gene in its original form as a cistron (a mixture of coding and non-coding DNA), as most biologists now do, then we can say that the original "dogma of molecular biology"—one gene codes for one protein—is wrong. The coding regions of a single cistron (gene) can be spliced together in different ways to make different proteins that will do different tasks at the cellular level.

It would seem that such details at the level of the molecular biology of gene expression are far, indeed, from the psychosocial level of meaningful human experience in general and therapeutic hypnosis, in particular. What brings these distant levels together is that Meshorer et al. (2002), as quoted above, are now reporting that *traumatic stress induces alternative splicing of mRNA in neurons.* From a psychosomatic perspective, this alternative splicing of mRNA by stress closes the so-called "gap between mind and body". It can now be understood how the holistic approaches of alternative and complementary medicine as well as therapeutic hypnosis can serve as a bridge from the psychosocial level of experience to the fundamental molecular dynamics of gene expression, protein synthesis and the possibilities of healing.

*3. Neuritic mRNA translocation during stress.* Meshorer et al. (2002) continue their exposition of the stress-induced modulations of gene transcription and translation as follows:

> Another stimulus-induced post-transcriptional process is dendritic mRNA translocation, which has been described for several transcripts. *Because psychological, physical, and chemical stressors all cause neuronal activation and hyper excitation, dendritic translocation of specific target mRNAs may follow...* Chemical, *psychological,* and physical stresses all shift splicing from the primary mRNA product that encodes the synaptic membrane AchE-S multimeric to the normally rare "readthrough" AchE-R transcript, which yields soluble monomers... Our findings suggest that neuronal hypersensitivity under stress involves neuritic replacement of AchE-S with AchE-R." (p. 508, italics added)

The neuroscience community has built a bridge from experiences of arousal ("hyperexcitation, hypersensitivity") commonly described as trauma, stress, and surprise on the psychological level to the loci of mind-body information transduction at the molecular-genomic and synaptic levels. Table 1 summarizes neuroscience research on the loci and functions of expectancy and surprise in mind-body communication and healing as it is currently understood. I hypothesize that surprise functions as an "implicit processing heuristic" that evokes and facilitates "unconscious (implicit) processing" that may generate creative responses at the molecular-genomic level that are by definition difficult to predict. Expectancy, by contrast, is a more focused effort to prime predictable responses at the level of synaptic neurotransmission. This is in keeping with the traditional ideals and methodologies of hypnosis. I have recently extended this mind-body bridge of neuroscience to include a wider variety of psychosocial experiences of arousal such as pain, novelty, and the REM stage of dream sleep, as well as creative moments of awe, wonder, and the numinous to facilitate rehabilitation and healing in the cultural and humanistic arts and sciences as well as therapeutic hypnosis (Rossi, 2000,2002).

How can the hypnotherapeutic community now extend this mind-body "communication link of neuroscience to the clinical practice of therapeutic hypnosis? Obviously this will require new levels of cooperation with neuroscientists who have the means of identifying changes in the molecular dynamics of gene expression that could be induced by therapeutic hypnosis (Rossi, 1999, 2000b). Notice that I use the term "therapeutic hypnosis" rather than simply hypnosis as defined by a score on an emotionally neutral scale of hypnotic susceptibility. A simple hypnotic susceptibility score may or may not be associated with changes in gene expression. Behavioral state changes involving psychobiological arousal, pain, relaxation, sleep and dreaming are associated with modulations of gene expression (Rossi, 2002), but emotionally neutral state changes induced by hypnosis have not been explored on the level of gene expression yet. The use of therapeutic hypnosis to induce behavioral state changes (e.g., changes in emotional arousal, stress, relaxation etc.), however, would be expected to modulate gene expression at the three levels described above (that is, changes in which genes are expressed, their alternative splicing, and the translocation of their protein products in neurons). In view of the recent success in finding experimental associations between hypnosis and modulations of the immune system under stress (Gruzelier, Smith, Nagy, & Henderson, 2001; Kiecolt-Glaser, Marucha, Atkinson, & Glaser, 2001), for example, it would require innovative extensions of the new methodologies of neuroscience to measure how hypnosis mediates these psycho-neuro-immune system changes by modulating gene expression and protein synthesis within cells of the immune system.

## *Expectancy and Surprise in Milton Erickson's Neuro-Psycho-Physiological Approaches*

I now propose some speculations about the utilization of the complementary dynamics of expectancy and surprise in the neuro-psycho-physiological approaches of Milton Erickson that may generate detectable changes in the dynamics of gene expression and neurotransmission.

Clinical examples of the utilization of both expectancy and surprise may be found by searching for these two key words in the indexes of a number of volumes of Ericksonian studies (Erickson, 1985, 1992; Erickson et al., 1976; Erickson &

**Fig. 5.3.** - The ultradian (less than 24 hours) and circadian time parameters of long-term changes in acetylcholinesterase activity following experiences of stress (forced swimming in rats) as measured in the cortex (cort), cerebellum (cer), and hippocampus (hipp). Note how the similarity of peak of activity in the hippocampus during stress corresponds to the 90-120 minute period (Kleitman's basic rest activity cycle) required for gene expression and protein synthesis at locus 3 in Figure 1; this supports the hypothesized association between stress, the chronobiology of psychosomatic dysfunctions and healing at the genomic level (Rossi, 2000c, 2002). Research is required to determine if such changes at the molecular level of synaptic neurotransmission and gene expression are associated with the complementary dynamics of expectancy and surprise respectively in therapeutic hypnosis. From Kaufer et al., 1998. Reprinted by permission from *Nature* 393:373-377 (1998), MacMillian Publishers Ltd.

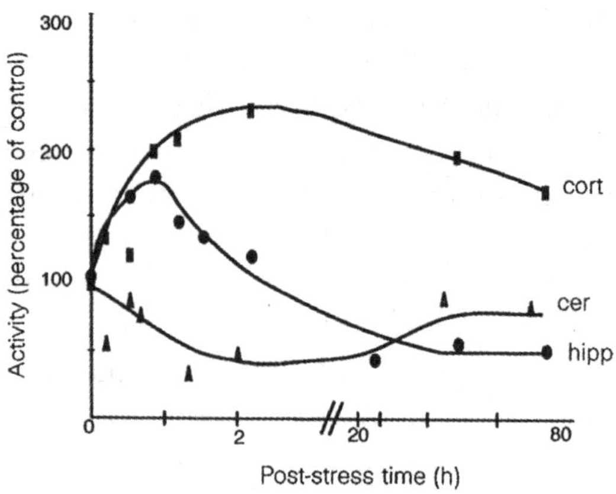

Rossi, 1979, p.288, 1981, 1989). Rossi (1973) published five cases that emphasize Erickson's utilization of psychological shock and surprise in breaking through resistance and rigid mental sets for the reorganization of personality via therapeutic hypnosis.

The integrated use of expectancy and surprise in a 5-stage paradigm of the micro-dynamics of trance and suggestion are outlined in Erickson & Rossi (1976). Zeig (1990, 1999) has reviewed and emphasized the important role of "seeding" concepts and metaphors to facilitate the priming of expectancy in the timeline of Ericksonian psychotherapy. Here are a few brief verbatim examples of Erickson's utilization of the complementary dynamics of expectancy and surprise in a single sentence or two to facilitate hypnotic induction and therapeutic effects.

1. ***Structuring Expectancy to Facilitate Hypnotic Responsiveness*** (Erickson et al., 1976)

> Erickson: (Quietly to Rossi but without altering his far away gaze past the subject during hypnotic induction.) "Notice the silent waiting, the *expectancy* in her. So far as the patient knows you are not pressuring, you're waiting. You are letting the patient *discover* (discovery is usually a *surprise*} how she can enter trance."

Erickson later commented on this by saying, "You can't wait for something without knowing it is going to happen." Rossi responded with, "Your waiting has *the hidden implication that trance will happen.* It is only an implication but it actually structures behavior without the patient being aware of it. You have structured an *expectancy* in her that may initiate her into a response attentativeness to any minimal changes in herself that can be the first signs of a new (novel, *surprising)* hypnotic experience" (pp. 87-88, italics added).

2. ***The Implied Directive*** (Erickson, et al., 1976).

"The implied directive is a label we are proposing for a fairly common type of indirect suggestion that is in current use in clinical hypnosis. The implied directive usually has three parts:

**Table 1:** The psychosocial genomics of expectancy and surprise that may mediate the dynamics of mind-body communication in stress, memory, problem solving, and healing.

| *Psychological Level* | *Molecular Level* | *Psychobiological Function* | *Response Predictability* | *Suggestion Style* |
|---|---|---|---|---|
| **Surprise** | 1. Cell Nucleus: Gene Expression<br>2. Ribosomes: Protein Synthesis | Implicit Processing Heuristics | ***Lower*** | Implicit, Creative |
| **Expectancy** | Synaptic Vesicles: Neurotransmission | Priming: Focusing Responses | ***Higher*** | Explicit, Deterministic |

(1) a time-binding introduction (that evokes *expectation),* (2) the implied (or assumed) suggestion, and (3) a behavioral response to signal when the implied suggestion has been accomplished (usually in a *surprisingly* easy manner). We may thus analyze an implied directive from this session as follows:

> Erickson: As soon as you know (pause)
> (1) A time-binding introduction that focuses the patient on (builds *expectation* for) the suggestion to follow.
> Erickson: only you or I, or only you and my voice are here (pause)
> (2) The implied (or assumed) suggestion (for a *surprising* dissociation) Erickson: your right hand will descend to your thigh.
> (3) The behavioral response signaling that the suggestion has been accomplished.

An implied directive frequently used by Rossi to end a hypnotherapeutic session is as follows:

> Rossi: "As soon as your unconscious knows (pause)"
> (1) A time-binding introduction that facilitates dissociation and reliance on the unconscious (that is, implicit processing to build *expectancy)*
> Rossi: "...it can again return to this state comfortably and easily to do constructive work the next time we are together, (pause)
> (2) The implied suggestion for easy reentry to trance phrased in a therapeutically motivating manner (that arouses *expectancy).*
> Rossi: ".. .you will find yourself awakening feeling refreshed and alert, (usually a pleasant *surprise* in a therapeutic con-

> text when discomfort and symptoms disappear).
> (3) The behavioral response signaling that the above suggestion has been accomplished.
> The implied directive is thus a way of facilitating an intense state of internal learning or problem solving. We may suppose that all of a subject's available mental resources (e. g., stored memories, sensory and verbal association patterns, various forms of previous learning, etc.) are marshaled toward a creative state of learning and problem solving. Since recent experiments in the neurophysiology of learning suggest that new proteins are actually synthesized in the appropriate brain cells during learning (Rossi, 1973), we may speculate the implied directive facilitates the internal synthesis of new protein structures that could function as the biological basis of new behavior and phenomenological experience in the patient. (Pp. 188-189)

In the generation since these words were written the psychosocial genomics of memory, learning, stress and healing have been investigated in great detail by neuroscientists. It will now require ingenious extensions of their new research methods on the levels of gene expression and synaptic neurotransmission to explore the utilization of the complementary dynamics of both surprise and expectancy in therapeutic hypnosis. Such research could utilize implicit processing heuristics as one type of therapeutic suggestion that may modulate mind-body communication and healing at the cellular-genomic level (Rossi, 2002).

* * * *

This conceptual review indicates how the leading edge of current neuroscience research is building a bridge between mind and body on the molecular level by identifying the cellular-genomic and synaptic loci of memory, learning and behavior change. Psychological experiences of trauma, stress, and surprise that evoke psychobiological arousal to engage the levels of gene expression and protein synthesis could also facilitate neurogenesis, creative problem solving and healing. Psychosocial experiences of expectancy, by contrast, may prime more focused and predictable hypnotic responses by modulating neurotransmitter release at the synaptic vesicle level between neurons. It would require ingenious extensions of current neuroscience methodology to measure how hypnosis mediates the psychosocial genomics of surprise and expectancy at the cellular-genomic and synaptic levels respectively. We have speculated how Milton Erickson's neuro-psycho-physiological approaches may engage the complementary dynamics of expectancy and surprise as a complex system of creative adaptation to facilitate rehabilitation and healing on many levels of human experience from mind to gene expression.

# VI
# STRESS-INDUCED ALTERNATIVE GENE EXPRESSION IN PSYCHOTHERAPY

## *Introduction*

Recent research documents how psychosocial stress can induce the alternative gene splicing in humans. Stress has been found to alter the expression of the acetylcholinesterase gene to generate at least 3 alternative proteins that are implicated in a wide variety of normal mind-body functions and a number of neuro-pathologies. Stress can alter the early embryological development of the brain in the fetus as well as plasticity of the brain in adulthood in the post-traumatic stress disorders (PTSD) and the stress-associated dysfunctions of the CNS, the endocrine and immune systems, as well as age-related neuropathologies. Such stress-induced alternative gene splicing is proposed here as a major mind-body pathway of psychosocial genomics – the modulation of gene expression by psychological, social, and cultural processes. We explore the types of research that are now needed to investigate how the stress-induced alternative splicing of the acetylcholinesterase gene may play a pivotal role in the deep psychobiology of creative psychotherapy, meditation, spiritual rituals, and the experiencing of positive humanistic values such as compassion, beneficence, serenity, forgiveness, and gratitude that have been associated with mind-body medicine and healing.

## Selye's visionary concept of stress

Hans Selye presented a visionary concept of stress in the 1956 edition of his famous book, *The Stress of Life*.

> Perhaps the most fascinating aspect of medical *research on stress* is its *fundamentally permanent value to man*... The study of stress differs essentially from research with artificial drugs because it deals with *the defensive mechanisms of our own body.* The immediate results of this budding new science are not yet as dramatic in their practical applications as are those of many drugs, but what we learn about nature's own self-protecting mechanisms can never lose its importance. Such defensive measures as the production of adaptive hormones by glands are built into the very texture of the body; we inherited them from our parents and transmit them to our children, who, in turn, must hand them on to their offspring, as long as the human race shall exist. *The significance of this kind of research is not limited to fighting this or that disease. It has bearing upon all diseases and indeed upon all human activities,* because it furnishes knowledge about the essence of THE STRESS OF LIFE. (Italics and capitals in the original, p. 304-5).

Selye's initial biological conception of stress focused on the activity of "adaptive hormones by glands [that] are built into the very texture of the body." This relationship between stress and hormones was the forerunner of our current recognition of the hypothalamic-pituitary-adrenal (HPA) axis in the psychobiological integration of mind-body medicine. Current research emerging from the human genome project traces this hormonal integration of mind and body via the HPA axis to the genomic level (Rossi, 2002). Sensory-perceptual cues of psychosocial and physical stress and trauma are communicated to the organs, tissues, and cells of the brain and body via the neural, endocrine, and immune systems. The ultimate level of adaptation to stress takes place when challenging signals from the environment are communicated to the cellular level to turn cascades of gene expression on and off. When genes are turned on by stress it means that the information in their DNA code is "transcribed" into messenger RNA (mRNA) that then serves as a template that is "translated" into proteins that make up the structure and adaptive functions of the brain and body. Proteins, for example, regulate the structural and energy dynamics of the cell as well as the neurotransmitters, hormones, and signals of the psychoimmune system in mind-body medicine.

The original "Dogma of Molecular Biology" presented by Watson and Crick (1953a & b) 50 years ago proposed that one or more signals from the environment and/or body turned on *one gene that made one protein that modulated one molecular structure, and one adaptive physiological function*. Since there are only about 35,000 genes in the human genome, however, this led to the question of how so few genes could account for the immense complexity of human adaptive behavior in health and illness. This question is now resolved in part by the general concept of stress-induced gene expression. As anticipated by Watson and Crick (1953a & b), their original dogma of *one gene – one protein – one structure, and one function* was an oversimplification. A gene does not exist as one piece of DNA on a chromosome, which codes for one protein. Rather, each gene is a mixture of "exons" that code for parts of a protein and "introns" that do not code for a protein. The introns, originally called "junk DNA" (before current research that suggests they have adaptive functions), first must be cut out and separated from the protein coding exons. These exons are then "spliced" together to transcribe the gene into its messenger ribonucleic acid (mRNA) that will be translated into a protein that will regulate a physiological function.

Currently it is estimated that about 30 % to 60 % of the genes in the human genome can be spliced together in different ways to generate alternative sets of proteins that modulate physiological adaptation in response to challenge, stress, and trauma on many levels ranging from the purely physical to the psychosocial (Lewis et al., 2003). It is not entirely clear, however, that the usage of the words, "adaptation, challenge, stress, and trauma" mean the same thing to molecular biologists and mind-body psychotherapists. Can the "oxidative stress" on a molecular level induced by physical trauma, for example, be equated with "psychosocial stress" and the post-traumatic disorders (PTSD) in mind-body medicine? Can activities on the cognitive-emotional-behavioral levels of human experience really modulate the molecular pathways that generate alternative gene expression in the brain and body in sickness and health? The "Cartesian divide" between mind and body is most obvious in the conceptual gap between genes and their biochemical pathways, on the one hand, and the apparently independent phenomenological activities of the mind and psychological experience, on the other. This chapter extends the emerging field of *psychosocial genomics* (Rossi, 2002a & b; 2003a, b, & c), to include the concept of *stress-induced alternative gene expression* as an empirical and scientifically verifiable link between mind and body. The surprising idea that experiences on the psychological, social, and cultural levels can modulate gene expression on the molecular level is an unexpected validation of Selye's visionary concept of "the stress of life" for understanding the integration of mind and body in human nature. Psychosocial genomics now seeks to explore research paradigms that can demonstrate how psychotherapy and related cultural processes and rituals such as meditation, prayer, and deeply meaningful humanistic experiences of art, drama, dance, music, poetry etc can modulate alternative gene expression to facilitate health, rehabilitation, and healing.

### Psychosocial stress alters cholinergic gene expression

In this conceptual review we will first outline the psychosocial genomics of stress-induced gene expression in the cholinergic system that mediates stimulation, excitation, and communication in the brain and the body. The cholinergic system consists of the acetylcholine gene, that codes for acetylcholine, which was the first neurotransmitter ever discovered, and its partner, the acetylcholinesterase gene, that codes for the enzyme that destroys (hydrolyzes) acetylcholine when it completes its job of excitatory neurotransmission in the central nervous system (at synapses connecting neurons) as well as the peripheral nervous system (at neuromuscular junctions). Such pairs (and sometimes families) of genes that turn physiological functions on and off in complex biofeedback loops of communication with the environment are typical of the *classical regulatory systems of body and mind involving neurotransmitters, hormones, psychoimmune system, and growth factors of the brain and body.*

In the past few decades, however, research teams lead by the molecular biologist, Hermona Soreq, and others have discovered and investigated many *non-classical* functions of the acetylcholinesterase gene induced by stress (Grisaru, et al., 1999; Kaufer, Friedman, & Soreq, 1999; Soreq & Seidman, 2001). In essence it has been found that a wide variety of stressors ranging from the psychosocial (e.g. crowding stress or the gentle handling of mice by humans) to the environmental (alternations in the circadian cycle), to the physical (forced swimming), and molecular (drug and chemical intoxication as in the Gulf War syndrome) can lead to the overproduction of the acetylcholinesterase (AChE) gene to modulate a wide range of mind-body processes as illustrated in figure one.

The wide range of stress-associated pathologies associated with massive amplification of AChE gene illustrated in figure 1 denotes a convergent outcome of acute psychosocial stress, chemical intoxication, and traumatic head injury. Chronic stress and a long term excess of AChE promotes cognitive decline and neurodeterioration as in Alzheimer disease (Sternfeld et al., 2000), Down's syndrome, and some symptoms of myasthenia gravis (Brenner et al., 2003). Age-related and late-onset dysfunctions that are particularly vulnerable to alternative splicing variations include age-related macular degeneration (AMD), amyotropic lateral sclerosis (ALS), and frontotemporal dementia with Parkinsonism (Meshorer & Soreq, 2002). Massive amplification of the human AChE gene has been associated with autoimmune disorders such as Grave's disease, the leukaemias (Grisaru et al., 1999), ovarian carcinomas, thrombocytopenia, and exposure to organophosphate insecticides (Soreq & Glick, 2000). AChE activity also has been found associated with meningioma, astrocytoma, and glioblastoma tumors (Razon et al.,1984; Gurwitz et al., 1984). In mice the accumulation of stress-associated AChE predicts male infertility (Mor et al., 2001). How such apparently divergent pathologies can be related has been investigated with more detailed investigations of the alternative splicing variants of the AChE gene produced under acute and chronic stress.

## Acute and chronic stress induce alternative gene splicing

Acute and chronic stress can induce a series of changes in cholinergic gene expression (or "splicing") that alter the normal balance of acetylcholine and acetylcholinesterase metabolism associated with the posttraumatic stress disorders (PTSD) and many related mind-body dysfunctions (Kaufer et al., 1998) that are well summarized by Sternfeld et al. (2000).

> Both chronic stress and acute stress promote neuroanatomic changes in brains of evolutionarily diverse species, including higher vertebrates and humans. Some of these changes likely reflect normal physiological adaptation to injury, environmental challenge, traumatic experience, or even standard maintenance conditions of laboratory animals. *However, stress may also precipitate delayed or prolonged neuropsychiatric dysfunction, the etiology of which is yet poorly defined. For example, up to 30% of individuals exposed to an acute traumatic experience develop posttraumatic stress disorder, a syndrome characterized by progressively worsening personality disturbances and cognitive impairments.* The cellular and molecular factors mediating the switch between physiological accommodation of stress and progressive disease are unknown but likely reflect complex interactions between the genetic background of the challenged individual and the nature of the stress insult. *The accepted notion is that physiological stress responses are beneficial in the short run but detrimental if overactivated or prolonged. This concept suggests the existence of stress modulators designed to regulate the extent, duration, and long-term impact of acute stress responses.* We recently reported massive induction of a unique mRNA species encoding the rare "read-through" variant of acetylcholinesterase (AChE-R) in brains of mice subjected to forced swimming stress. AChE-R differs from the dominant "synaptic" variant, AChE-S, in the composition of its C-terminal sequence. . . In hippocampal brain slices, induced AChE-R seemed to play a role in delimiting a state of enhanced neuronal excitation observed after acute cholinergic stimulation. *This observation suggested that AChE-R acts as a stress modulator in mammalian brain.* (p. 8647). . .Our current findings therefore demonstrate that *AChE-R, most likely with another modulator or modulators, may be beneficial in the response to acute stress at two levels: (i) by dampening the acute cholinergic hyperactivation*

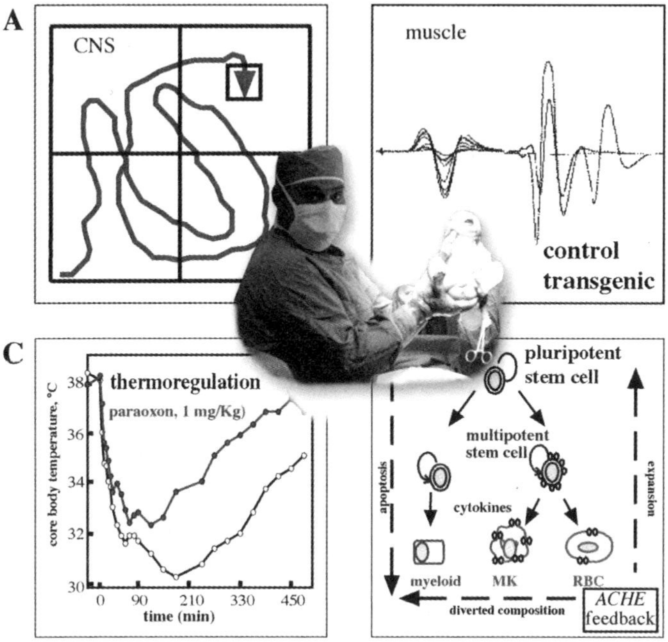


**Figura 6.1.** Images of a wide range of the physiological connections and consequences of stress induced gene expression of acetylcholinesterase (AChE) in the cholinergic system of mind-body medicine. A. Over expression of AChE in brain neurons results in cognitive impairments in the CNS as evident in failures in the Morris water maze. B. Progressive deterioration at the neuromuscular junction as evident in rapid fatigue in the electromyographic response. C. Failure in thermoregulation on exposure to drugs such as paraoxon. D. Enhanced expansion of multipotent stem cells in the blood (primary haematopoistic cells) and diversion of their differentiation toward myeloid lineages. (With permission from Soreq & Glick, 2000).

It would be really proper to study in the newborn the differences in gene expression of the acetylcholinesterase (AChE) stress induced by birth in the cholinergic system. Try to imagine the stress to which the fetus is submitted during delivery, its progression through the birth-canal, the abrupt expansion of the lungs clinically evidenced by the first strong cry, the delicate obstetric manipulation and cutaneous stimulation, the removal of the oro-pharingeal mucus, not to mention the cutting of the umbilical cord as well as the circumcision practiced in various cultures and religions. It would be very interesting to study the possible effect on gene expression of the fetus when completely separated from the mother by the umbilical cord cutting. With these studies we could have a better understanding of the fetal-neonatal stressful conditions and devise preventive methods to give the unborn child a "***better and more serene birth***"(Iannotti, 2004)

> *that accompanies stress and (ii) by protecting the brain from entering a downward spiral into progressive neurodegeneration through an as-yet unidentified mechanism, which could involve noncatalytic activities and/or direct competition with AChE-S.* In that case, the *diversion of up-regulated AChE expression after insults to the central nervous system from production of the usual AChE-S to the unusual AChE-R isoform [by stress-induced alternative gene expression or splicing] would reflect an elegant evolutionary mechanism to avoid the dangers of over-expressed AChE-S.* These findings imply that mutations conferring heritable up-regulation of AChE-R would protect the mammalian central nervous system from some age-dependent neuropathologies. The definitive role of AChE-R after transient stress or drug-induced over expression remains to be examined in additional animal models permitting conditional regulation of AChE-R expression. (p. 8652. Italics added here.)

Figure 6.2 summarizes a major feedback loops between AChE-S (synaptic AChE) and AChE-R (readthrough AChE) in what has been called "the vicious circle of stress and anti-cholinesterase responses" (Kaufer, Friedman, & Soreq, 1999). AChE-S, the principle form of acetylcholinesterase, is found in the synapse where it regulates classical excitatory function of acetylcholine neurotransmission between neurons in the brain mediating visual, motor, and emotional experiences in normal and highly aroused states of consciousness as well as REM dream sleep (Hobson & Leonard, 2001) and at the neuromuscular junctions in the body (Soreq & Seidman, 2001). AChE-R (R is for *readthrough* because of the way it is spliced together), the second most common form of acetylcholinesterase, accumulates in the brain and the blood in response to acute psychological stress as well as toxins from the environment (Soreq & Glick, 2000).

In the next section we will frame a number of open questions about the possible associations between the immediate-early gene c-fos, stress-induced alternative gene expression illustrated in figure 2, and the psychosocial genomics of mind-body psychotherapy that are now ripe for experimental investigation.

## The 4-Stage Creative Process of psychosocial genomics

The central role of the immediate-early genes such as *c-fos* in psychophysiology has been described from many points of view: 1) A marker for neuronal activity and states of consciousness (Bentivolglio & Grassi-Zucconi (1999); 2) A modulator of acute stress signals (Kaufer, Friedman, & Soreq, 1999); 3) A mediator of mind-body communication and healing as we already seen back in Chapter 2, Figure 2.4(Rossi, 2002a & b). Figure 3 illustrates the typical 90-120 minute activity cycle of *c-fos* and 10 other genes in what has been called "Single-Cell Gene Expression Profiling" (Levsky et al., 2002). A key goal of the Human Genome Project is to relate the expression of specific genes to the proteins they code for and the physiological pathways they regulate, often in response to environmental factors such as stress, in real time. Levsky et al. summarize their work in this area.

> "The instantaneous transcriptional activity of genes in single cells allows observation of causes and effects of expression. *Eventually, the physiological state of cells within tissues will become synonymous with a pattern of gene expression.* This will provide a quantitative approach to factors influencing gene expression patterns, such as occur in cytopathology, development and cell differentiation, infectious disease, or response to drug treatment. Investigation of functional genomics may now be approached at the cellular level. *We expect that the enormous information inherent to the expression of many genes in large cell populations will aid the understanding of relationships among genes in single nuclei and their cooperative and cumulative roles in physiology and disease.* " (p. 836, Italics added.)

This chapter now presents the case for extending the relationship between gene expression and the classical physiology of Kleitman's (1969) Basic Rest-Activity Cycle (BRAC) that has an identical 90-129 minute periodicity as described previously in chapter 3, to the psychosocial genomic foundations of the classical 4-stage creative process applied to psychotherapy (Rossi, 1996, 2002a). The wave-like gene expression profile of *c-fos* over 120 minutes in figure 6.3c is strikingly similar to the

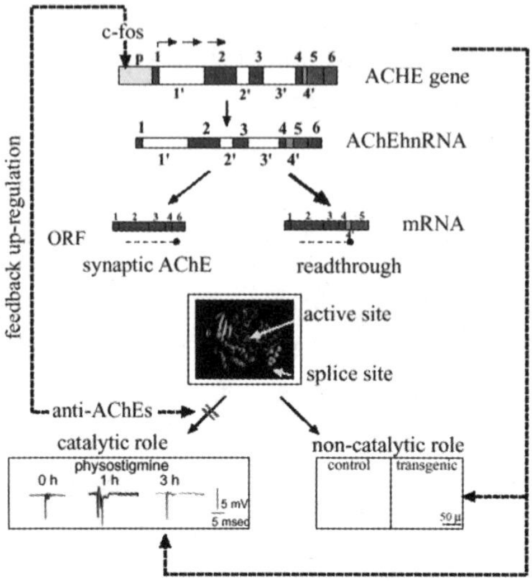


**Fig. 6.2** An outline of the vicious biofeedback circle of stress and anticholinesterase responses.

Psychosocial stress, physical head trauma, and chemical intoxication use a similar biofeedback loop (dashed lines) in which the up-regulation of the immediate-early gene, *c-fos* induces transcription of the acetylcholinesterase gene (AChE) by binding to its promoter (p). Alternative gene splicing takes place when AChE hnRNA (nascent heteronuclear RNA) is preferentially processed (heavy arrow) into AChE-R mRNA (readthrough) that can protect the CNS from hyperactivity and neurodeterioration via over-activated AChE-S (synaptic AChE). Open reading frames (ORF) of the alternatively spliced genes (AChE-S & AChE-R) that are eventually translated into a protein with physiological functions are shown as short dashed lines with a dot. In its *catalytic role* AChE modulates acetylcholine availability and cholinergic excitation that may have survival value in acute short term life emergencies. Exposure to anticholinesterases (such as *physostigmine*) increases acetylcholine levels increasing cholinergic excitation in the hippocampus of the brain. Within 3 hours there is a delayed suppression of this initial excitation (downward right broken arrow) due to trauma and stress. The *non-catalytic role* of AChE modulates the growth of neurites (the extensions on the dendrites of neurons where synapses are located). In *transgenic* mice (where a gene is inserted experimentally), in comparison with *controls,* the overexpression of AChE, however, attenuates dendritic branching on brain neurons, which results in a deterioration of memory, learning, and behavior due to the accumulation of excess AChE in this vicious circular biofeedback response. A third form of the acetylcholinesterase gene, AChE-E (erythrocyte AChE), that can arise from alternative splicing associated with a range of physiological functions and pathologies of the blood and bone is not illustrated here. (With permission from Kaufer et al., 1999).

wave-like profile of the 4-stage creative cycle of growth oriented psychotherapy long described as the "Breakout Heuristic" (Rossi, 1967, 1968, 1972/2000, 1996, 2002a) that usually takes place over 90-120 minutes as illustrated in figure 6.4.

At this point we do not know whether the similarity between 1) the wave-like activity profile of the immediate-early gene *c-fos* (fig. 6.3) in its role in mediating the acute adaptive and chronic maladaptive dynamics of alternative cholinergic gene expression, and 2) the wave-like 4-stage creative cycle of psychotherapy (fig. 6.4) is simply a coincidental observation or a deeply meaningful psychobiological connection that is now ripe for more direct experimental investigation with DNA microarrays (Whitney et al., 2003). We can summarize the open questions that need to be answered to establish the case for a causal albeit complex non-linear association between gene splicing, expression and psychotherapy in mind-body medicine.

## Chronic Stress In Neurological, Autoimmune, & Age-Related Dysfunctions

The basic idea that the positive adaptive functions of brain plasticity can be facilitated by the initial excitatory cholinergic dynamics of acetylcholine in acute stress while chronic stress overloads the CNS, endocrine, and immune systems to the point of dysfunction now requires more direct documentation via gene expression profiling of AChE-S, AChE-E, and AChE-R in real time throughout the entire 24 hour circadian cycle in normal and clinical populations. The shifts between dominance of AChE-S, AChE-E, and AChE-R may be a common denominator in the critical transitions between the initial positive adaptations of acute but brief stress versus the pathologies associated with chronic stress that are the mission of mind-body medicine to understand and heal.

## Time Parameters of Stress, Gene Expression, and Psychotherapy

The hypothesized association between 1) the normative 90-120 minute profile of gene expression, 2) the cholinergic dynamics of the acute and chronic stress response, 3) Kleitman's Basic-Rest Activity Cycle, and 4) the 4-stage creative response in mind-body psychotherapy now require direct investigation via coordinated studies with brain imaging and DNA microarray technology (Gibson, 2003; Rossi, 2002a). Note in fig. 6.4 how the first two stages of the creative process in psychotherapy evoke a state of psychobiological *Arousal* that corresponds to the state of excitement of the CNS mediated at least in part by the neurotransmitter acetylcholine. In the older psychiatric literature of 100 years ago this state of arousal sometimes leads to an emotionally driven state of *conflict, negative affect,* and *Crisis* at the peak of stage 2 when a patient apparently lost emotional control in what was described as hysteria. Recently, emotionally driven fear and *conflict* behavior have been associated with stress-induced alternative splicing and AChE-R (Birikh et al., 2003). Depth psychotherapy began when this *Crisis* was channeled into a thera-

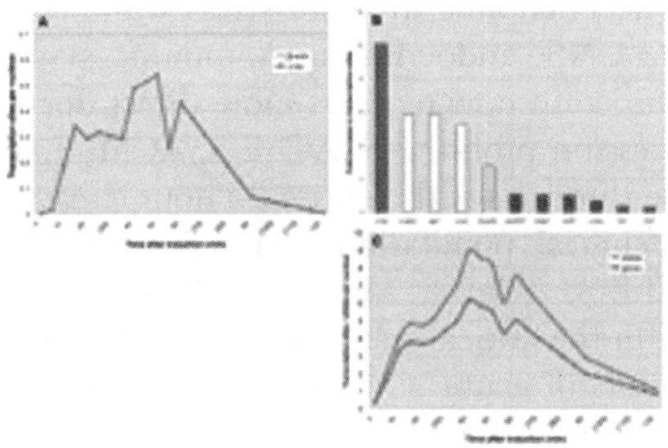

**Fig. 6.3.** The gene expression profiles of the immediate-early gene *c-fos* and 10 other genes (alleles) from normal human fibroblasts. All genes showed measurable activation within 5 or 10 minutes. By 40 minutes a peak of activation was reached and by 90-120 minutes gene expression had returned to the baseline. (With permission, Levsky, et al. 2002)

peutic catharsis that evoked traumatic early memories that were the putative source of PTSD and "neuroticism" in the early work of Pierre Janet, Sigmund Freud, and others (Ellenberger,1970; Zilboorg, & Henry, 1941). With a bit of luck and good guidance this state of hyperarousal (acetylcholine excitation) sometimes led to a breakthrough to new insight, problem solving, positive affect, and the resolution of stress-induced mind-body symptoms in stage 3 of the creative cycle. When the patient was encouraged to appreciate the value and meaning of this apparently spontaneous insight and positive emotions the *Crisis* was resolved, symptoms were reportedly healed, and *Relaxation* was experienced spontaneously in the second half of the creative cycle (stages 3 and 4 as illustrated in figure 6.4). While his psychobiological shift from arousal to crisis, insight, and symptom resolution with relaxation and a sense of well-being is characteristic of all modern cognitive-behavioral approaches to psychotherapy, its relationship to Kleitman's Basic Rest-Activity Cycle and the classical 4-stage creative process is rarely noted (Rossi, 1996, 2002a).

Note the two small parallel lines indicating a break just after the peak or crisis between stages 2 and 3 of the creative cycle in figure four. This "break" was used to indicate the unknown process by which the *Arousal* of stages 1 and 2 was shifted to the Relaxation of stages 3 and 4 of the creative process. It is now hypothesized that this break is mediated by alternative gene splicing leading to a dominance of AChE-R over AChE-S that is summarized above by Sternfeld et al., "Our current findings therefore demonstrate that *AChE-R, most likely with another modulator or modulators, may be beneficial in the response to acute stress at two levels: (i) by dampening the acute cholinergic hyperactivation that accompanies stress and (ii) by protecting the brain from entering a downward spiral into progressive neurodegeneration through an as-yet unidentified mechanism, which could involve noncatalytic activities and/or direct competition with AChE-S*" (p. 8652). We now need more direct tests of this psychobiological hypothesis of mind-body therapy by assessing the time parameters of the relationships between *AChE-S* (more present in stages 1 and 2 of the creative cycle) and *AChE-R* (more evident in stages 3 and 4 of the creative cycle).

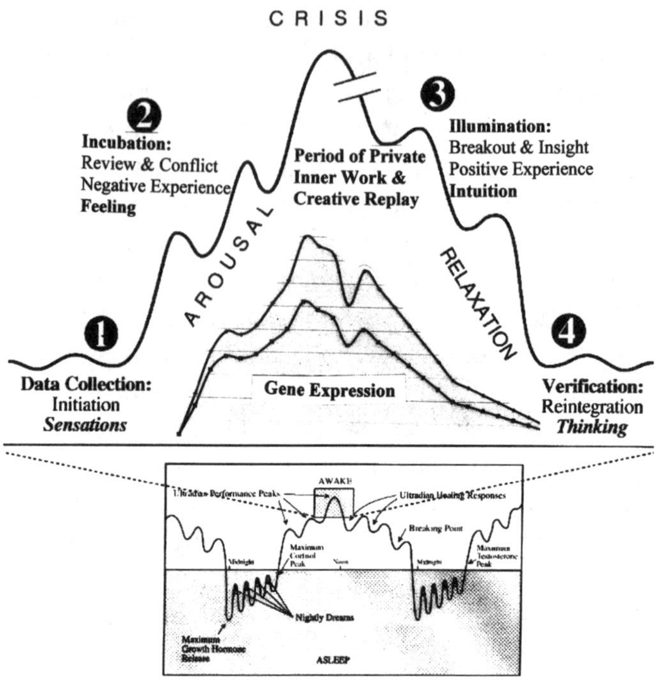


Fig. 6.4. An illustration of our basic psychosocial genomics hypothesis: Gene expression is the ultimate psychophysiological foundation of Kleitman's Basic Rest-Activity Cycle and the 4-stage creative process. The lower diagram summarizes the normal circadian (~ 24 hours) profile of the alternating 90–120 minute ultradian (> 20 hours) rhythms of waking and sleeping characteristic of Kleitman's Basic Rest-Activity Cycle (BRAC) in a simplified manner. The ascending peaks of rapid eye movement (REM) sleep typical of nightly dreams every 90–120 minutes or so are illustrated along with the more variable BRAC rhythms of activity, adaptation, and rest in the daytime. This lower figure also illustrates how many hormonal messenger molecules of the endocrine system such, as *growth hormone*, the activating and stress hormone *cortisol,* and the sexual hormone *testosterone,* which all modulate gene expression, have BRAC peaks at different times throughout the 24-hour cycle (Kleitman & Rossi, 1992; Lloyd & Rossi, 1992).

The upper diagram outlines the basic psychobiological unit of psychotherapy as the creative utilization of one natural 90–120 BRAC ultradian rhythm of arousal and relaxation illustrated in the lower diagram. The classical four stages of the creative process: 1) Data collection; 2) Incubation; 3) Illumination; 4) Verification have been well documented by Wallas (1926) and others (Smith, 1995). In general the four basic psychological functions of sensations, feeling, intuition, and thinking as originally described by Jung (1923;1943/1966) appear to be related to the 4 stages of the creative process (Rossi, 2002a & b). The Levsky, et al. (2002) image in fig. 3c of the expression profiles of the immediate-early gene *c-fos* and 10 other genes (alleles) over the same ultradian time period of 90-120 minutes is included here to illustrate our basic psychosocial genomics hypothesis that gene expression is the ultimate psychophysiological foundation of the classical 4-stage creative process in psychotherapy as well as Kleitman's Basic Rest-Activity Cycle.

## Brain Plasticity and Alternative Gene Splicing in the Creative Cycle

Brain plasticity refers to the growth and transformations in the physical structure, connections, and functions of the cells of the brain generally described as *synaptogenesis* (Cohen-Cory, 2002) and *neurogenesis* (Kempermann & Neumann, 2003) during 1) normal embryological development and the entire life cycle and during 2) the normal, everyday life changes in the neural networks associated with psychological experiences of novelty, creativity, significant memory, learning, and behavior. Acetylcholinesterase (AChE) is expressed during the embryological development of the brain where it is involved in the *growth of neurites*, the extensions on the axons of neurons where the synapses are located (Grisaru et al., 1999).

In the well functioning adult the alternative splicing of the AChE gene presumably is involved in brain plasticity in regulating the classical excitatory functions of acetylcholine neurotransmission in mediating visual, motor, and emotional experiences in normal and highly aroused states of creative consciousness as well as REM dreaming sleep (Hobson & Leonard, 2001).

We now need DNA microarray studies of the relationships between the 3 alternative splicing variants of the acetylcholinesterase gene, AChE-S, AChE-R, and AChE-E, with working and declarative memory (Blank et al., In Press; Cohen et al., In Press) during the 4-stage creative process while awake during psychotherapy as well as in the various states of sleep (particularly REM dream versus stage 4 deep sleep) in normal and clinical populations throughout the entire life cycle.

A recent example from mathematics illustrates the phenomenology of the 4-stage creative process (Aczel, 1997) that may correspond to the psychobiological dynamics of the growth of neurites and synaptogenesis in brain plasticity. I hypothesize that the shift from a dominance of AChE-S to AChE-R is manifest in this example during the emotional transition from the peak of creative excitation, conflict, and crisis typical of the state of psychobiological arousal of stages 1 and 2 to the relaxation of stages 3 and 4 when the new creative insight is finally integrated and experienced as real (Rossi, 2002a).

Research on the cognitive and emotional aspects of relationship between *intuition* in stage 3 and *thinking* in stage 4 has been recognized by the 2002 Nobel Prize for Economics awarded to Kahneman (2003).

Everything new seems to be unreal on first exposure to it. A creative experience sometimes seems unreal precisely because it is still so new that it does not yet have a well-established place in our familiar contexts and ways of understanding. Those odd sensations and qualities of seeming distortion and grotesqueries in dreams are actually new patterns of awareness breaking through to consciousness as they shatter, reframe, and replace previous contexts and meanings in a seemingly spontaneous manner.

***That which seems absurd, bizarre, or meaningless in dreams only seems so from the older more established attitudes and points of view that still dominate the conscious mind***. Consider this poignant description of the initial sense of unreality and disbelief that mathematician Andrew Wiles experienced at the critical moment of correcting an error to finally solve a 300-year-old puzzle by proving Fermat's Last Theorem (Aczel, 1996, pp. 132-33).

'Wiles studied the papers in front of him, concentrating very hard for about 20 minutes. And then he saw exactly why he was unable to make the system work.

'It was the most important moment in my entire working life.' he later described the feeling. 'Suddenly, totally unexpectedly, I had this incredible revelation. Nothing I'll ever do again will.' At that moment tears welled up and Wiles was choking with emotion.

What Wiles realized at that fateful moment was so indescribably beautiful, it was "so simple and so elegant–and I just stared in disbelief." Wiles walked around the department for *several hours*. [Acetylcholinesterase (AChE) accumulates in the hippocampus of the brain for *several hours* to modulate a peak in cholinergic hyperarousal (Kaufer et al., 1999). ] He didn't know whether he was awake or dreaming. Every once and a while, he would return to his desk to see if his fantastic finding was still there–and it was.'

Many similar experiences are reported by creative workers in literature and the arts (Hadamard, 1954; Zeki, 1999, 2001). Such creative moments are the basis of much art, drama and literature in fantasy and fiction.

What, for example, are the magical moments so common in fairytales and myth supposed to mean? Why do children, as well as adults, continue to find delight in reading such fictions? Well, you might respond such delights are normal experiences of "enjoyment."

What precisely is this positive experience of stage 3 of the creative process that we call enjoyment?

We can understand why nature builds in the perception of danger and the sensation of pain as an aid to survival but what good is the enjoyment of the fictions of drama and tales of fantasy and wonder?

I propose that *the many experiences of so-called "enjoyment" on an explicit conscious level are aspects of the numinosum (fascination, mysterious, tremendous) that signal the occurrence of gene expression, synaptogenesis, and/or healing at an implicate level.*

*This is the essence of what I proposed as the "Novelty-Numinosum-Neurogenesis Hypothesis" characteristic of mind-body healing in chapter three* (Rossi, 2002a)

## The Wide Range of Mind-Body Therapies and Cultural Rituals

We can now appreciate the broader and deeper implications of Hans Selye's vision when he wrote, "Perhaps the most fascinating aspect of medical *research on stress* is its *fundamentally permanent value to man*." (Selye, 1956, 304). We need to investigate how the stress-induced alternative splicing of the acetylcholinesterase gene and related psychosocial genomic systems may play a pivotal role in the deep psychobiology of the creative process in the arts and sciences, psychotherapy, meditation, spiritual rituals, and the experiencing positive humanistic values such as compassion, beneficence, serenity, forgiveness, and gratitude that have been associated with mind-body medicine (Rossi, 2002a). The receptivity of the neuroscience community to investigating the therapeutic potentials of such positive values associated with meditation, prayer, and spiritual practice as well as psychotherapy suggests progress will be made in this area of mind-body medicine in the immediate future (Barinaga, 2003; Davidson et al., 2003).

* * * *

The basic concept of psychosocial genomics – that psychological states, creative experience, social stressors, and deeply meaningful cultural rituals can modulate alternative gene expression and psychophysiology – is proposed here as an emerging scientific foundation for mind-body medicine and psychotherapy. Research documents how psychosocial stress can alter the expression of the acetylcholinesterase gene to generate 3 alternative proteins that are implicated in a wide variety of normal and adaptive mind-body functions ranging from brain plasticity in acute but brief challenges - to the chronic stress and age-related pathologies of the central nervous system, the endocrine and immune systems. We have extended the possible applications of stress-related alternative splicing of the acetylcholinesterase gene to the deep psychobiology of Kleitman's Basic Rest-Activity Cycle and the 4-stage creative process in brain plasticity, psychotherapy, and significant developments of human consciousness in the arts, sciences, and humanities. Consistent with Hans Selye's broad vision of the significance of stress in human experience, we outlined how positive humanistic values and cultural rituals generally associated with mind-body medicine may be investigated with current neuroscience methods of brain imaging and DNA microarray technology.

# VII
# SYSTEMS BIOLOGY AND THE PSYCHOSOCIAL GENOMICS OF THERAPEUTIC HYPNOSIS

## Introduction

This chapter is a heuristic effort to integrate therapeutic hypnosis and psychotherapy with the emerging concept of systems biology. Therapeutic hypnosis is identified as a neuro-psycho-physiological work function related to the bi-modal circadian profiles of hypnotic susceptibility, body temperature, and gene expression. A bioinformatic approach to computer data mining is used to compare the *profile of circadian hypnotic susceptibility* to the *profiles of gene expression while awake and asleep* to explore the psychosocial genomic foundations of therapeutic hypnosis. Such research may determine what profiles of gene expression related to health, stress, and illness could be modulated by suggestion and thereby establish a broad scientific approach for facilitating mind-body healing via the systems biology of therapeutic hypnosis and psychotherapy.

## Therapeutic Hypnosis as a Neuro-Psycho-Physiological Work Function

Researchers as well as clinicians in hypnosis have noted that subjects can be understood as being engaged in varying types of psychophysiological work as they carry out hypnotic suggestions. Clinicians utilizing therapeutic hypnosis, for example, have occasionally commented on the physiological responses of warmth, heat, and sweating that accompanies the intense emotional arousal characteristic of deep catharsis and the re-experiencing of post-traumatic stress symptoms in hypnotherapeutic work. Milton H. Erickson (1948/1980) described these responses as "neuro-psycho-physiological" and his early student, David Cheek, used them to assess the validity of Ideodynamic finger signaling in therapeutic hypnosis

## Temperature and Psychobiological Work in Therapeutic Hypnosis

"Milton H. Erickson's patients often became emotionally excited and aroused: they might weep or rage or became *hot and actually sweat during the psychobiological work of their hypnotherapeutic sessions* (Erickson 1958/1980)." (Rossi, 2002, p.143, italics added here). David Cheek, used perspiration as a physiological index in what he called the 3-stage sequence for assessing the validity of the hypnotic recall of emotionally meaningful memories of stress and trauma with ideomotor finger signals as follows.

> "*True unconscious ideodynamic signals are always repetitive and often barely visible*. Sometimes we must rely on the slight vibratory movements shown by the tendon leading to a designated finger. With recall of stressful experiences, it is sometimes possible to see an accelerated release of droplets of perspiration around the tip of a finger that eventually will lift. This is a physiological response preceding the skeletal muscle lifting that finger. My basic clinical hypothesis is that there is a definite three-stage sequence involved in the valid recall of meaningful material." (Rossi & Cheek, 1988, p. 21-22)

## 3-Stage Criteria for Assessing Validity of Ideodynamic Signaling

1. *Emotional and physiological memory* can be seen first through changes in respiration, pulse rate, and emotional reactions [such as feeling hot, facial and/or finger flushing, and perspiration]. These occur very rapidly and must occur *before* a designated finger lifts to show an inner orientation to the time of an important experience.

2. *Ideodynamic finger signals* indicate the accessing of memory at an unconscious level. They usually occur a few seconds after the appearance of physiological memory. At the moment the finger lifts signaling this second, higher level memory, the

patient still does not have a verbal level of awareness of the experience; there are only feelings of anticipation, vague unrest, or discomfort.

3. *Verbal reporting* of the experience follows these physiological and ideomotor indications of the inner accessing of meaningful material. To reach this conscious horizon of verbal thought, the entire experience may have to be reviewed repeatedly. The patient is told that one finger will lift to signal the beginning of an experience and another finger to signal its ending. The number of required repetitions to elevate the memory from deep unconscious zones of memory storage depends upon the gravity of the experience.

While the therapeutic applications of hypnosis have focused on relaxation or what is sometimes called "low phase hypnosis", research by Hautkappe & Bongartz (1992) and Unterweger et al. (1992) suggests that hypnosis engages a significant *"work function"* that operates differently in high and low hypnotic susceptibility subjects. Consistent with Cheek's recognition of heart and pulse changes as an index of responsiveness in therapeutic ideomotor signaling, Hautkappe & Bongartz (1992) found that heart rate variability is a useful physiological index for discriminating high and low hypnotic susceptibility. High susceptible hypnotic subjects have less heart rate variability. "High susceptible subjects do not have to work as hard on passing a suggestion as do low susceptibles" (Unterweger et al., 1992, p. 87). Milton Erickson often described good hypnotic subjects as having higher *"response attentiveness"* or focus of attention so their mind-body system does not require an indiscriminate massive arousal to do certain tasks (Erickson and Rossi, 1979). Erickson actually used psychological shocks and creative moments to focus attention in what we would now call "high phase" hypnosis (Rossi, 2002). We may hypothesize that high hypnotic susceptibility is associated with a more efficient psychobiological use of information and energy. More recently Barabasz and Barabasz (1996) have documented how "alert hypnosis" can facilitate neural biofeedback in children with attention deficit hyperactivity disorder (ADHD). From a historical perspective this clinical and experimental research may be interpreted as consistent with Braid's conception of hypnosis as a state of *monoideism* in focusing attention on various types of inner work performed during the classical 4-stage creative process (Rossi, 2000a, 2002).

While it is generally believed that that the molecular-genomic revolution initiated by Watson & Crick (1953) and others fifty years ago eventually will serve as a foundation for integrating all the medical and psychological disciplines into systems biology (Hood & Galas, 2003), it has had relatively little impact on the field of clinical and experimental hypnosis at this time (Rossi, 2002, 2003a & b). In this chapter I outline evidence indicating how the psychobiological foundations of therapeutic hypnosis may be extended to the genomic level by adopting the research strategies of current systems biology in neuroscience, functional genomics, and chronobiology (Rossi, 2000 a, b, c).

## A Systems Biology Approach to the Psychosocial Genomics of Therapeutic Hypnosis

Following the author's assembled evidence for an hypothesized association between therapeutic hypnosis, circadian physiology, and behavior (Rossi,1990, 1992, 1996, 2002), a number of pilot studies reported tentative support (Aldrich & Bernstein, 1987; Sanders & Mann, 1995). The experimental documentation of a circadian profile of hypnotic susceptibility by Aldrich & Bernstein (1987) appears, in retrospect, to be the most seminal in relation to current research on circadian profiles of behavior, physiology, and gene expression (Panda et al., 2002 a & b; Storch et al., 2002; Ueda et al., 2002). Panda et al. (2002a) conceptualized this new systems approach embracing all levels from the molecular and genomic to the physiological and complex cognitive-behavioral as follows.

> Although the molecular mechanism by which the central oscillator controls timekeeping is becoming increasingly clear, knowledge of how this timing information is transmitted to regulate behavior and physiology is only just

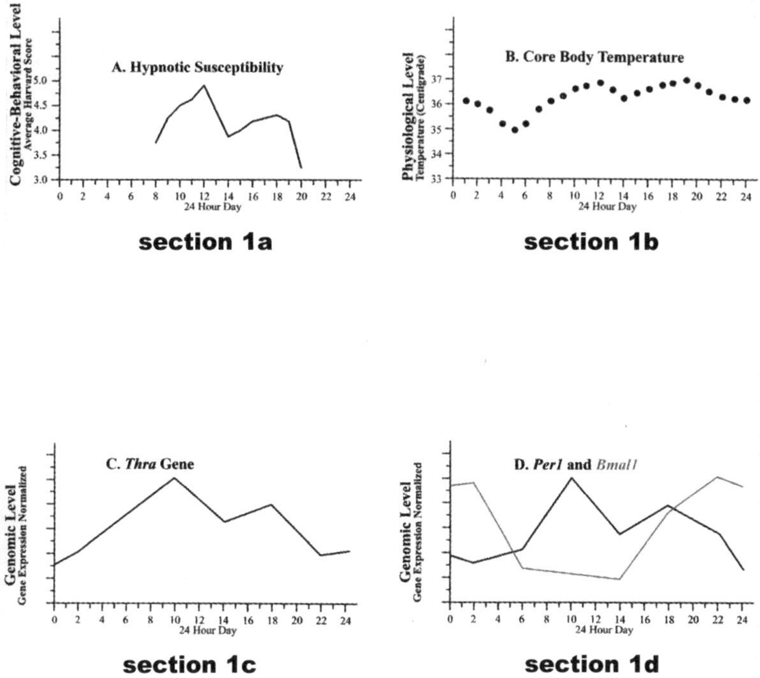


**Fig. 7.1.** - Matching the apparently similar circadian profiles of hypnotic susceptibility, body temperature, and gene expression. This may be an unusually clear illustration of the psychosocial genomic relationships between the cognitive-behavioral level of hypnosis in (1a), the physiological level of core body temperature in (1b), and expression of the *Thra* gene in (1c) associated with rate of metabolism and body temperature. Figure (1d) illustrates how the circadian profile of the *Per1* gene, typical for the awake state, is similar to the *Thra* gene in (1c) having a peak of expression about 90-120 minutes before the peak of hypnotic susceptibility and core body temperature around noon. By contrast notice how the circadian profile of the *Bmal1* gene in (1d), which is a marker for the sleep state, is in *antiphase* (the opposite of) the awake profiles of *Per1* and *Thra* gene expression associated with peaks of core body temperature and hypnotic susceptibility.

> emerging. A common theme in connecting the clock to physiological outputs has been the identification of cycling components, followed by molecular genetic and histological tests to establish a connection. (p. 333)... *The exciting possibility that complex behaviors can be described at the molecular level, and are well conserved across species, underscores the importance of the use of model organisms and comparative behavioral genomics*. (p. 334, italics added)

Figure 7.1 illustrates how this new systems biology approach to psychosocial genomics can be applied by juxtaposing Aldrich & Bernstein's (1987) circadian profile of hypnotic susceptibility (the cognitive-behavioral level), with a typical circadian profile of core body temperature (the physiological level that Aldrich & Bernstein hypothesize as underlying hypnotic susceptibility) in humans with the circadian profile of the *Thra* gene (the genomic level) in tissues of the heart and liver of the murine rodent (Storch et al., 2002). Aldrich & Bernstein (1987) summarize their results as follows.

> Figure 7.1a shows the distribution of the mean HGSHS: A scores for each hour at which groups were hypnotized. The *distribution is bimodal, with peaks at 12:00 noon and 4:00 – 6:00 p.m.* and a local minimum at 2:00 p.m. (p. 143, italics added)... The results provide preliminary evidence that hypnotizability may be related to the circadian rhythm of body temperature. (p. 144).

As may be seen, the circadian profile of core body temperature in Figure 1b is also bimodal and closely approximates the circadian profile of hypnotic susceptibility in Figure 1a. Figure 1c illustrates the circadian expression profile of the *Thra* gene, which is also bimodal and resembles the circadian profiles of hypnotic susceptibility and body temperature. The *Thra* gene, coding for the thyroid hormone receptor-, is itself induced by the thyroid hormones T3 and T4, which are fundamental in regulating the physiological work of metabolism and body temperature (Storch et al., 2002). Figure 1d illustrates the circadian expression profile of the clock gene *Period (Per 1),* which is associated with many daytime activities in humans, and resembles the circadian profiles of hypnotic susceptibility and body temperature even more closely than the *Thra* gene. Notice how the circadian profiles of the *Per1* and *Thra* gene are similar in having a peak of expression about 90-120 minutes *before* the peaks of core body temperature and hypnotic susceptibility around noon. This is consistent with the fact that about 90-120 minutes is required for many (but certainly not all) genes to be expressed via gene transcription and translation into the proteins that ultimately generate physiological and cognitive-behavioral effects (Rossi, 2002).

Figure 1d illustrates the circadian profile of the *Bmal1* gene associated with the sleep state *(the opposite of the Per1* and *Thra* gene profiles associated with being awake). Storch et al. (2002) research on the circadian modulation of *Thra* gene expression related to body temperature, psychosocial stress (the glucocorticoids), and the immune system (tumor necrosis factor alpha) are of great interest for a psychobiologically oriented approach to therapeutic hypnosis and psychoneuroimmunology at the genomic level. Note the striking similarity of their description of the bimodal circadian profile of gene expression with the bimodal distribution of hypnotic susceptibility scores reported above by Aldrich & Bernstein.

> Many mammalian peripheral tissues have circadian clocks; endogenous oscillators that generate transcriptional rhythms thought to be important for the daily timing of physiological processes. (p., 78) ...The *distribution of phases is essentially bimodal, with most genes showing peak expression between circadian time 6 h and 14 h, and a smaller group peaking in phase at about circadian time 20 h* (p. 81).

Innovative researchers investigating circadian profiles of gene expression in the brain, heart, liver, and immune system such as Storch et al. (2002) and others (Panda et al. 2002 b; Ueda et al., 2002; Rosbash & Takahashi, 2002), however, do not discuss the implications of their findings for the therapeutic applications of hypnosis. The assemblage of matched bimodal circadian profiles of Fig. 7.1 is consistent with *but certainly do not yet prove* that there are causal and reciprocal relationships in the complex interactions between the cognitive-behavioral level of hypnotic susceptibility, physiology, and gene expression (Rossi, 1996, 2002, 2003a & b). Such proof would require many novel types of integrative systems biology research by interdisciplinary research teams. The value of such research is that it may be able to resolve many of the highly controversial questions about the fundamental nature of therapeutic hypnosis that are still unresolved today - to which we will now turn our attention.

## Is Therapeutic Hypnosis a Psychogenomic State More like Sleep or Awake?

Figure 7.2 is a phase histogram that illustrates a more global type of circadian bi-modal profile of gene expression in the heart and liver that suggests how the new data and methods of systems biology could be adopted for exploring associations between gene expression, physiology, and complex behavior such as hypnosis. In support of this approach Panda et al. (2002 b) have stated,

> "The circadian control of transcription in higher organisms is integrated with the spatial control of *gene expression to target rate-limiting steps in major pathways in their relevant organs, resulting in a systems-level temporal orchestration of behavior and physiology for optimal adaptation of the organism to its environment.*" (p. 317, italics added)

Global circadian profiles of gene expression such as figure 7.2 may enable us to get a precise answer to centuries of controversy regarding the fundamental psychobiological basis of the hypnotic state: Will the circadian profiles of gene expression found during hypnosis be more closely associated with sleep as proposed by Pavlov (1927) or the awake state as maintained by Hull (1933/1968)? This question can be answered on the genomic level by investigating the implications of research by Ueda et al. (2002), which they summarize, "Here we demonstrate the role of the Rev-ErbA/ROR response element in gene expression during circadian night, which is in phase with *Bmal1* [during sleep, which Pavlov would predict as associated with hypnosis], and in antiphase to *Per2* oscillations [while awake, which Hull would predict as associated with hypnosis]." (p. 534).

A more direct assessment of the circadian correlates of hypnosis would be to continuously measure core body temperature and gene expression in high hypnotic susceptible subjects during their sleep. When their temperature is at its lowest point during sleep we could induce hypnosis without awakening the subject (*sleep induced hypnosis*) and keep them occupied with personally relevant inner work during hypnosis for an ultradian period of 90-120 minutes. If Pavlov is correct (hypnosis is like sleep) their core body temperature should remain low while their "Rev-ErbA/ROR response element in gene

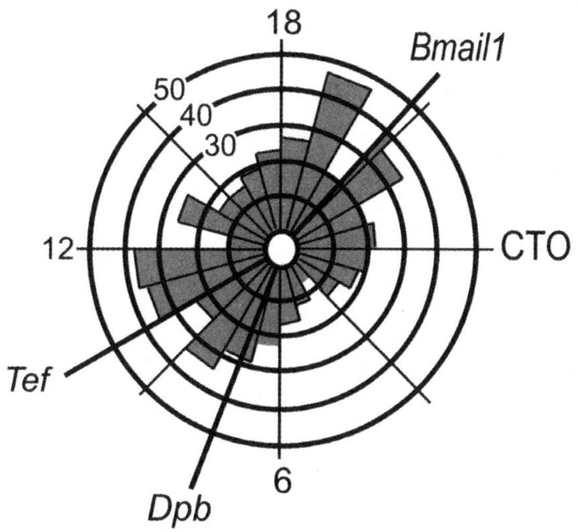


**Fig. 7.2.** - A phase histogram of circadian gene expression in the murine heart and liver. *Bmal1* is a gene with a known robust circadian expression during sleep while the *Dbp* and *Tef* genes serve as phase markers associated with being awake. This circular histogram resembles a circadian 24 hour clock face but with CT0 (Circadian Time Zero) on the right representing awakening (or the standardized initial point of data collection such as when movement and physical activity occurs). The nested numbers from the periphery toward the center (50, 40, 30) are the number of genes expressed in each circadian phase of the histogram during this particular experiment. (With permission from Storch et al. 2002).

expression" and *Bmal1* gene expression remains at a peak illustrated in figure 1d (indicating that the circadian night profile of gene expression prevails during sleep and hypnosis). If, on the other hand, Hull is correct we would expect the reverse: core body temperature should rise during the hypnosis and hypnotically induced inner work while subjects were sleeping; on the genomic level their Rev-ErbA/ROR response element and *Bmal1* gene expression should go down in antiphase with their *per1* gene expression should go up toward a peak as is characteristic of the awake state. Such research would be an important step in answering on the genomic level an issue first raised by the author 20 years ago regarding the degree to which therapeutic hypnosis may *modulate, entrain* or *shift* the *phase response curves* (Johnson, 1999) of circadian and ultradian rhythms during health, stress, and illness (Rossi, 2002).

Extending this type of thought experiment even further, it may be possible to differentiate between waking consciousness and hypnosis on the level of gene expression. If a difference is found, for example, between the profiles of gene expression in *sleep induced hypnosis* versus a control group (where subjects are simply awakened from their sleep and kept awake for a similar 90-120 minute ultradian period), these *circadian difference gene expression profiles* may be useful in distinguishing between gene expression during hypnosis, normal waking consciousness, and the spectrum of consciousness in creative states and meditation (Wilber, 1993) as well as during stress and illness. This may enable us to determine precisely which profiles of gene expression could be modulated by hypnosis during health, stress, and illness to eventually establish a firm systematic scientific foundation for the psychosocial genomics of therapeutic hypnosis.

## Prospects for therapeutic hypnosis today

Throughout its colorful history hypnosis has expressed its creatively protean nature by continually updating itself by assimilating the scientific methods and healing metaphors of its prevailing culture (Carrer, 2002).

Today the emerging scientific methods of systems biology that are being developed to organize the data of The Human Genome Project is the prevailing culture that hypnosis needs to assimilate. It is now possible to assess the entire human genome from a few drops of blood with DNA microarray technology (Rossi, 2000b) and the ultra sensitive nano technologies that will permit the screening of gene expression during psychotherapy in real time (Goho, 2004).

It is no longer visionary to believe we can assess gene expression and brain plasticity within the time frame of a typical session of therapeutic hypnosis (Rossi, 2000c, 2002, 2004). The community of therapeutic hypnosis and psychotherapy needs to summon its collective intelligence, will, and resources to meet this challenge for its current and future development.

# VIII
# CREATIVE REPLAY IN THERAPEUTIC HYPNOSIS AND PSYCHOTHERAPY

In this chapter we review current research in neuroscience and the psychosocial genomics of memory, learning, and behavior have important implications for the theory and practice of therapeutic hypnosis and psychotherapy.

It is proposed that many phenomena associated with therapeutic suggestion and hypnosis – typically explained by the ideodynamic action hypothesis – actually describe the phenotypic or observable cognitive-behavioral manifestations of activity-dependent gene expression, brain plasticity, and mind-body healing. This chapter outlines how the *neuroscience trace reactivation theory of the construction and reconstruction of consciousness, memory, and behavior* is consistent with an update of the classical ideodynamic action hypothesis of therapeutic suggestion: therapeutic hypnosis can facilitate brain plasticity and mind-body healing by replaying the activity-dependent gene expression/protein synthesis cycle in the reconstruction of fear, stress, and post-traumatic memories and symptoms. A new psychosocial genomic paradigm of mind-body research is proposed for assessing the possible role of therapeutic hypnosis and related psychotherapeutic processes in modulating the various classes of gene expression associated with memory, learning, dreaming, performance, psychoimmunology, stress and healing.

## The Classical Ideodynamic Action Hypothesis of Hypnosis and Suggestion

Since its inception by Mesmer over 200 years ago, no general agreement has been achieved about the essential nature of hypnosis and therapeutic suggestion (Bongartz, 1992).

Recently, for example, an eminent, international authority on the history of research in hypnosis cogently summed up the current theoretical picture (Weitzenhoffer, 2001).

My position today (Weitzenhoffer, 2000) is that any theorizing regarding hypnosis has been and continues to be premature. There is still much groundwork to be done before anything fruitful of the sort can be accomplished. Today I have little in the line of a theory – just a few hypotheses which are insufficient to account for all the facts that have been satisfactorily established (p. 157).

In his historical and critical effort to assemble, describe, and clarify "the facts that have been satisfactorily established," regarding suggestibility, however, Weitzenhoffer (2000) describes the *ideodynamic action hypothesis* as coming closest to a theory of "hypnotic effects" as follows.

> Few formulations regarding what the suggestion process is, exist that can be called a theory. The most widely accepted and influential so-called theory, still really a hypothesis, is known as the *ideodynamic action theory,* often being improperly referred to as the "ideomotor theory" and as a theory of hypnosis. Strictly speaking, it pertains directly only to suggested behavior. It has nothing to do with hypnosis, but of course, indirectly it does. Of all the hypotheses that have been proposed regarding the production of hypnotic effects (understood as suggested effects), it is the one that comes closest to being a theory and more workers in the field have ascribed to it than any other hypothesis (p. 123). If the responses involved here [e.g. *automatisms,* postural sway suggestion, Chevreul explorer pendulum etc.] are to be viewed as being ideodynamic in their entirety, they call for a more complex picture of the neuromotoric mechanisms involved (130) . . . Furthermore, *many automatisms are self-correcting processes in which feedback plays an important part.* This last is important because this, combined with the capacity for self-termination, probably makes some complex automatisms appear like intentional behavior (p. 128). . . Last, *far more complex responses than the one originally suggested might be expected to take place through ideodynamic action because one idea often leads to another idea, indeed to a whole chain of ideas, with loops and side branches. Not only is it reasonable to posit that a number of associated and interlinked*

> *ideodynamic responses can thus take place, but also that the more complex composite "idea" thus formed may have the capacity to produce its own specific ideodynamic effect.* Indeed, the overall response to a suggestion has the potential for being something quite different than what was originally proposed (p. 129, italics added).

What, precisely, is the *action* of the ideodynamic *action* hypothesis of suggestion that Weitzenhoffer describes here? Historically the *ideodynamic*, a word that does not appear in most English dictionaries, referred to an idea – a "*mental*" process – that generates a *dynamic* – "an energy, relating to or tending toward a change or productive *activity*" (New World Dictionary, 2ed College Edition, 1986). More than a century ago Bernheim (1886/1957) offered a charming description of ideodynamic action (with ideomotor, ideo-sensory, ideo-reflexive components) that gave rise to hypnotic *automatisms* (involuntary action) such as the Chevreul pendulum and "table turning" in séances.

"A well-known experiment shows the influence of an *idea* upon the *act*. I hold the end of my watch chain between two fingers at the height of my forehead; the watch hangs vertically suspended and moves to the right or left, backwards or forwards, or in a circle, according as I conceive the idea of these successive movements. I try in vain not to interfere voluntarily, but am unconscious of the motion, which my hand imparts to the chain. *Simply the idea of motion is enough to occasion it.* Is this not the secret of table turning, which has turned so many heads for 30 years? Involuntarily and unconsciously everyone imparts a slight movement to the table, and the sum of all these unconscious movements, results in the tipping of the table." (p. 133, italics added).

While the Ideodynamic action hypothesis of suggestion has played an important role toward the development of a theory of hypnosis for over 100 years, current neuroscience does not recognize this historical priority. Neuroscientists are now rediscovering many phenomena previously subsumed under the *ideodynamic action hypothesis* and, without any regard for the history of hypnosis, give these phenomena different names derived from recent experimental research on *activity-dependent gene expression, behavior state-related gene expression, and brain plasticity* in the molecular dynamics of memory, learning, behavioral adaptation, sensation, perception, emotions, dreaming, stress, trauma, and healing. There has been little or no communication between the three disciplines of therapeutic hypnosis, neuroscience, and psychosocial genomics (Rossi, 1972 – 2004, in press). This paper is a conceptual review integrating the current neuroscience "trace reactivation theory" of memory and learning with the molecular-genomic mechanisms of the ideodynamic action hypothesis of therapeutic suggestion, hypnosis, and psychotherapy (Rossi, 2000 a, 2002 a, 2004, in press).

## The Neuroscience Trace Reactivation Theory of Memory Consolidation

The central issue of the current neuroscience trace reactivation theory of memory consolidation is to determine how new experiences are converted into stable, long-term memory, while still being available for change and updating with later experience. Experimental evidence documents that after a salient, new, surprising, unusual, or unexpected life experience many brain circuits replay the *memorable event during the "offline" periods of rest, recovery, quiet time, sleep, and dreaming.* This reactivation and replay of the novel experience integrates and consolidates the memory (Sutherland & McNaughton, 2000; Wittenberg et al., 2002).

New memories initially consist of separate parts, which are stored in different regions of the cortex of the brain where they were first received and encoded. These separate parts of the new memory initially are linked indirectly through the common connections they share with the hippocampus. The hippocampus is the focal area of the brain that converts short memories to long term. Later, when the new memory is *"replayed offline during rest,"* direct links form among the individual parts in the cortex that initially encoded the memory. The new short term memory becomes consolidated into a long term memory that is now independent of its initial links with the hippocampus. The hippocampus is then free to operate on further new experiences that need to be linked and converted into long term memory. Hoffman and McNaughton (2002) describe how the *concurrent reactivation or replay between brain cells encoding different parts of memory is essential for linking the correct pieces of memory together into a coherent whole* as follows.

> "Neural ensembles in the rat hippocampus & neocortex show *memory trace reactivation during "offline periods" of quiet wakefulness, slow-wave sleep, and in some cases REM sleep. Reactivation of recent memory traces is also observed during sleep in motor areas of the zebra finch brain...neuro-imaging in humans reveals that brain areas with increased signal during a task have continued or reappearing activity after the task is completed.* (p. 2070, italics added)

The important implication for therapeutic hypnosis and psychotherapy is that the access, reactivation, and replaying of any part of a memory tends to reactivate a possible therapeutic reconstruction of the whole original experience. Humans often are able to recall distant past memories in vivid detail. Recent memories are susceptible to disruption and/or change, however, especially in the first minutes to days after a memorable event - before the hippocampus is able to convert the short term memories to long term during an appropriate rest period. This may be one the source of so-called "hypnotic amnesia," that is a standard phenomenon that is used to assess "hypnotic susceptibility" on standardized scales commonly used in assessing the reality of the "trance state" in hypnosis research (Hilgard, 1965, 1991) and clinical practice (Erickson, 1980). Some forms of *"hypnotic amnesia" may occur when the hippocampus has not had an appropriate opportunity to reactive and replay new experiences to consolidate them into long term memories encoded in the cortex.* Research is now required to assess this hypothesis.

Repetition is a cardinal process in the historical accounts of hypnotic induction and therapeutic suggestion as well as the modern standardized hypnotic susceptibility scales. Why is repetition needed? What ideodynamic action is facilitated by repetition? The trace reactivation theory of memory consolidation implies that "ideodynamic action" on the neuroanatomical level may be the replay between the cortex and hippocampus that is well described by Lisman & Morris (2001).

> "Newly acquired sensory information is funneled through the cortex to the hippocampus. Surprisingly, only the hippocampus actually learns at this time - it is said to be online. Later, when the hippocampus is offline (probably during sleep), it replays stored information, transmitting it to the cortex. The cortex is considered to be a slow learner, capable of lasting memory storage only as a result of this repeated replaying of information by the hippocampus. In some views, the hippocampus is only a temporary memory store — once memory traces become stabilized in the cortex, memories can be accessed even if the hippocampus is removed. There is now direct evidence that some form of hippocampal replay occurs... These results support the idea that the hippocampus is the fast online learner that "teaches" the slower cortex offline." (p. 248-249, italics added)

Research by Shimizu et al., (2000) provides more detail about how repetition, recall, creative replay required for the transformations of consciousness, memory, and behavior on the cellular level. They found that the NMDA (N-Methyl-D-Aspartate) receptor on brain cells in the CA1 region of the hippocampus serves as a "gating switch" in the construction and reconstruction of memory.

> "Our results indicate that memory consolidation may require multiple rounds of site-specific synaptic modifications, possibly to reinforce plastic changes initiated during learning, thereby making memory traces stronger and more stable. Recent studies report that the learning-induced correlation states among CA1 neurons are reactivated spontaneously in a post-learning period. Such a co-activation of these neurons might suggest the existence of the natural condition within the hippocampus by which recurrent synaptic strengthening can occur during memory consolidation. We hypothesize that such synaptic re-entry reinforcement (SRR) process can also be applied to explain how

the hippocampus transfers newly created memories to the cortex for permanent storage. As the hippocampus undergoes reactivation during consolidation, it may also act as a coincidence regenerator for activating neurons in the cortical area such as the association cortex. This would allow cortical neurons previously corresponding to the different sensory modalities to be reactivated together, leading to the strengthening of the connections between them through SRR. Indeed, such a coordinated reactivation of hippocampal-cortical neurons after learning has been observed recently ... Once these cortical connections are fully consolidated and stabilized, the hippocampus itself becomes dispensable for the retrieval of the 'old memory'... Therefore, we postulate that the hippocampus, by serving as a coincidence regenerator, may induce the reinforcement of synaptic connection within the cortex during memory consolidation as the cellular means to convert short-term memories into long-term memories." (Pp. 1172-1173, italics added)

## Ultradian Time Frame of Brain Plasticity, Healing and the Basic Rest Activity Cycle

A recent review of these psychobiological dynamics of the "*multiple rounds of site-specific synaptic modifications, possibly to reinforce plastic changes initiated during learning*" described above by Shimizu et al. (2000) was illustrated by Cohen-Cory (2002) who also provides more detail about the time parameters of *activity-dependent synaptogenesis* and brain plasticity in the central nervous system as well as the body.

"During development, more synapses are established than ultimately will be retained. Therefore, the elimination of excess synaptic inputs is a critical step in synaptic circuit maturation. Synapse elimination is a competitive process that involves interactions between pre- and postsynaptic partners. The dynamics of synapse formation and of synapse elimination may be much more rapid in the CNS than at the NMJ [Neuro Muscular Junction], where synapse elimination has been well characterized. At the vertebrate NMJ, a single muscle cell is initially innervated by multiple motor axons. *The transition from multiple innervations to innervation by a single motor*

> *axon occurs gradually as some terminal branches retract from each muscle fiber before others, a process requiring about 24 hours for withdrawal of the presynaptic terminal...* In the CNS, as with the NMJ, a developmental, activity-dependent remodeling of synaptic circuits takes place by a process that may involve the selective stabilization of coactive inputs and the elimination of inputs with uncorrelated activity. *The anatomical refinement of synaptic circuits occurs at the level of individual axons and dendrites by a dynamic process that involves rapid elimination of synapses. As axons branch and remodel, synapses form and dismantle with synapse elimination occurring rapidly, in less than two hours... hippocampal neurons in which glutamate receptor function was altered demonstrated that synapse disassembly in the CNS occurs rapidly, within 1.5 hours after synapses are no longer functional* (p. 771)... *Studies investigating the effects of long-term synaptic plasticity have generally used experimental paradigms in which repetitive, high-frequency stimulation gives rise to synaptic potentiation [called long-term potentiation, LTP] that is accompanied by structural and molecular changes at the level of single synapses...* Recent imaging experiments reveal that both NMDA and AMPA receptor activation are indeed involved in synapse formation and maturation." (p. 773, italics added).

Notice how the time frame of 1.5 to 2 hours required for brain plasticity via synaptogenesis, as reviewed above by Cohen-Cory (2002), appears to be identical to Kleitman's 1.5 to 2 hour Basic Rest-Activity Cycle (BRAC), which is the fundamental time parameter of the REM dream cycle where it was originally discovered (Aserinsky & Kleitman, 1953; Kleitman & Rossi, 1992). Most of the basic chronobiological life processes of homeostasis, adaptation (Lloyd and Rossi, 1992, 1993; Rossi, 1982, 1986,1986/1993, 1996, 2002a; Rossi & Nimmons, 1991), stress and trauma (Kaufer et al., 1998), memory, learning, and neurogenesis (Kandel, 2000) as well as the dynamics of neuroendocrinology and psychoimmunology are similar. The time parameters of this broad psychobiological perspective suggests that Milton Erickson's typical 90-120 sessions of therapeutic hypnosis may have been efficacious, at least in part, because of their association and utilization with the natural chronobiology of Kleitman's BRAC (Rossi, 1982, 1986, 1992, 2002a). From this neuroscience perspective, brain plasticity in general and synaptogenesis, in particular, is the most recent addition to the list of *complex adaptive systems of the BRAC* that is evident on all levels from the molecular-genomic to the cognitive-behavioral that we facilitate via the psychosocial genomics of activity-dependent therapeutic hypnosis and psychotherapy (Rossi, 1996, 2002a, 2004 in press).

## The Shaky Memory Trace: The Activation and Therapeutic Resynthesis of Fear, Stress, and Post-traumatic Memories

Current neuroscience research documents how the classical process of Pavlovian fear conditioning requires the recall and re-activation of a conditioned memory before it can be extinguished and/or reconstructed at the gene expression and protein synthesis level as demonstrated experimentally by Nader et al. (2000 a & b). Nader et al. (2000a) summarize their research with these words,

> *"Our data show that consolidated fear memories, when reactivated, return to a liable state that requires de novo [gene expression] and protein synthesis for reconsolidation.* These findings are not predicted by traditional theories of memory consolidation (p.723, italics added).

Dudai comments on the potentially therapeutic implications of this finding in a paper titled, "The Shaky Trace," (2000).

> "The current textbook version, in a nutshell, goes like this. Training modifies proteins at synapses in the neuronal circuit that acquires the new memory. This alters synaptic efficacy and thus the encoding of information in that circuit. *But protein molecules survive only for periods of minutes to weeks, whereas many memories are destined to live longer. It seems that at least part of the immunity of memory to this molecular turnover is achieved by training-induced modulation of gene expression in the modified neurons. The new gene products promote long-lasting re- modeling of the activated synapses, in a process that involves cross talk between the synapses and neuronal cell bodies. It takes a few hours for the new pattern of gene expression and the synaptic change to be consolidated.* During this time, the

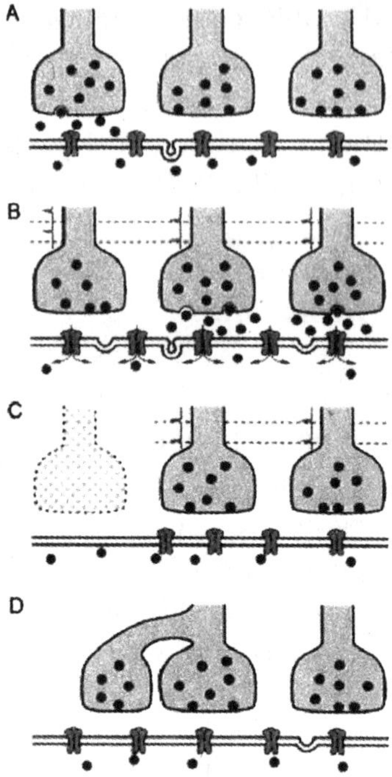


**Fig. 8.1.** - Synapse formation and stabilization in the vertebrate central nervous system is a dynamic process, requiring bi-directional communication between pre- and postsynaptic partners. Numerous mechanisms coordinate where and when synapses are made in the developing brain. This review discusses cellular and activity-dependent mechanisms that control the development of synaptic connectivity.

process can be halted by inhibitors of protein synthesis..." "So it seems that *fear-associated memories become temporarily labile on retrieval.* Why should the brain invest so much energy in the original consolidation and then risk losing the trace by interference each time it is used? One can come up with teleological explanations — for example, that the brain prefers plasticity at the expense of stability — or mechanistic ones, suggesting in-built constraints on the *synaptic machinery*. . .

More generally, might these results apply to different types of memory? Previous studies hinted that Pavlovian fear conditioning may not be unique in being shaky on retrieval. But even if just a few types of memory must reconsolidate after use, the implications of the results of Nader *et al.* (2000a) are remarkable. *Consider, for example, the prospect of intentionally recalling the memory of a traumatic experience and then selectively erasing it. What such a possibility would mean for psychoanalysts on the one hand, and poets on the other, is quite a different matter."* (p. 686, italics added).

These experimental findings may have important implications for understanding the long and controversial tradition of using hypnosis to facilitate memory recall and emotional re-experiencing in the so-called "cathartic cure." The implication is that *"intentionally recalling the memory of a traumatic experience and then selectively erasing it"* is precisely what took place during many historical approaches to reactivating traumatic memory recall and their "Mental Liquidation" (Pierre Janet's original words, 1925/1976, pp. 589) via therapeutic hypnosis. Extending the implications of Dudai's remarks, *I hypothesize that this activity-dependent process of reactivating a fear memory in order to extinguish and re-construct it on the level of gene expression, protein synthesis, and brain plasticity (including synaptogenesis, neurogenesis) is the psychobiological essence of creative replay in the practice of therapeutic hypnosis and psychotherapy.* We can generalize this mind-body essence of therapeutic suggestion to the many alternative and complementary approaches to medicine as well as the creative process in cultural rituals (Greenfield, 1994, 2000) and the humanistic arts in general. They all typically engage the ideodynamic recall and replay of memory in the therapeutic reconstruction of human learning and behavior (Rossi, 2000a, b; 2002a, 2004 in press). This appears to be the neurobiological substance of the popular psychotherapeutic metaphor, "Every Replay is a Reframe."

The concept of *positive, creative, therapeutic replay during offline psychological states as potentially important periods for the transformation of psychological experience and mind-body healing* finds further support in the research of Lisman & Morris (2001).

> "...newly acquired sensory information is funneled through the cortex to the hippocampus. Surprisingly, only the hippocampus actually learns at this time — it is said to be online. *Later, when the hippocampus is offline (probably during sleep), it replays stored information, transmitting it to the cortex. The cortex is considered to be a slow learner, capable of lasting memory storage only as a result of this repeated replaying of information by the hippocampus.* In some views, the hippocampus is only a temporary memory store — once memory traces become stabilized in the cortex, memories can be accessed even if the hippocampus is removed. *There is now direct evidence that*

> *some form of hippocampal replay occurs... These results support the idea that the hippocampus is the fast online learner that "teaches" the slower cortex offline."* (p. 248-249, italics added)

The creative replay of problems and traumatic memories was a typical hypnotherapeutic approach practiced frequently by Milton H. Erickson when I studied with him during the last 8 years of his life (1973a & b). Erickson's (1970/1980) described his use of trauma reactivation for therapeutic replay in dreams in a series of hypnotherapeutic sessions with suggestions such as these.

> "Dream the same dream with the same meaning, the same emotional significance, but with a different cast of characters. This time maybe it won't be so dark. Maybe you can see a bit more clearly. It won't be pleasant, but maybe it won't hurt so much. So go ahead as soon as you can and have your dream.' Within four minutes the dream developed; 20 minutes later, streaming with perspiration, Edward said, 'It was bad. It was awful bad. But it didn't hurt so much..." (pp. 62-63)
>
> Such intense states of psychobiological arousal were characteristic of many of Erickson's case histories (Rossi, 1973a). Erickson (1948/1980) maintained that *'therapy results from an inner resynthesis of the patient's behavior achieved by the patient himself...* this experience of re-associating and reorganizing his own experiential life that eventuates in a cure, not the manifestation of responsive behavior, which can, at best, satisfy only the observe... Not until sometime later did the therapist [Erickson] learn by what train of thought he had initiated the *neuro-psycho-physiological process...*" (pp. 38-39, italics added).

One of the most interesting lines of research on the natural dynamics of replay during dreaming indicates a possible psychosocial genomic mechanism for mind-body healing by replaying traumatic memories with a more creative and therapeutic script as illustrated above by Erickson. It has been found that when experimental animals experience novelty, environmental enrichment and physical exercise, the *zif-268 gene* is expressed during their REM sleep (Ribeiro, 2003;

Ribeiro et al., 1999, 2002, 2003). *Zif-268* is an *immediate-early gene* and *behavioral-state related gene* that is associated with the generation of proteins and growth factors that facilitate brain plasticity. Ribeiro et al (2004) recently summarized the complementary role of "*neural replay*" during the stage of REM dreaming versus deep slow wave sleep in the consolidation of new memories as follows.

> "The discovery of experience-dependent brain reactivation during both slow-wave (SW) and rapid eye-movement (REM) sleep led to the notion that the consolidation of recently acquired *memory traces requires neural replay during sleep*. . . Based on our current and previous results, we propose that the 2 major periods of sleep play distinct and complementary roles in memory consolidation: pretranscriptional recall during SW sleep and transcriptional storage during REM sleep. . .In conclusion, *sustained neuronal reverberation during SW sleep, immediately followed by plasticity-related gene expression during REM sleep, may be sufficient to explain the beneficial role of sleep* on the consolidation of new memories." (p. 126 – 135, italics added.)

Two recent papers provide new details of how the reactivation of fear, stress, and traumatically encoded memories in a therapeutic context can be the first step in initiating a molecular-genomic of reconstructing them at the levels of gene expression, brain plasticity, and behavior. Lee et al. (2004) summarize their research as follows.

> "The idea that new memories undergo a time-dependent consolidation process after acquisition has received considerable experimental support. More controversial has been the demonstration that established memories, once recalled, become labile and sensitive to disruption, requiring "reconsolidation" to become permanent... *We show that consolidation and reconsolidation are doubly dissociable component processes of memory. Consolidation involves [gene transcription and expression of] brain-derived neurotrophic factor (BDNF) but not the [gene] transcription factor Zif-268, whereas reconsolidation recruits Zif-268 but not BDNF. These findings confirm a requirement for BDNF specifically in memory consolidation and also resolve the role of Zif-268 in brain plasticity, learning, and memory. " (pp. 839, italics added).*

A related paper by Franklin et al. (2004) extends these findings by imaging activity-dependent gene expression in the anterior cingulated cortex during the activation of fear memories. These finding have particular significance for therapeutic hypnosis because *the anterior cingulated cortex has been implicated in hypnotic susceptibility* (Rainville et al. 1997, 1999)

> "Although the molecular, cellular, and systems mechanisms required for initial memory processing have been intensively investigated, those underlying permanent memory storage remain elusive. *We present neuroanatomical, pharmacological, and genetic results demonstrating that the anterior cingulate cortex plays a critical role in remote memory for contextual fear conditioning. Imaging of activity-dependent genes shows that the anterior cingulate is activated by remote memory* and that this activation is impaired by a null - CaMKII mutation that blocks remote memory. *Accordingly, reversible inactivation of this structure in normal mice disrupts remote memory without affecting recent memory."* (p. 881, italics added).

From the neuroscience perspective, *the (1) apparently simple process of recalling and (2) creatively replaying fear, stress, and traumatic memories within a new positive therapeutic perspective can (3) initiate the molecular-genomic dynamics of Erickson's neuro-psycho-physiological process of healing. I propose that the psychosocial genomics of gene expression and brain plasticity is operative within the natural 4-stage ultradian creative process of deconstructing the old neural networks that a encoded posttraumatic stress disorder (PTSD) and re-synthesizing new neural networks capable of effective problem solving and symptom resolution* (Rossi, 2002a, 2004 in press). An outline of this 4-stage psychosocial genomics core of brain plasticity via the "ideodynamic action" of therapeutic hypnosis is illustrated in figure 8.2.

**The Neuro-Physics of Brain Plasticity**

The concept of brain plasticity has been reviewed from a formal information, probabilistic, and energetic perspective that extends to the quantum level (Rossi, 1972/1985/2000). Laughlin & Sejnowski (2003) outline some of the basic neuro-physics as follows.

"A typical cortical neuron receives on the order of 10,000 synapses, but the probability that a synapse fails to release neurotransmitter in response to an incoming signal is remarkably high, between 0.5 and 0.9. Synaptic failures halve the energy consumption of gray matter, but because there are so many synapses, the failures do not necessarily lose information. . . Does the energy-efficient cortical neuron. . . select signals from sites that are most informative? This question draws energy efficiency into one of the most active and important areas of neuroscience: synaptic plasticity.

> "The initiation of action potentials in the cortex can occur with millisecond precision but... the communication at cortical synapses is probabilistic. On a short time scale of milliseconds to seconds, presynaptic mechanisms briefly increase or decrease the probability of transmission at cortical synapses over a wide range, depending on the previous patterns of activity... A new view of the cortical network is emerging from these discoveries. Rather than being a vast, fixed network whose connection strengths change slowly, the effective cortical connectivity is highly dynamic, changing on fast as well as slow time scales. This allows the cortex to be rapidly reconfigured to meet changing computational and communications needs... The more we learn about the structure and function of brains, the more we come to appreciate the great precision of their construction and the high efficiency of their operations. Neurons, circuits, and neural codes are designed to conserve space, materials, time, and energy... To an extent yet to be determined, the noise and variability observed among neurons is compensated by plasticity—the ability of neurons to modify their signaling properties. Neural plasticity also has the potential to direct the brain's scarce resources to where they will be of greatest benefit." (p. 1873, *italics added*)

Case examples illustrating how psychotherapists may engage these therapeutic and reconstructive neurobiological dynamics on all levels from the quantum to the molecular, genomic, and psychological have been presented elsewhere (Rossi, 1972/1985/2000, 1986/1993, 1996, 2002a, 2004 in press).

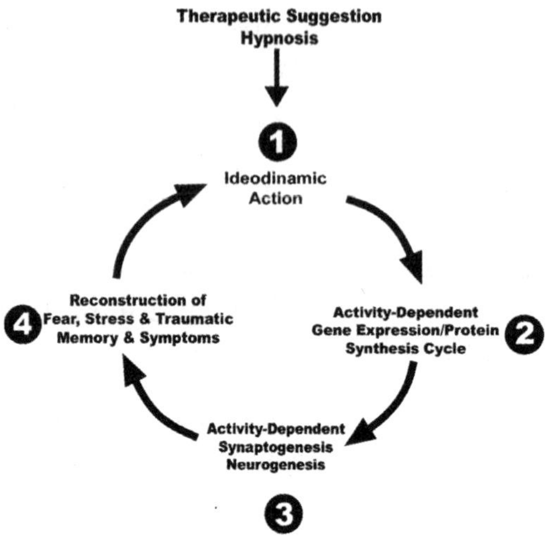


**Figure 8.2.** - The proposed 4-stage psychosocial genomics core of the ideodynamic action hypothesis of therapeutic hypnosis and related psychotherapeutic processes. Note how this creative cycle engages feedback loops whereby ideodynamic action generates "far more complex responses than the one originally called for" as described by Weitzenhoffer (2000, p. 128).

## A Psychosocial Genomic Research Paradigm for Therapeutic Hypnosis

In a pioneering paper Whitney et al. (2003) have documented how individuality and variation in gene expression patterns in the blood can be assessed with DNA microarray technology to investigate questions about states of behavior, consciousness, stress, and human condition in general.

> "The extent, nature, and sources of variation in gene expression among healthy individuals are a fundamental, yet largely unexplored, aspect of human biology. *Future investigations of human gene expression programs associated with disease, and their potential application to the detection and diagnosis, will depend upon an understanding of normal variation within and between individuals, over time, and with age, gender, and other aspects of the human condition* (p.1896, italics added)

This suggests that DNA microarrays (often called "gene chips") could be used as a safe and relatively non-invasive approach for the assessment of the states of therapeutic hypnosis (Rossi, 1999, 2000 a &b). Table 8.1 is a brief sampling

Table 8.1. - **A brief sampling of gene candidates for assessing the possible role of therapeutic hypnosis and related psychotherapeutic processes in modulating brain plasticity and mind-body healing via DNA microarray technology (From Rossi, 2002a, 2004 in press).**

| ***Brain Plasticity in Consciousness, Memory, Learning and Behavior Change*** | |
|---|---|
| c-fos, c-Jun, krox, NGFI-A & B | Bentivoglio & Grassi-Zucconi, 1999 |
| CREB | Kandel et al., 2000 |
| BNDF | Spedding et al., 2003 |
| CYP-17 | Ridley, 1999 |
| **Replay in the Reconstruction of Fear, Stress and traumatic Memories** | |
| Zif-268 | Ribeiro et al., 2002, 2004 |
| **Chronic Psychosocial Stress** | |
| Nerve Growth Factor (NGF) | Alfonso et al., 2004 |
| Membrane Glycoprotein 6a (M6a) | |
| CDC-like Kinase 1 (CLK-1) | |
| G-protein alpha q (GNAQ) | |
| CRE- dependent reporter gene | Alejel et al., 2002 |
| Acetylcholinesterase (AChE-S & AChE-R) | Soreq & Seidman, 2001 |
| **Psychoneuroimmunology** | |
| Interleukin 1, 2, 1ß Cox-2 | Castes et al., 1999; Glaser et al., 1990 |
| **Clock Genes & Behavior State-Related Genes** | |
| About 100 sleep related genes | Cirelli et al., 2004 |
| Clock, Period 1, BMAL | Rossi, 2004 |
| Period 2 | Rosbash & Takakshi, 2003 |
| **Maternal behavior and therapeutic touch** | |
| ODC gene | |

of gene candidates for assessing the possible role of therapeutic hypnosis and related psychotherapeutic processes in modulating gene expression and brain plasticity associated with memory, learning, dreaming, performance, psychoimmunology, stress and mind-body healing.

DNA microarrays could provide a more sensitive, comprehensive, and reliable measure of the varying psychological states, brain plasticity and mind-body healing as a bioinformatics approach to the psychosocial genomics of therapeutic hypnosis (Rossi, 2004).

* * * *

The classical *ideodynamic action hypothesis* of therapeutic suggestion has been updated with the current neuroscience theory of memory consolidation via trace reactivation and replay. It is proposed that the ideodynamic action hypothesis of therapeutic suggestion describes the phenotypic or observable cognitive-behavioral manifestations of activity-dependent gene expression, brain plasticity, and mind-body healing. The activity-dependent process of reactivating an old traumatic memory in order to re-construct it on the level of gene expression, protein synthesis, and brain plasticity is proposed as a new psychosocial genomic perspective on the role of creative replay in the humanistic and cultural arts as well as therapeutic hypnosis and related psychotherapeutic processes. Research on the psychosocial genomics of creative replay with DNA microarray would be an important step toward a general neuroscience theory of the efficacy of therapeutic hypnosis on all levels from the cognitive-behavioral to the genomic. A variety of research projects in this area are briefly outlined in the appendix 2.

# IX
# WORKSHOP: ACTIVITY-DEPENDENT APPROACHES TO CREATIVE WORK

Our activity-dependent approaches to creative work have a long history. Introductory college courses in the humanities remind us that the earliest roots of theater, drama, art and literature over the millennia had their origins in bringing the community together for mutual experiences of sharing, bonding, and healing. Indigenous societies around the globe throughout human history have always had hubs of social identity mediated by rituals of spiritual communion. Current research in the psychobiology of gene expression and brain plasticity expands the meaning and significance of these deep cultural traditions with a profound view of how mind and molecule co-evolve together creatively on multiple levels during the activities and silences of group experience.

Over the past hundred years schools of psychotherapy, art, drama, and the theater have tiptoed around each other with mutual wariness about how they facilitate the creativity and healing. The original theater school of Stanaslavski in Russia and its offshoots in the teaching of Strasberg and the Group Theater in America, for example, were adapted as a new approach to psychotherapy in psychodrama of Moreno. The Gestalt approach to psychotherapy of Fritz Pearls and many of its stepchildren in Transactional Analysis and Re-decision therapy echo the numinous dramas of ancient myth, poetry, and song regarding human choice, destiny, and fate. The source of therapeutic hypnosis, meanwhile, had its origin in the *baquet* of Mesmer, a kind of dramatic group therapy of altered experiential states that were believed to facilitate healing somehow. Today deeply engaging experiential states that have much in common with drama are used in the training of clinicians learning how to facilitate the process of psychotherapy (Zeig, 1997). The late Carl Rogers puzzled with me over the seemingly unreasonable effectiveness of "Demonstration Therapy" wherein a brief 20 minute experience of client centered counseling in front of a large group of our peers often became a "memorable, life transforming" event for the client. After centuries of pursuing every-which-way-theory to explain these therapeutic effects, it is now becoming evident that science is exploring genuinely new research paradigms searching for a new common ground to these questions.

*What is a creative experience?*
*What is the connection between art, science, and the humanities?*
*What is the relationship between drama, psychotherapy, and healing?*
*What is the essence of the mind-body connection?*
*What do spiritual beliefs, rituals, and practices have to do with life?*
*How do we recognize, facilitate, and measure optimal human experience?*
*Why is psychotherapy NOT about the prediction and control of behavior?*

The answer to all these questions is the same from the perspective of neuroscience and psychosocial genomics. As we have seen from previous chapters, neuroscience research documents how *novelty, enrichening experience*, and *exercise* can turn on activity-dependent gene expression, brain growth (synaptogenesis, neurogenesis etc.) and healing at the molecular level. I have proposed that creative experience in the arts, sciences, and humanities are certainly *novel* and *enrichening* kinds of *exercise* that can turn on the activity-dependent dynamics of gene expression, brain plasticity, and healing. I speculate that scholars who have described the essence of the spiritual experience as the *numinous* (*fascinating, mysterious, and tremendous*) are likewise referring to *novel* and *enrichening* kinds of *exercise* that turn on activity-dependent gene expression and its natural sequelae of brain plasticity and healing. *Notice that this creative perspective is the opposite of the typical definition of psychology as the science of the prediction and control of behavior.* The history of therapeutic suggestion reminds us that the experience of "fascination" was an early term for what later became know as "hypnosis." Every novel and fascinating science discovery, art form or style, new school of psychotherapy or spiritual practice enjoys a period of "grace" wherein its *surprise and unexpectedness evokes a state of psychobiological arousal* that within minutes evokes (1) immediate-early gene expression that turns on (2) alternative and behavior state-related gene expression that is associated with (3) activity-dependent gene expression, that (4) we ultimately recognize as "creative experience" on the phenotypic or phenomenological level. Such shifts in gene expression generate the synthesis of the proteins that drive the energy, matter, and metabolism of all life processes. When these effects are only short-lived we tend to call them merely "placebo effects." When these effects are startling, unusually beneficial, and seemingly inexplicable in terms of well understood science we tend to call them "miracles." When they are undesirable, chronic, and associated with stress and trauma we call them "psychosomatic." If they are more desirable and associated with psychotherapy and positive psychological experience we call them "mind-body" healing. *New, unexpected, and deeply meaningful psychosocial experiences stimulate consciousness to evoke psychobiological arousal that turns on the gene expression/protein synthesis cycle, which is the essence of creativity.* Ultimately, we hope to assess the psychosocial genomics of human creativity by actually measuring changes in gene expression in the blood (Whitney et al., 2003). Meanwhile, this chapter outlines some of the practical details of how to facilitate novel activity-dependent approaches to creative work and healing so we will be ready to make these objective measurements at the genomic level in our therapeutic workshops in the future.

## A 4-Stage Outline of the Creative Psychosocial Genomics Workshop

I typically present my therapeutic workshops over 2 or 3 days to combine an optimal balance of new intellectual content and education along with opportunities for personal growth and professional training for students and licensed professionals in the healing arts. The daily plan has four basic activities to optimize the possibilities of activity-dependent gene expression, brain growth and mind-body healing in all the participants. In the future, as measurements of gene expression and brain plasticity become more available, we would like to actually measure the facilitation of optimal gene expression, brain plasticity and mind-body healing on the molecular level as they take place in real time in all the workshop participants.

As we accumulate research findings on the deep psychobiological integration and healing that is facilitated on all levels from mind to gene in our Creative Activity-Dependent Psychosocial Genomics Workshops, they could become a model for facilitating all types of personal relationships and group processes in the future. These range from personal relationships between couples and family members as well as interactions between therapists and individuals in psychotherapy. The so-called "transference" between people will be reified in the actual assessment of what happens on level of gene expression to their stress, immune, and endocrine systems as people interact together in real time. Will we find that we can learn to modulate through education and good will the deep psychosocial genomics that can optimize (or hinder!) relationships between people?

It certainly may seem visionary at this difficult point in the current human condition where power politics and war are considered to be effective strategies in human relationships – to believe that group interactions and relationships can be facilitated by assessments at the deep psychobiological levels of gene expression, brain growth, and healing. This vision, however, is the ultimate goal that inspires us to explore the practical integration of biology and psychology at the levels of gene expression, brain plasticity, and healing in our therapeutic workshops. We hope to develop a new methodology for assessing education and facilitating human communication on all levels ranging from local community groups like the parent teacher association, the political caucus on the national level, and congresses on international relations.

To realize our vision it is important that our therapeutic

workshops be an open society available to all people and interested guests that may wish to be participant-observers. Friends, family, and patients of all workshop participants are strongly encouraged to attend.

Outside community, education, civic, media, political, and spiritual leaders should be encouraged to attend for their perspectives and contributions. I have led such workshops where the attendance has ranged from dozens to hundreds and occasionally thousands.

My experience suggests it is optimal to have between dozens and about 100 participants. Numbers beyond that require numerous professional staff to monitor and facilitate the workshop.

Let us now focus on how we can optimize our Activity-Dependent Psychosocial Genomics Workshops with 4 classes of positive learning experiences such as these.

1. Lecture on the New Sciences of Gene Expression, Brain Plasticity & Healing

2. Group Experiences of Heightened Self Sensitivity and Creativity

3. Group and Individual Training Experiences for Professional Development

4. Individual Experiences of Personal Growth and Optimal Creativity

### 1. Lecture on the New Sciences of Gene Expression, Brain Plasticity, Creativity, and Healing

Each chapter of this book presents a unique and currently interesting, novel, numinous, enrichening experience of learning that hopefully will evoke and optimize gene expression, brain plasticity, and mind-body healing in all the workshop participants. Perhaps Edwin Schrödinger (1944) said it best years ago in his little book on *Mind and Matter.*

> "One might say, metaphorically, that consciousness is the tutor who supervises the education of the living substance, but leaves its pupil alone to deal with all those tasks for which it is already sufficiently trained. But I wish to under-

> line three times in red ink that I mean this only as a metaphor. The fact is only this, that *new situations and the new responses they prompt are kept in the light of consciousness*; old and well practiced ones are no longer so." (pp. 103, italics added)

Simply being curious and hearing about the ever new advances in neuroscience and their implications for our everyday life and creativity can be a wondrous and numinious inspiration for most workshop participants. These up-to-date presentations are interesting "new that everyone can use" rather than a critical analysis and review of data more appropriate to scientific conferences by working scientists. Our therapeutic workshops seek to evoke Leonardo da Vinci's curiosità (Gelb, 1998) and Rudolph Otto's numinous experience together with their attendant facilitation of gene expression, brain plasticity, and healing by exploring the ever expanding frontier of human possibilities.

From the perspective of the history of therapeutic hypnosis Pierre Janet described the essence of our creative therapeutic workshop with his prescient concept of the educational gymnasium of attention and emotion.

> "The most important phenomena that have been adduced to substantiate the alleged power of hypnotic suggestion are the *physiological modifications which suggestion is said to have caused*. Normally the power we can exercise over our visceral organs and our physiological functions greatly restricted, and is usually indirect. Hypnotic suggestion, we are told, can influence these organs markedly and directly... A precise verification of these phenomena, and the discovery of the exact mechanisms of their causation, would do a great deal to demonstrate the power of suggestion. [Pp. 308-309] This historical summary, though cursory and incomplete, is an attempt to display the remarkable evolution of the methods of treatment by gymnastics and education. At the onset, the endeavor was to treat by these means various elementary and circumscribed disorders affecting well-known motor-functions. Then the claims of the gymnastics were enlarged, and they were applied as remedial measures to functions of a more lofty and complicated character. The aim was to practice the gymnastics of emotion and atten-

> tion just as at an earlier stage there had been practiced the gymnastics of walking and speech... The treatments we have just been discussing were considered under the caption 'Education and Reeducation.' [Pp. 732]

After an introductory lecture on the new developments in the sciences of gene expression, brain plasticity, and healing the workshop audience is usually ready to take a break with the promise that they will all have an opportunity to actually experience these new creative developments in group exercises when they return.

## 2. Group Experiences of Heightened Self Sensitivity and Creativity: How NOT to Predict and Control Behavior

After the initial lecture on the current edge of numinous neuroscience research, most workshop participants are ready to have a personal experience wherein they will have an opportunity to explore new possibilities of extending their horizons. All who wish to have a personal experience within a group process are asked to place their books and notes aside so they can place their hands in one of the initial novel activity-dependent approaches as illustrated below. Audience members who do not wish to participate are encouraged to watch their neighbors quietly but not disturb them in any way. As indicated by the words and suggestions offered in the 4-stage outlines of the creative process presented in this chapter, the so-called *"locus of control" of everyone's experience always remains within each workshop participant rather than the therapist who is guiding the group process. The therapist is only asking questions about possible human experiences. Hypnosis is not being used in the popular but misguided sense of programming, manipulation or the presumption of taking control of anyone's mind or behavior.* Because of this it would be more appropriate to speak of "locus of *creativity*" rather than the more familiar psychotherapeutic concept of "locus of *control*." The therapist/guide is simply asking the audience a series of questions to facilitate each participant's highly personal and entirely private exploration of their own creative process. In effect, the therapist is asking every individual in the audience to take full responsibility for their own personal exploration and experience. *Participants experience, explore and guide their own creative process by operating simultaneously on three levels.*

***Level One.*** Participants are empowered to *privately experience* their own thinking, emotions, and personal responses to the therapist's questions. They are instructed to quietly but zealously protect their privacy in every way! They participate in their own experience in this workshop situation just as they do in all other activities of their everyday life. *The locus of control (creativity) is always within the participants' private experience. This locus of control (creativity) does not reside in the therapist (as is commonly believed) who is facilitating the group process by simply offering a series of novel and hopefully numinous questions that function as implicit processing heuristics.*

***Level Two.*** Participants are empowered to carefully observe and monitor the therapist's words, questions, and behavior for appropriateness and acceptability to the participant's moral and value system. The participants are enjoined to take full responsibility for ignoring anything that is not appropriate and, instead, substitute something that is better for them.

***Level Three.*** Participants are encouraged to receive and guide their own creative and entirely unique inner process in a wise and self-compassionate manner so that they are fully responsible for taking good care of themselves. *This, in fact, is the essence of the 4-stage creative cycle of psychosocial genomic therapy – each participant learns how to recognize and guide their unique creativity in an appropriate manner.* The therapist can provide helpful hints in the form of generalized implicit processing heuristics that are helpful to people in general, but no therapist actually lives inside the participant with access to all the numinous living experiences that are needed to turn on the gene expression/protein synthesis cycle to reframe neural networks (via synaptogenesis) into new patterns of self-creation within the participant. In the final analysis, neither the therapist nor the participant (patient, client, subject) knows how to predict and control the creative process. ***Creativity***, by its very nature, ***is an open ended adventure***. Sometimes we can learn how to facilitate it, however, with *implicit processing heuristics.*

## Implicit Processing Heuristics to Facilitate the 4-Stage Creative Cycle

The 4-stage creative cycle as presented here is an activity-dependent process to access, replay, and resynthesize memory, learning and behavioral systems that encode significant life experiences. This 4-stage creative cycle, facilitated with activity-dependent self-exploration, opens up infinite possibilities of never-before realized experiences even when an attempt is made to standardize the therapist's manner, verbalizations, and implicit processing heuristics. Recall that this is not an effort to program, prescribe or condition people in a predictable, behaviorist manner. *We are using heuristics to engage creative processes whose outcome cannot be predicted in advance.* In the initial preparation stage, for example, it is of no concern whether the person actually experiences any of the sensations that are offered as implicit processing heuristics (Is one of you hands warmer or cooler than the other?). The intent of the preparation stage is simply to focus the subject inward with heightened self-sensitivity. There are implicit, complex adaptive systems operating within the person that continually evaluate, respond to, and modulate what the therapist is saying. The therapist's words are *heuristics* —not suggestions, programming, directives, covert demands, commands or interpretations in the conventional sense. Heuristics are defined by the Second College Edition of Webster's New World Dictionary as "to invent or discover: see EUREKA; helping to discover or learn…using rules of thumb to find solutions or answers."

The therapist offers novel, enriching, and interesting implicit processing heuristics to help people self-reflect and co-create within the safe privacy of *ideodynamic* discovery within. "*Ideodynamic,*" a word that does not appear in most dictionaries, means that an *idea* can generate *dynamic* transformations of psychological experience (e.g. memories, sensations, images, emotions, cognitions, behaviors, and psychobiological states) that are surprising to consciousness. When it is going well the person typically experiences these novel and surprising ideodynamic transformations as insights, intuitions, creativity, solutions, and healing. Future research will be needed to assess how these surprising experiences are associated with altered patterns of gene expression, brain plasticity, creativity, and healing. Detailed verbatim models of typical sets of implicit processing heuristics that are used to facilitate each stage of the 4-stage creative cycle in therapeutic workshops are presented in the appendix.

After about 20 minutes of experiencing The Four-Stage Creative Cycle in (Group Exercise 1 in the appendix) the audience is ready for some modest sharing of experiences. *Recall from earlier chapters that psychosocial enrichment is one of the most significant life experiences (along with novelty and exercise) that facilitate gene expression, brain plasticity, creativity, and healing at a molecular level.* I typically begin this important stage of mutual sharing and enrichment by asking the audience for a show of hands for everyone who actually responded to my heuristic questions in the first preparation stage with an actual experience of warmth or coolness, lightness or heaviness etc in their hands. I encourage the audience members to keep their hands up while looking around the room to witness how many others had the same subject responses to my exploratory questions. Sometimes I will actually count and calculate what percentage of the audience experienced each subjective sensory response. This is a most convincing demonstration to students, observing visitors and professional that it is okay and even desirable to explore one's own heightened personal responsiveness to therapeutic and exploratory questions. To explore novel and numinous experiences in our creative workshop is safe and good!

I then proceed in a similar manner asking for how many audience members responded to each of the other 3 stages of the 4-stage creative process. I begin to encourage individual audience members to share with the group a few sentences describing *any aspect of their experience that they found interesting, surprising, unusual, unexpected, novel, or fascinating*. I emphasize they are to share only what they feel is safe and appropriate to be shared in public. I am thereby emphasizing, one again, that each individual in the group is responsible for their own experiences, sharing whatever they wish, and at all times taking appropriate care for guarding their own privacy and facilitating their own self-care and self-management of their own creative process.

Experience shows that the people in the audience who initially are willing and even eager to share some personal experiences in this manner are those who feel natural and comfortable talking about themselves in public. They often serve as models for others who have less experience and skill in sharing the subjective aspects of their creative process. They enrich those who need help in learning how to experience themselves in a safe and therapeutic manner. *This simple and forthright sharing of interesting, numinous, and unusual experiences tends to open the entire audience to the novelty-numinosum-neurogenesis effect to facilitate gene expression, brain plasticity, creativity, and healing on a molecular level.*

After about 20 minutes of such sharing individuals are now encouraged to describe how they actually experienced a successful 4-stage creative process by actually solving a personal problem or resolved a genuine mind-body symptom of stress and/or trauma. The more articulate audience members are encouraged to describe how they discovered a way of facilitating their own "Period of Private Creative Inner Work" so they experienced a rewarding shift form the difficulties of the dark night of the soul in stage 2 of the creative process to the joy of stage 3 insights and healing. *These creative teaching stories that happened in reality here-and-now are the most powerful models of the therapeutic process for the entire audience. They create the positive healing atmosphere of our therapeutic workshop.* They set the stage for the next and usually most dramatic live demonstrations of personal growth, optimal creativity, problem-solving, and healing in the workshop.

## 3. Individual experiences of Personal Growth and Optimal Creativity Healing

Members of the workshop who did not have creative and healing experiences in the initial group demonstration of the 4-stage creative process now have an opportunity to learn, perhaps for the first time in their lives, how to access and facilitate their own inner resources and manage their own personal development and self-care. We witness how some people in the audience, who were initially unsuccessful in the group process, will learn how to make dramatic breakthroughs by volunteering for an individual demonstration to work one-on-one with the therapist/guide in front of the entire workshop audience. How volunteers select themselves for an individual experience of personal growth, optimal creativity, and healing within the time constraints of highly private experience during a public demonstration psychotherapy is a very delicate art worthy of careful study.

Most people usually do not recognize how they are continually engaged in a creative process in normal everyday life. Our therapeutic workshop will teach them how to recognize and facilitate their own creative process for problem solving and healing in everyday life as well as psychotherapy. In any large group (20 or more), people will be in various stages of creative work as they usually are in normal everyday life. It is not necessary to look for motivated volunteers for these individual demonstrations ahead of time before the workshop begins. It is usually not wise to promise anyone that they will be chosen for a public demonstration before the workshop begins. We do not yet know if and when the delicate 4-stage creative process will manifest itself in an appropriate manner for a therapeutic demonstration with any individual. To try to choose volunteers ahead of time is a misunderstanding of the psychosocial genomics of psychotherapy which is always an essentially creative (and therefore unpredictable) process.

From 40 years of experience, however, I know that in every audience there will always be a few people who will not be happy with the initial therapeutic exercises of the 4-stage creative cycle with the entire workshop audience. They will not have a successful therapeutic experience as initially reported by many others in the audience. These unhappy people will be fairly easy to recognize, however, because they will be quiet, anxious, perhaps tearful or even resentful and angry. They believe they have failed. It is from this unhappy group that the best individual volunteers will come from the audience.

This small but unhappy group has not actually failed, as they believe. Their problem simply is that they only reached stage 2 of the creative cycle. The therapist's job is to find these audience members who remain momentarily stuck in stage 2 (the incubation stage – the dark night of the soul - in which the state dependent encoding of problems and symptoms are re-experienced but not yet therapeutically reframed, reconstructed, and re-synthesized as they soon will be in stage 3 which comes next). The therapist can usually locate these audience members who are temporarily stuck in stage 2 by asking if anyone in the audience is unhappy with their initial experience. Is anyone willing to share a sentence or 2 about their difficulty while keeping private everything that they do not wish to share with the public. If no unhappy person volunteers - I will sometimes look searchingly at the obviously unhappy people in the audience. Or perhaps other audience members or the staff sponsoring the workshop will look at them and point them out to me.

It is very rare but sometimes at this point an unhappy audience member may try to leave the room. This is a potentially difficult situation but it is easy to correct. The person is allowed to leave if they really wish (hopefully with a friend that can stay with them). But they are usually encouraged to a least watch the next individual demonstration from the back of the room in a safe manner. It is important that they witness a successful individual demonstration. Later, usually after a break, such a disturbed audience member can be encouraged to volunteer for a brief individual therapeutic demonstration after they have seen how I have worked successfully with others who also thought they were failures.

I will then ask the unhappy audience member if he or she believes they can be helped with a very brief personal demonstration in front of the entire audience where their privacy will be respected. It is at this delicate moment that the unhappy person either decides to either volunteer - or not - for an individual demonstration with the therapist in front of the audience. If they decide not to volunteer - that is graciously accepted because it means the unhappy person does not yet believe and understand how she or he can be helped. I simply tell the unhappy person to watch how the next volunteer learns how to have a therapeutic experience. These are usually the best and often most dramatic therapeutic demonstrations that everyone talks about later and wishes they had a videotape of for more detailed study. One example of a therapeutic demonstration that was videotaped at a professional teaching workshop sponsored by the Milton H. Erickson Foundation. This videotape is available to licensed professionals and researchers from the Erickson Foundation, 3606 North 24th street, Phoenix, Arizona, 85016; www.erickson-foundation.org; Tel. 602-956-6196 (Identified as Videotape Code E 297-V8, Erickson Foundation, 1992).

A complete word by word detailed analysis of this videotape is presented in chapters 7 and 8 of *The Psychobiology of Gene Expression* (Rossi, 2002). The summary of this demonstration presented here illustrates Milton H. Erickson's naturalistic and utilization of to a patient's spontaneous behavior in hypnotic induction and psychotherapy. I have outlined this demonstration to emphasize ***Theory and Research*** as well as the ***Therapeutic Practice*** of each of the 4 stages of the creative cycle.

### Stage One. - Preparation: Immediate-Early Gene Expression in The Therapeutic Alliance

***Theory and Research.*** Many psychobiological states of arousal in everyday life are associated with the initiation of immediate-early gene expression (Rossi, 1996, 2002). The most fundamental experimental evidence that immediate-early gene (IEG) expression is associated with behavioral states of arousal, waking, sleeping, and dreaming (REM sleep) in the ultradian time frames (90-20 minutes) comes from the observation of limbic-hypothalamic-pituitary oscillations during the normal circadian replay of the sleep-wake cycle (Kelner & Bloom, 2000; Lloyd & Rossi, 1992). Figure 9.1 emphasizes how many processes of psychobiological arousal — such as pain, stress, novelty, the basic rest-activity cycle (BRAC),

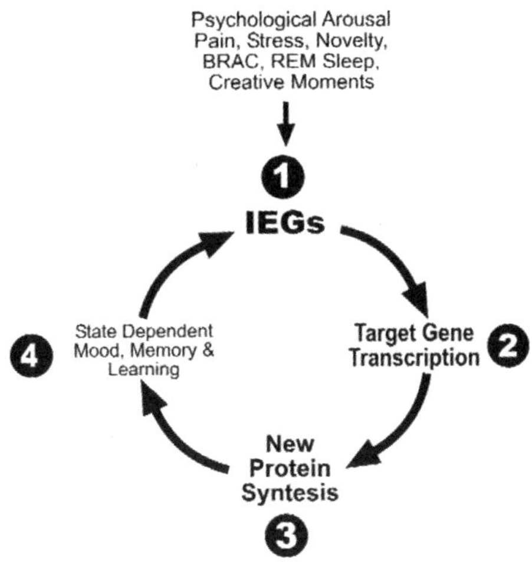


**Figure 9.1.** A psychogenomic view of the essential dynamics of psychosocial genomics in therapeutic hypnosis, psychotherapy, and the holistic healing arts. Mind-body communication via psychobiological arousal can initiate (1) immediate-early gene expression (IEGs) that, in turn, leads to (2) the expression of specific target genes, which (3) code for new protein synthesis that is the molecular basis of (4) state-dependent memory, learning, behavior (SDMLB) that is replayed on conscious and unconscious levels in the creation and recreation of human consciousness and experience.

dreaming (REM sleep), and creative moments — can initiate (1) immediate-early gene expression (IEGs) that, in turn, lead to (2) the expression of specific target genes which (3) code for new protein synthesis that is the molecular basis of (4) state-dependent memory, learning and behavior (SDMLB). As reviewed previously (Rossi 1999a, 2000a), there are many immediate-early genes such as c-fos and c-jun that are turned on within a minute or two by any strong psychobiological stimulus to activate target genes associated with the sleep-wake cycle (Bentivoglio & Grassi-Zucconi, 1999; Rossi, 2000a, b). These include target genes associated with maternal behavior (Schanberg, 1995), the construction and reconstruction of memory and learning (Kempermann & Gage, 1999), stress and emotions (Autelitano,1998), sleep and dreams (Ribeiro et al., 1999; Toppila et al., 1995), pain and reward (Tölle et al., 1995), psychosomatic disorders (Senba & Ueyama, 1997), addictions (Azar, 1999), psychosis (O'Neil et al., 1998), and many of the target genes whose transcription and translation into proteins generate the neurotransmitters and messenger molecules (eg. hormones, cytokines) of mind-body regulation and healing (Rossi, 1986/1993, 2002; Rossi & Cheek, 1988).

***Therapeutic Practice.*** The author has proposed how immediate-early gene expression begins with the varying degrees of emotional arousal that usually takes place in the typical history taking at the beginning of any psychotherapeutic process (Rossi, 2002). More than mere words and talk are involved in any form of therapeutic hypnosis or psychotherapy from the deep perspective of psychosocial genomics. The typical tears and distress in an initial interview, for example, indicate that patients are already accessing and *replaying* state-dependent memory and the emotional dynamics that signals how they are embarking on a potentially healing adventure. *The patient's task in this initial stage is to simply receive with courage and compassion the state-dependent memories, emotions, and experiences associated with their problems. The therapist's task at this point is to recognize and facilitate the natural ultradian creative cycle of gene expression, neurogenesis, and healing has already begun. Implicit processing heuristics* (that is, open ended therapeutic suggestions that enable patients to create their own response rather than authoritarian suggestion that attempts to direct the patient in an exact manner as in programming a computer) are employed to optimize the entire psychogenomic process (Rossi, 2002). The psychotherapeutic process may begin with a *symptom scaling* of the patient's currently experienced symptom, emotions, or attitudes. A subjective scale from one to ten scale (where ten being the worst the problem was ever experienced, five about average, and zero a completely satisfactory state) may be used to assess the patient's experiential state before, during, and after psychotherapeutic process to assess and validate it.

Figure 9.2 illustrates the first stage of the creative cycle in a demonstration of therapeutic hypnosis at an Ericksonian Congress (Rossi, 1992) with a young woman volunteer who has severe rheumatoid arthritis in her hands. The therapist models a delicately balanced and symmetrical hand position a few inches above the lap to initiate an activity-dependent approach to the induction of therapeutic hypnosis and the creative cycle (Rossi, 1996, 2002). The thought bubble notes the psychogenomic concerns of the therapist during this first stage. The therapist is interested in knowing what stage of the basic rest activity cycle (BRAC) the patient may be experiencing initially in order to facilitate further progress in whatever direction is appropriate (Rossi & Kleitman, 1992). He wonders whether CYP17 — the "social gene" associated with the synthesis of arousal and sexual hormones — is becoming engaged as a natural manifestation of the psychotherapeutic transference. The psychosocial genomics of CYP17 gene expression is well described by Ridley (1999).

**Figura 9.2.** Stage One: The therapist models a delicately balanced and symmetrical hand position a few inches above the lap to initiate a hand levitation approach to the induction of therapeutic hypnosis. The therapist wonders what stage of the basic rest-activity cycle (BRAC) the patient may be experiencing, whether CYP17 — the social gene — is becoming engaged as a natural manifestation of the psychotherapeutic transference, and to what extent immediate-early genes (IEGs) such as c-fos and c-jun — associated with a creative state of psychobiological arousal, problem solving, and healing — are becoming engaged.

> "The brain, the body and the genome are locked, all three, in a dance. The genome is as much under the control of the other two as they are controlled by it. That is partly why genetic determinism is a myth. The switching on and off of human genes can be influenced by conscious or unconscious external action [p.148] ... genes need to be switched on, and external events—or free-willed behavior—can switch on genes [p.153]...Social influences upon behavior work through the switching on and off of genes [p.172]...The psychological precedes the physical. The mind drives the body, which drives the genome" [p.157].

The therapist wonders to what extent immediate-early genes (IEGs) such as c-fos and c-jun — associated with a creative state of psychobiological arousal, problem solving, and healing — will become engaged to evoke the target genes that will be essential to optimize protein synthesis and therapeutic transformations in state dependent memory, learning, and behavior that will be reviewed and replayed in stage 2.

***Theory and Research.*** Important advances in research on gene expression and the molecular biology of waking, sleeping, and dreaming have profound implications for a new view of the deep psychobiological dynamics of therapeutic hypnosis (Rossi, 1999, 2000a). *Behavioral state-related gene expression* (BSGE) refers to research documenting how a wide range of behavioral states, such as sleeping, rest, dreaming, consciousness, vigilance, stress, emotional arousal, and depression is associated with different patterns of gene expression (Bentivoglio & Grassi-Zucconi 1999; Lydic 1998; Lydic & Baghdoyan, 1999). This wide range of states associated with behavioral state-related gene expression resolves one of the apparent mysteries of historical hypnosis: How can the so-called "miraculous healing" of therapeutic hypnosis can be facilitated with apparently opposite states such as *rest and relaxation* versus *activity and excitation* (Erickson, 1958/1980, 1959/1980; Janet, 1925/1976)? Our answer is that behavioral state-related gene expression modulates the entire range of psychobiological states ranging from rest to high activity (Lloyd & Rossi, 1992; 1993). This leads us to predict that behavioral state-related gene expression will be important in future experimental research defining the psychosocial genomics of the "hypnotic state" (Rossi, 2000a, b, 2002). Behavioral state-related gene expression is a fundamental link between psychology and biology and it is of essence in the relationships between unconsciousness and consciousness.

***Therapeutic Practice.*** Figure 9.3 illustrates how the patient experiences a mild state of confusion in stage two of the creative cycle as the therapist points out evidence of psychobiological arousal and (by implication, behavioral state-related gene expression) in the very fine vibration she begins to experience involuntarily in her hands and fingers. She is surprised by these movements and she spontaneously comments that her "right hand is doing some stuff" (unusual sensations and involuntary movements *that were not suggested by the therapist).* Unexpectedly her hands also become hot. These are typical manifestations of psychobiological arousal in stage two of the creative cycle that we hope to channel into a therapeutic process of salient self-engagement and healing (Rossi & Cheek, 1988; Rossi, 1986/1993).

**Figura 9.3**. Stage Two: The patient experiences psychobiological arousal (associated with behavioral state-related gene expression (BSGE). She evidences surprise and confusion about her unusual sensations and involuntary movements that were not suggested by the therapist. The therapist wonders how to facilitate the psychogenomics of immunological variables such as interleukin-1, 2, and 1? associated with Cox2 that has been implicated in rheumatoid arthritis which is the patient's presenting symptom.

The therapist now wonders about how to engage the psychogenomic dynamics of immunological variables such as interleukin-1 & 2 that are messenger molecules of the immune system that could facilitate healing. He also wonders about the activity of IL-1? that is associated with Cox2, which has been implicated in rheumatoid arthritis (the patient's presenting symptom). While there are clinical tests for gene expression leading to the formation of these molecules that would be important to measure in any real therapeutics at the molecular level, they are not yet available to clinicians engaged in therapeutic hypnosis (Rossi, 2000b). This is unfortunate, indeed! It means that at the present time therapists and patients are, for the most part, blind to the results of their therapeutic work on the genomic level as it is taking place in real time. Until we develop practical approaches to assessing gene expression, protein synthesis, neurogenesis, and healing during the hypnotherapeutic process, we will remain limited to the phenotypic level of observations on cognitive-behavioral levels somewhat as follows.

The second stage of the creative cycle may be experienced on the phenotypic level by patients as periods of incubation, meditation, or inner search for the source of their problem. They typically encounter depression, anger, tears, and the valley of shadow and doubt ("the dark night of the soul," "the storm before the light") that is portrayed in poetry, song, and dance in the rituals of all cultures during important psychosocial transitions. These rituals — including those of therapeutic hypnosis, psychotherapy, and virtually all the healing arts — initiate behavioral state-related gene expression, neurogenesis, and the possibility of problem solving and healing. As patients review the origin and history of their problems the natural dynamics of Darwinian variation are experienced on deep psychobiological levels: as they attempt to recall the details of their traumatic experiences they are, without quite realizing it, replaying their past with new Darwinian variations. As Heraclitus said, "you cannot step into the same river twice." *The patient's main task in this second stage is to simply receive and appreciate what come up spontaneously regarding their problems in the manner of Freud's technique of free association. The therapist's tasks during this second stage are (1) to offer open-ended therapeutic questions that may function as implicit processing heuristics (that is, open ended questions) to access and replay the state-dependent memory and behavior encoding the patient's symptoms and (2) to support the patient through the sometimes painful arousal of their natural ultradian cycle of creativity as they make a courageous effort to recall the problems of the past.* Less is often more at this stage of respectful listening rather than giving advice.

What is actually happening on the level of gene expression

during the recall and replay of problematic past experience during this second stage of the creative cycle? Recent research by Nader et al. (2000a, b) illustrates how recall and the reconstructive aspect of memory can engage the dynamics of psychosocial genomics in a manner that supports the growth paradigm of psychotherapy (Rossi, 1967, 1968, 2000a, b, c, 2002). Nader et al. demonstrated how fear memories that are stored in the amygdala of the brain could be disrupted and made to disappear when they are reactivated under special experimental conditions. *Their research illustrates how consolidated fear memories return to a liable state during recall so that gene expression and protein synthesis cycle are reactivated to resynthesize the memory in a new way!* When the rat brain is infused with anisomycin (an inhibitor of protein synthesis) shortly after the reactivation of a long consolidated memory (from one to 14 days after conditioning), the memory is extinguished. The same treatment of the brain with anisomycin but without reactivating the consolidated memory leaves the memory intact. *This means that the gene expression and protein synthesis cycle is reactivated when important emotional memories are recalled and replayed.* Nader et al., (2000a) describe their research in this way. "Our data show that consolidated fear memories, when reactivated, return to a liable state that requires *de novo* protein synthesis [via gene expression] for reconsolidation. These findings are not predicted by traditional theories of memory consolidation." (P.723).

While Nader et al.'s research is conducted on an animal model at the present time, it does imply that similar molecular mechanisms of creativity and the transformation of consciousness take place in humans (that psychotherapists attempt to describe when they say, "every recall is a reframe.") That is, every recall of a memory automatically replays, re-synthesizes, and changes it in some way.

Further support for this recall-replay-resynthesize principle comes from more recent research on the molecular-genetic dynamics of novel learning and memory versus relearning with familiar cues. Berman & Dudai (2001) describe their research in this way.

> Experimental extinction is the decline in the frequency or intensity of a conditioned response following the withdrawal of reinforcement. It does not reflect forgetting due to the obliteration of the original engram, but rather "relearning," in which the new association of the conditioned stimulus with the absence of the original reinforcer comes to control behavior... *This is in accordance with the suggestion that the cortex contains molecular "novelty" switches, which are turned on only on the first highly salient [numinous] encounter with the stimulus... This means, among other things, that the cortex honors at the molecular level the distinction between learning anew and learning the new.* (Pp. 2417–2419, italics added).

We propose that the research of Nader et al. (2000a, b) and Berman & Dudai (2001) describes essential mechanisms of psychosocial genomics in therapeutic hypnosis, the cognitive-behavioral psychotherapies, and the holistic healing arts in general: *The recall and replaying of traumatic and problematic memories opens the possibility of reconstructing and re-synthesizing them on the levels of gene expression, protein synthesis, neurogenesis, and healing* (Rossi, 2001, 2002).

The skeptic could well challenge this view by asking: Why don't patients with post-traumatic stress syndromes cure themselves when they involuntarily replay their traumatic and painful memories in everyday life? Aye, this is a fundamental tragedy of the human condition. Most people do not know how to utilize the recall and replay of traumatic memories during stage two of the creative cycle to move on to stage three wherein they can experience new insights and a satisfactory resolution of their problems. Rather, when they experience the spontaneous recall of traumatic memories during the normal states of arousal in everyday life, they quickly try to suppress their painful memories. This self-suppression blocks their incipient creative process; patients now feel stuck and do not know how to help themselves. They believe they have a problem they do not know how to deal with. They do not realize that the apparently autonomous upwelling of traumatic memories actually means they are already engaged in stage two of the creative cycle. That's precisely why they need a psychotherapist to support them going through their "dark night of the soul." The therapist encourages the patient's spontaneous anamnesis by encouraging them to have the courage to allow the painful memories to continue for "just a little while longer" in the safety of the therapeutic situation. This replay of memory continues until the entirely natural process of Darwinian variation iterates new possibilities that move the patient's implicate associative dynamics closer and closer to appropriate options for problem solving and healing. Healing finally takes place when patient's consciousness discovers interesting options and possibilities that reframe their past with new insights that generate the therapeutic reconstruction of memory, meaning, and personal identity in state 3 of the creative cycle.

**Stage Three. - Illumination:**
**Activity-dependent Gene Expression and Conscious Selection of the Novel and Numinous.**

***Theory and Research.*** The basic working assumption of neuroscience is that the psychological experiences of mind, memory, emotions, learning, and behavior are encoded in the neural networks of the brain (Gould et al., 1999; Van Praag et. al., 1999, 2000). A generation of research by the Nobel Laureate Eric Kandel (2001), and others, has found that messenger molecules can facilitate the molecular-genetic basis of memory, learning, and behavior within neurons of the brain. Laboratory researchers identified three factors relating psychological experience to neurogenesis and brain growth: 1. novelty (Eriksson et al., 1998; Kempermann & Gage, 1999); 2. environmental enrichment (Kempermann et al., 1997; Van Praag et al. 2000); 3. physical exercise (Van Praag et al., 1999). It has been found, for example, that activity-dependent gene expression in response to novelty and exercise can double the number of neurons and connections between them generated in the hippocampus leading to the construction of new memory, learning, and behavior (Gould et al., 1999). In a series of remarkable drawings Lüscher et al. (2000) have shown how the number of synapses can double during the growth process over a period of 60 minutes. This is the approximate time frame of therapeutic hypnosis and most forms of psychotherapy.

In a series of pioneering papers, Kandel (1983, 1989, 2000, 2001) proposed how these molecular-genetic mechanisms could account for many of the classical phenomena of memory and learning as well as clinical phenomena of psychopathology such as chronic anxiety, neurosis, and schizophrenia. Kandel's theoretical and experimental work is the clearest expression of the possibilities of activity-dependent gene expression as the molecular basis of psychosomatic medicine, psychotherapy, and the healing arts. The role of activity-dependent gene expression and neurogenesis in the way we express ourselves with language (using measures of "idea density" and grammatical complexity) from an early age is evident in the quality of human experience as well as organic brain dysfunctions such as Altzheimer's disease. A major finding of Snowdon's (2001) research in his study of nuns suffering from Alzheimer's disease is that exercising whatever mental capacity one has with *positively* stimulating intellectual activities and emotions offers some protection against age-related mental decline.

***Therapeutic Practice.*** At its best, the patient's experience in stage 3 may be described as the famous cry of "Ah-ha" or "Eureka" that is celebrated in ancient and modern literature when the creative process is described in the arts and sciences (Sternberg, 2000, 2001; Sternberg & Davidson, 1995). People are usually surprised when they receive a creative intuition. Many people automatically dismiss their own originality as worthless since it has never been reinforced in their early life experience. *The patient's task is to consciously recognize the potential value of the new that appears in stage 3. Likewise, the therapist's task is to help the patient recognize and appreciate the value of the novel and the numinous that usually emerges spontaneously and unheralded after the period of inner struggle in stage 2.* Often the patient may have already thought of the options that come up for problem solving at this stage but dismissed them since they were never validated. Figure 9.4 illustrates just such a creative moment between patient and therapist in stage 3 of the creative cycle. After the patient reviewed the history of her rheumatoid arthritis in stage 2 she gained new insights during stage 3 regarding the emotional dynamics of her suppressive life experiences with her employer and boyfriend. The reactive anger she experienced was channeled into playful shadow boxing that was supported by the therapist. This playful shadow boxing could be interpreted as an activity-dependent exercise that could generate activity-dependent gene expression, protein synthesis, and healing. During this playful shadow boxing of stage 3 the patient experienced a creative breakout of her typically restrained hand and finger movements associated with her rheumatoid arthritis. She did report some pain associated with her newly flexible hand and finger movement, but she was obviously exhilarated that she now could move them well enough to make a clenched fist.

**Figura 9.4.** Stage 3 of the creative cycle wherein the patient experiences the playful activity-dependent exercise of shadow boxing as a creative breakout of her typically restrained hand and finger movements associated with her rheumatoid arthritis. Future research will be needed to determine if activity-dependent gene expression (ADGE) — such as the CREB genes associated with new memory and learning — as well as the ODC and BDNF genes associated with physical growth and neurogenesis are actually being engaged during such creative moments.

The therapist wonders at this high point if activity-dependent gene expression (ADGE) — such as the CREB genes associated with new memory and learning — as well as the ODC gene (associated with physical growth) and the BDNF (brain derived neurotropic factor) gene associated with neurogenesis and physical growth are actually being engaged (Schanberg, 1995; Rossi, 2002). Further research assessing the molecular dynamics of the gene expression-protein synthesis cycle during therapeutic hypnosis and psychotherapy as described by the methodology of genomic neuroscience will be needed to know for sure (Chin & Moldin, 2001).

**Stage Four. - Verification:**
**Consolidating Target Gene Expression, Neurogenesis, Problem Solving, and Healing**

***Theory and Research.*** The recall of a memory in everyday life invariably involves a mingling of the old representations of the past with the new perceptions of the present. This mingling of past and present in everyday recall results in new associations that lead to a natural Darwinian process of variation and conscious selection in the change and transformations of memory over time. Current research documents how the classical process of Pavlovian stimulus-response conditioning requires the activation of a conditioned response before it can be extinguished (Dudai, 2000). As note above, this process of reactivating a memory in order to extinguish, reframe, or resynthesize it has profoundly important implications for the practice of psychotherapy and, by extension, many of the therapeutic arts.

The problem in stage 4 of the creative cycle, however, is that the new therapeutic insights and symptom resolutions experienced in stage 3 are themselves fragile — sometimes they are initially experienced only momentarily in the therapeutic situation and are immediately lost when patients walk out the door and move back into their usual everyday life situations that tend to re-evoke their old traumas, symptoms, and problems. How do we consolidate the nascent therapeutic process so that it is not immediately lost?

Erickson dealt with this problem by using posthypnotic suggestions to support therapeutic advances (Erickson & Erickson, 1941) and posthypnotic amnesia to protect the fragile new memory reconstructions from the negativity of the patient's own psychodynamics as well as that of their psychosocial milieu (Erickson & Rossi, 1974). Current neuroscience research indicates how the consolidation process can be facilitated with *multiple rounds of recall and replay of past life experiences for the reconstruction of memory.* Research by Shimizu et al., (2000), for example, clarifies how a particular region of the hippocampus called "CA1" is crucial for converting new memories into long-term memories. This psychobiological process may continue for weeks after the initial learning and memory event. They found that *multiple rounds of recall and replay of past life experiences are required for the reconstruction, resynthesis, and reconsolidation of memory.*

> Our results indicate that *memory consolidation may require multiple rounds of site-specific synaptic modifications, possibly to reinforce plastic changes initiated during learning*, thereby making memory traces stronger and more stable. Recent studies report that *the learning-induced correlation states among CA1 neurons are reactivated spontaneously in a post-learning period.* Such a *co-activation* of these neurons might suggest the existence of the natural condition within the hippocampus by which recurrent synaptic strengthening can occur during memory consolidation. We hypothesize that such a *synaptic re-entry reinforcement (SRR)* process can also be applied to explain how the hippocampus transfers newly created memories to the cortex for permanent storage. As the hippocampus undergoes *reactivation during consolidation, it may also act as a coincidence regenerator for activating neurons* in the cortical area such as the association cortex. *This would allow cortical neurons previously corresponding to the different sensory modalities to be reactivated together,* leading to the strengthening of the connections between them through SRR. Indeed, such a *coordinated reactivation of hippocampal-cortical neurons after learning* has been observed recently... Once these cortical connections are fully consolidated and stabilized, the hippocampus itself becomes dispensable for the retrieval of the 'old memory'... Therefore, we postulate that the hippocampus, by serving as a *coincidence regenerator,* may induce the reinforcement of synaptic connection within the cortex during memory consolidation as the cellular means to convert short-term memories into long-term memories. (Pp. 1172-1173, italics added)

This outlines a neuroscience approach to dealing with many of the mysteries of how to consolidate the new insight, problem solving, and symptom resolution that are usually experienced as taking place spontaneously in stage 3 of the creative process.

***Therapeutic Practice.*** As indicated, the main task of patient and therapist in stage 4 is to consolidate the therapeutic work of stage 3. The first step in this consolidation process can be facilitated by the therapist offering a posthypnotic suggestion when it is apparent that patients have navigated stage 3 well and are beginning to make spontaneous shifts in breathing, facial gestures, body position, and minimal movements that indicate they are already experiencing a natural ending to their therapeutic trance. A posthypnotic suggestion that supports the patient's behavior at this point together with a continuation and consolidation of the therapeutic work during their natural ultradian rhythms of homeostasis and adaptation in everyday life runs somewhat as follows.

"*When your unconscious mind knows it can continue this healing work all by itself* [pause] *whenever it's entirely appropriate* [pause], *and when your conscious mind knows it can cooperate by helping you recognize those moments throughout the day when it is right to take a rest* [pause], *will you find yourself awakening feeling refreshed, alert, and as aware as you need to be of the meaning of your experience here today?*"

With this implicit processing heuristic in the form of an open ended question, awakening from therapeutic trance is made contingent on the healing process continuing on an inner, implicate, unconscious level. Simultaneously, the conscious mind is focused on cooperating by learning recognizing those ultradian mind-body cues that can be used to break the stress cycle to facilitate our natural "ultradian healing response" (Rossi, 1986/1993, 1996, 2002; Rossi & Nimmons, 1990). The dynamics of many variations of this type of posthypnotic implicit processing heuristic with a variety of problems and symptoms in clinical practice have been described in detail (Rossi, 2002; Rossi et al. 1994, 1995). From our current perspective of psychosocial genomics, it becomes apparent how accessing and utilizing the patient's natural ultradian rhythms of adaptation and healing in everyday life in this manner may be facilitating the molecular mechanisms of activity-dependent and behavior state-related gene expression, protein synthesis, neurogenesis, and healing.

After the patient has awakened, the process of consolidating therapeutic progress can be facilitated by reviewing and discussing insights and symptom resolution with a sense of wonder and appreciation — a positive reinforcement of the new! Sometimes patients will affirm the therapeutic change by considering how they now can voluntarily change their thinking, feeling, and behavior to maintain their progress in everyday life. If not, the therapist can facilitate the process of consolidation by offering implicit processing heuristics such as these.

**Figura 9.5.** Stage 4 of the creative cycle when the patient received a standing ovation from the audience. The therapist speculates that the zif-268 gene will certainly be expressed in her REM dream states tonight to encode her new therapeutic experiences with this unusually strong show of psychosocial support.

*"Having had this wonderful experience here today – how will your behavior change in everyday life?*

*What do you suppose you can do differently now to support the progress you have made today?*

*What will you be saying to yourself and others in your everyday life to continue this therapeutic work you have begun here today?"*

In constructing a response to such questions the patient will, in effect, give themselves a cognitive-behavioral prescription whereby they utilize their own resources in their own way to synthesize a psychosocial genomic bridge between their new therapeutic progress and the real everyday circumstances of their present and future life. As illustrated in Figure 9.5 the patient experiences a satisfying self-empowerment with the support of a standing ovation from the therapeutic community in the audience. The therapist hopes this encounter with the experiential theater of demonstration therapy will be sufficiently numinous to activate zif-268 gene expression in her REM dreaming tonight to optimize the therapeutic reorganization and reconstruction of her mind and memories in keeping with the psychobiological dynamics of current neuroscience research. Recent research (Ribeiro et al., 1999) has documented how a day rich in memorable experiences leads to creative replays of gene expression, protein synthesis, and neurogenesis (literally brain growth) during dreaming (REM sleep).

Most people are not aware of how they need to facilitate and consolidate gene expression, neurogenesis, problem solving, and healing in everyday life as well as psychotherapy. This fourth stage of the creative process requires cooperation between the implicit, unconscious dynamics of target gene expression (the specific genes engaged in memory, learning, and healing) and the explicit, conscious dynamics of ratifying the reality of the new. Patients participate in this cooperative process but they do *not* direct it. This is the greatest source of misunderstanding made about the creative process by both patients and therapists: patients and therapists are co-creators, not directors of the natural dynamics of gene expression, neurogenesis, and healing. *The patient's co-creative task is to recognize and value the new that arises in stage 3 and then plan in stage 4 how the new can be practiced in real life. The therapist's co-creative task in stage 4 is to (1) facilitate the patient's creative experiences to validate the value of their psychotherapeutic process and (2) help reframe and resynthesize symptoms into signals and psychological problems into inner resources* (Rossi, 1986/1993, 1996, 2002). Symptom scaling may be used to validate the therapeutic experience and what may need to be done in future sessions.

This completes my outline of the theory, research, and practice illustrated in videotape Code E 297-V8, Erickson Foundation, 1992). The leading edge of neuroscience is tracing the pathways of mind-body communication, neurogenesis, and healing on all levels from the psychosocial to the cellular-genetic in a manner that makes a genuine science and practice of therapeutic hypnosis, psychotherapy, and the healing arts possible. Many of the problems and paradoxes of historical hypnosis can be resolved by a deeper understanding of the deep psychobiological parameters of mind-body communication and healing in the emerging field of psychosocial genomics. We propose that the psychosocial genomics of therapeutic hypnosis and psychotherapy can be modeled as a 4-stage creative cycle that facilitates the replay and resynthesis of problematic memory, learning, and behavior in an adaptive manner. While most current research on the genomic neuroscience of memory and learning is still being conducted with animal models in the laboratory, it is likely that new methods of assessing gene expression, neurogenesis, and healing on the molecular level will become available to guide clinical practice in the future. Meanwhile we need to practice more highly structured approaches to as outlined in the next section.

## Small Group Practice for Professional Development

After observing a few individual demonstrations as outlined in the previous section most participants in our therapeutic workshops are eager to practice themselves. I ask the entire workshop to break-up into small groups of 4 to 6 participants to practice together. They are to select among themselves who wants to volunteer as a patient and who wants to practice as the therapist – the other 2 or 3 group members act as observers. The most interesting challenge is for this small group practice to proceed "blind." That is the "patient" does not tell anyone in the group anything about their personal history, the issue, problem or symptom that is to be explored and resolved therapeutically. The "patient" simply says, "I want to work on myself" and the "therapist" immediately proceeds with one of the verbatim outlines presented here. This so-called "blind" approach, of course, is not really blind; rather it immediately focuses both patient and therapist on the essence of the 4-stage creative experience. It is a *novel, enrichening, and numinous exercise* that radically shift everyone's conventional preconceptions about what "talk therapy" is supposed to be.

I usually ask the participants to study the structured outlines presented here as guides to their first efforts. This series of four structured approaches are graded in their level of difficulty and the amount of previous student and professional experience recommended before they are explored in our therapeutic workshops.

## THE 4-STAGE CREATIVE PROCESS

**Stage One: Initiation. Symptom Scaling and Privacy.** The initiation of the natural 4-stage creative cycle begins with the initial questions and typical history taking at the beginning of any form of psychotherapy. More than mere words are involved. The typical tears and distress in an initial interview indicate that patients are already accessing state dependent-memories and emotional arousal that signals they are embarking on a potentially healing adventure. *The therapist's main job at this point is to recognize that therapy has already begun and simply facilitate it by letting people to tell their own story.* **Implicit Processing Heuristics** are open ended questions that can facilitate the creative process without the therapist even knowing the nature of the subject's problem. The therapeutic process formally begins with a subjective **Symptom Scaling** of the subject's current emotional state. A one to ten scale (ten being the worst the problem was ever experienced, five about average, and zero a completely satisfactory state) evaluates the creative process.

**Stage Two: Incubation. The Dark Night of the Soul.** This is the valley of shadow and doubt, or "**The Storm Before the Light**" that is portrayed in Poetry, Myth and Song in many cultures. The creative paths outlined here facilitate *precognitive implicate processing* and *information transduction* so that the subject's inner nature can signal when autonomous inner processes of mind-body communication & healing are taking place. *The therapist's main job during this second stage is (1) to offer open-ended therapeutic_questions designed to access state dependent memory encoding symptoms and (2) to support the subject through the sometimes painful arousal that makes them abort their natural ultradian cycle of creativity, problem solving and healing in everyday life.* Less is often more at this stage of respectful listening rather than the therapist attempting suggestion, analysis, counseling, programming or advice giving.

**Stage Three: Illumination. The Novelty-Numinosum-Neurogenesis Effect.** Creative Insight, Problem Solving & Healing seems to happen spontaneously. This the famous "Ah-ha" or "Eureka" experience celebrated in ancient and modern literature when the creative process is described in the arts and sciences. Subjects are usually surprised when they receive a creative thought and many people automatically dismiss their own originality as worthless since it has never been reinforced in their early life experience. *The therapist's main job at this stage is to help the subject recognize and appreciate the value of the "new" that usually emerges spontaneously and unheralded.* Often the subject may have already thought of the options that come up for problem solving at this stage but dismissed them since they were never validated.

**Stage Four: Verification.** *The therapist's job here is to (1) facilitate a follow-up discussion to validate the value of the psychotherapeutic process and (2) help the subject reframe symptoms into signals and psychological problems into inner resources.* The symptom scaling of the subject's state of being before and after the psychotherapy is a measure of progress, problem solving and healing that is used to validate the value of the therapeutic experience and what may need to be done in future sessions.

## 1
## Creative Process Via Activity-Dependent Hand Signaling (All Students)

INITIAL SYMPTOM SCORE (0 IS BEST...10 IS WORST) ______.

**1. Induction:** "Place your hands up with the palms facing each other about 6 to 8 inches apart [Therapist demonstrates]... With great sensitivity notice what you begin to experience... Do the arms *feel the same or different?...* One may be *lighter or heavier?...* One hand may be *warmer or cooler?...* Something like *a magnetic force or energy pulling those hands togeth-*

*er or pushing them apart?*...Or perhaps they seem to take on *a life of their own moving in a special way?... Different feelings in one part of your body or other?... Heat, sweating, vibration, shaking,* or *tingling?...* Whatever you are experiencing is to be received and explored. Mind-body signals of getting ready to do inner work."

**2. Incubation & Inner Work:** "Will just one of those hands now begin to drift down slowly to signal that your inner nature will now explore some *private...* even *secret emotions and memories about you earliest experiences of learning?...* The *Courage* to simply *wonder and receive* everything *interesting to you as a child?...* Continue experiencing that for another moment or two...only long enough to *learn what it is about so you can receive what come next?...* Feeling as much or as little as you want *privately...*Then letting it go... And welcoming what comes next about your *most enjoyable, earliest creative experience...* One part of you experiencing that as fully as you need to... while *another part guides your personal and deeply meaningful experience in a safe way.*

**3. Illumination:** "Will the other hand now drift down slowly as you explore new possibilities of creativity, learning, and problem solving today?... Will that hand go down slowly signaling when you are ready to begin *to experience something new?... Interesting?... Curious?* ... Working well with yourself?... Something *pleasantly surprising you can look forward to?...*What *you really need to learn something new about what is personally most interesting and important to you*?...Simply receiving and continuing to explore the *sources of strength for dealing successfully* with that issue... Yes, *experiencing as fully as you need to while taking good care of yourself* as that hand finally comes to rest in your lap.

**4. Verification:** Review the entire session. Recognize issues & insights about the inner struggle to recognize and value one's own experiences, perspectives, and develop further. "*On a scale of 0 to 100%, how confident are you of this new understanding*?... When your inner nature knows you can continue these positive developments... & when you really know *you can take-a-break for about 15 or 20 minutes several times a day to simply rest* *and recognize what your inner nature is up to...* Will those eyes close for a minute or so to confirm your inner resolve?... On a scale of one to ten...where 10 is the most difficult you ever felt about yourself and 0 is *a completely satisfactory state...* What number comes up?... ____.

**Repeats If Necessary:** If a number greater than one is reported, facilitate with, *"If your inner nature knows it can do another unit of creative work right now so you can reach a completely satisfactory state, will those eyes close for a few moments so you can fully receive everything you need at this time?"* If a number greater than one continues after two or three repeats reassure the subject, "You know that your mind and body go through *a natural cycle of ultradian healing and problem solving every couple of hours throughout the day and at night when you are asleep & dreaming.* Notice how your progress continues all by itself and we will pick it up from there the next time we meet."

**End Symptom Score (0-10)____.**
**Confidence (100% is Best) _____.**

## 2
## Problem Solving Via Activity-Dependent Hand Signaling (Clinical Training)

INITIAL SYMPTOM SCORE (0 IS BEST...10 IS WORST) ______.

**1. Induction:** "Place your hands up with the palms facing each other about 6 to 8 inches apart [Therapist demonstrates]... With great sensitivity notice what you begin to experience... Do the arms *feel the same or different?...* One may be *lighter or heavier?...* One hand may be *warmer or cooler?...* Something like *a magnetic force or energy pulling those hands together or pushing them apart?...*Or perhaps they seem to take on *a life of their own moving in a special way?... Different feelings in one part of your body or other?... Heat, sweating, vibration, shaking, or tingling?...* Whatever you are experiencing is perfectly okay..

Mind-body signals of getting ready to do inner work."

**2. Incubation & Inner Work:** "Will just one of those hands now begin to drift down slowly to signal that your inner nature will now explore some private... even *secret emotions and memories?*... The *Courage* to simply wonder and receive everything that is *necessary to help you?*... Continue experiencing that for another moment or two...only long enough to *learn what it is about so you can receive what come next*?... Feeling as much or as little as you want *privately*...Then letting it go... And welcoming what comes next... One part of you experiencing that as fully as you need to... while *another part guides your personal and deeply meaningful experience in a safe way.*

**3. Illumination:** "Will the other hand now drift down slowly as you explore possibilities of healing and problem solving?... Will that hand go down slowly signaling when you are ready to begin *to experience something new*?... *Interesting*?... *Curious*? ... Working well with yourself?... Something *pleasantly surprising you can look forward to*?...What *you really need for healing and problem solving?*...Simply receiving and continuing to explore the *sources of strength for dealing successfully with that problem*... Yes, experiencing, healing and problem solving as that hand finally comes to rest.

**4. Verification:** Review the entire session. Reframe symptoms & problems into signals of the inner struggle to heal one's self and develop further. "*On a scale of 0 to 100%, how confident are you of this new understanding*?... When your inner nature knows you can continue these positive developments... & when you really know *you can take-a-break for about 20 minutes several times a day to transform Symptoms into Signals & Problems into Resources*... Will those eyes close for a minute or so to confirm your inner resolve?... On a scale of one to ten...where 10 is the worst you ever felt about yourself and 0 is a *completely satisfactory state*... What number comes up?... ____.

**Repeats If Necessary:** If a number greater than one is reported, facilitate with, *"If your inner nature knows it can do another unit of healing right now so you can reach a completely satisfactory state, will those eyes close for a few moments so you can fully receive everything you need at this time?"* If a number greater than one continues after two or three repeats reassure the subject, "You know that your mind and body go through *a natural cycle of ultradian healing and problem solving every couple of hours throughout the day and at night when you are asleep & dreaming.* Notice how your progress continues all by itself and we will pick it up from there the next time we meet."

**End Symptom Score (0-10)____.**
**Confidence (100% is Best) _____.**

## 3
## Conflict & Problem Solving via Creative Replay (Psychotherapy License)

INITIAL SYMPTOM SCORE (0 IS BEST...10 IS WORST) _____.

1. **Induction:** "If you are ready do some important inner creative work on that conflict ...Will you place your hands with the palms up as when you are ready to receive something important? [Therapist demonstrates: Leave arms free to move!] "As you focus on those hands in a sensitive manner...I wonder *if you can begin by letting me know which hand seems to experience or express that fear* (anxiety, anger, bad habit or whatever the negative side of the subject's conflict may be) more than the other? When subject indicates which hand is the problem side reinforce with, "*Wonderful!*"

2. **Incubation & Inner Work:** "As you continue with that I wonder *what you begin to experience in your other hand by* ***contrast*** *at the same time*? Reinforce whatever the subject responds with, "Good! Really going with that...*Allowing yourself to experience* ***both sides*** *of that conflict* (problem, issue, your personality etc.) *at the same time!*...Privately?...Or telling me what begins next?...

I don't know if those hands or fingers will begin to *move more or less all by themselves as both sides of your nature each express themselves in their own way*?...Together?...Apart?...An *inner drama*?... Day Dreaming?...Up?...Down?...Near?...Far?... A life of their own? *Experiencing that as deeply as you need to while another part of you takes good care of you.*

3. **Illumination:** "And is that going well? [If subject says No or gives negative facial gestures continue with] The strength to continue for a while until?...Sometimes darkness, a storm & struggle before the light... Yes...Allowing that to continue more or less all by itself...Both sides having their say...expressing what they need... Something changing? [as subject makes subtle changes of breathing and movements]...Exploring?...Receiving?...Surprising?...And is that really possible?...One way or another?...*Yes, the courage to really go with that*! [as subject smiles or makes more positive facial gestures & body language]...Very fine!...Really appreciating that!"

4. **Verification:** Review the entire session: Reframing symptoms & problems into signals of the inner struggle to heal one's self and develop further]..."*On a scale of 0 to 100%, how confident are you of this understanding*?... When your inner nature knows you can continue these positive developments... & when you really know *you can take-a-break for about 20 minutes several times a day to transform Symptoms into Signals & Problems into Resources*... Will those eyes close for a minute or so to confirm your inner resolve?... On a scale of one to ten...where 10 is the worst you ever felt about yourself and 0 is a *completely satisfactory state*...What number comes up?"____.

**Repeats If Necessary:** If a number greater than one is reported, facilitate with, "*If your inner nature knows it can do another unit of healing right now so you can reach a completely satisfactory state, will those eyes close for a few moments so you can fully receive everything you need at this time?*" If a number greater than one continues after two or three repeats reassure the subject, "You know that your mind and body go through *a natural cycle* *of ultradian healing and problem solving every couple of hours throughout the day and even at night when you are dreaming.* Notice how your progress continues all by itself and we will pick it up from there the next time we meet."

**End Symptom Score (0-10) ____.**
**Confidence (100% is Best) _____.**

## 4
## Merry Symptom Chase & Mind-Body Healing (Licensed Medical Doctors)

INITIAL SYMPTOM SCORE (0 IS BEST...10 IS WORST) ______.

1. **Induction:** "Do you have the courage to allow yourself to honestly receive just what you are really feeling right now so you can fully experience what comes next all by itself?...[Accepting whatever symptom, sensory, emotional shifts or doubts] Good, now going with that so you can receive what comes next!... Yes, and with great sensitivity noticing how even that will change all by itself...Wonderful, the curiosity to stay with these spontaneous changes until..." [And so the merry symptom chase continues with the therapist expressing approval of each shift and full expectation of further change.]

2. **Incubation & Inner Work:** "Yes, difficult as it may be, do you have the strength to experience [whatever symptom, pain or discomfort the subject has just mentioned] just enough so that some part of you can get the message?...Allowing yourself to explore *private*...even *secret emotions and memories you know you would like to deal with*?... Only long enough *so you can receive what come next*?... Feeling and expressing as much or as little as you need to so you can fully experience...Then letting it go... And welcoming what comes next... *One part of you experiencing this as deeply... as much as is necessary... while another part of you safely guides this meaningful therapeutic experience.*" Therapist facilitates the natural alternations of catharsis and quietness for 5 to 10 minutes.]

**3. Illumination:** "And is that going well? [If subject says No or gives negative facial gestures continue with] The strength to continue for a while until?...Sometimes darkness, a storm & struggle before the light... Yes...Allowing that to continue more or less all by itself... Both sides having their say...expressing what they need... Something changing? [as subject makes subtle changes of breathing and movements]... Exploring?... Receiving?... Surprising?... And is that really possible?... One way or another?... *Yes, the courage to really go with that*! [as subject smiles or makes more positive facial gestures & body language]... Very fine!... Really appreciating that!

**4. Verification:** [Review the entire session: Reframing symptoms & problems into signals of the inner struggle to heal one's self and develop further]..."*On a scale of 0 to 100%, how confident are you of this understanding?*... When your inner nature knows you can continue these positive developments... & when you really know *you can take-a-break for about 20 minutes several times a day to transform Symptoms into Signals & Problems into Resources*... Will those eyes close for a minute or so to confirm your inner resolve?... On a scale of one to ten...where 10 is the worst you ever felt about yourself and 0 is a *completely satisfactory state*...What number comes up?... ____.

**Repeats If Necessary:** [If a number greater than one is reported, facilitate with] *"If your inner nature knows it can do another unit of healing right now so you can reach a completely satisfactory state, will those eyes close for a few moments so you can fully receive everything you need at this time?"* [If a number greater than one continues after two or three repeats reassure the subject] "You know that your mind and body go through *a natural cycle of ultradian healing and problem solving every couple of hours throughout the day and even at night when you are dreaming.* Notice how your progress continues all by itself and we will pick it up from there the next time we meet."

**End Symptom Score (0-10) ____.**
**Confidence (100% is Best) _____.**

* * * *

A discussion period wherein each small group has an opportunity to explore the salient issues that came up during this new learning experience provides another level of enrichment for the entire workshop. The novelty of working "blind" in these workshops (but not in actual clinical practice) usually has an exhilarating effect on everyone when they witness how easy it is for people to learn to resolve their issues and problems privately in their own ways. A major source of therapist stress is resolved when it is fully recognized that the therapist does not have to be omniscient. Psychotherapists can facilitate the creative process on many levels, but it is only the patient who can learn to apply it appropriately in real life.

# X
# HOW TO ENJOY LIVING A CREATIVE AND HEALTHY DAY

The writer Aldous Huxley, author of *The Perennial Philosophy*, certainly knew how to enjoy living a creative day. After an early morning breakfast he would let himself sink into a state of what he called "Deep Reflection" in his favorite armchair in his study. After about an hour, or so, he would awaken spontaneously with words, sentences and paragraphs teeming through his mind for expression. He would then write easily and steadily throughout the morning almost as if he were taking down dictation from a boundless source of creativity.

What is Deep Reflection and what is the creative source? Huxley used these words to describe the nature of his Deep Reflection when he explored it with the help of Milton H. Erickson (1965/1980). "I use Deep Reflection to summon my memories, to put into order all of my thinking, to explore the range, the extent of my mental existence, but I do it solely to let those realizations, the thinking, the understandings, the memories seep into the work I'm planning to do without my conscious awareness of them. Fascinating ... never stopped to realize that my Deep Reflection always preceded a period of intensive work wherein I was completely absorbed..."

After a morning of writing Huxley would have lunch, rest a bit in the afternoon and then enjoy some social life and light exercise like a walk. After supper he reentered the creative mode in quite another way. He would relax in his study browsing a bit here and there from his vast home library. He was vaguely aware that he was searching for material for the next mornings writing, but he was not working actively on it with his conscious mind. He was simply enjoying himself with comfort and relaxation as a prelude to sleep and dreaming so that he could awaken the next morning primed for another creative day.

Recent research supports this view of Huxley's creative life style and more. A basic idea emerging from a variety of fields ranging from ecology and engineering to molecular genetics and neuroscience is that there is a natural Complex Adaptive System of psychobiological rhythms of creative life. Creative life processes are organized around natural complex rhythms of growth, self-reflection and co-creation. Howard Odum (1988), calls this the "Pulsing Growth Paradigm" of life.

This complex adaptive pattern of life underlies what has been called "the wave nature of consciousness" (Rossi & Nimmons, 1991). Huxley was successful in his creative work at least in part because he followed this natural complex rhythm in his daily life style. The biographies of many creative workers throughout the ages confirm that there is a highly individualized natural pacing of creativity and rest throughout the day. Let us now explore emerging research in this area by looking at a typical day's 90–120 minute Basic Activity Rest Rhythms. We will discover many hidden connections between these complex adaptive circadian (daily) and ultradian (many times a day) rhythms and our energy levels, moods, stress, exercise, diet, accidents, addictions and even the natural pacing of our sexuality and relationships.

## MIDNIGHT to 2:OO A.M.: Deep Sleep and Growth Hormone

One of the most remarkable advances in current science is our ability to trace how our genes are turned on and off throughout the day by the "circadian clock." The main circadian clock, which is regulated by a central pacemaker in the suprachiasmatic nucleus (SCN) of the hypothalamus of the brain, modulates many of the most obvious psychological processes in our daily rhythms of behavior. As we reviewed previously in chapter 3, many of these deep psychobiological rhythms can be modulated by psychosocial genomic cues in the interaction between psychological, social, and cultural experiences and the expression of our genes. Computer animations illustrating the basic rhythms of gene expression throughout the day are now easily available in great detail (Van Gelder, 2003). In this chapter we will outline how some of these rhythms of gene expression are manifest when we cooperate with nature in learning how to enjoy a creative day.

The new day really begins during the first two hours of sleep when growth hormone is released to facilitate the growth and repair of tissues throughout the body and mind. As we sleep the pineal gland in our brain releases melatonin, a major hormone that modulates growth as well as the rhythms of our sleep, dreams, immune system, and the aging process. Meanwhile the liver is synthesizing cholesterol. Although high levels of cholesterol can be dangerous, it is a vital substance that forms the basic building block for the daily production of a variety of hormonal messenger molecules that mediate mind-body communication, our sex drive, moods, energy levels, consciousness and productivity.

Sleep is not simply rest and recovery. Sleep is also an active process of creatively replaying and integrating the most novel, enrichening, and deeply meaningful experiences of the day to update our brain (Strickgold et al., 2002a & b). Dreams are not simply a review of the past, however. Dreams are actively replaying and reintegrating the past with the present and future possibilities on the levels of gene expression and synaptogenesis (brain growth) if we are sensitive and astute enough to pay attention to them. Carl Jung (1943/1966) described his realization of the *synthetic nature of sleep, dreaming and psychotherapy* when he wrote, "I had first to come to the fundamental realization that *analysis, in so far as it is reduction and nothing more, must necessarily be followed by synthesis…*" (,p. 81Italics added). With this statement of the importance of synthesis Jung emphasized in his break away from Freud's focus on psycho-*analysis.*

Some of the most dramatic scientific data that supports the idea that the *creative replay of novel and numinous psychological experience from mind to gene is the essence of all the psychotherapies and cultural arts* comes from the research of the research team of Sidarta Ribeiro at the Department of Neurobiology in Duke

University Medical Center. Ribeiro et al. (1999) first documented experimentally how a day rich in memorable experiences leads to expression of the zif-268 gene during dreaming (REM sleep). If we do not have novel and memorable experiences during the day, we still dream but the zif-268 gene is not expressed. The zif-268 gene is also known as *nerve growth factor-induced A* (as well as egr-1, krox 24, tis-8). This leads Ribeiro, et al. (1999) to conclude that the activation and transcription of the zif-268 gene leads to the formation of new proteins that can encode new memory and learning as follows.

> "Whereas non-exposed controls show generalized *zif-268 down-regulation* during SW and REM sleep, *zif-268* is *up regulated* during REM sleep in the cerebral cortex and the hippocampus of exposed animals. *We suggest that this phenomenon represents a window of increased neuronal plasticity during REM sleep that follows enriched waking experience.*" (p. 500, italics added)

In more recent updates of this research Ribeiro et al. (2002, 2004) present more details about the replay of gene expression and brain plasticity (growth) during the deepest stages of slow wave (SW) sleep and dream sleep (REM) as follows.

> "The present findings and the current literature suggest... that SW and REM sleep play separate roles on memory consolidation with *memory recall* occurring during SW sleep and *memory storage* taking place during REM sleep. . . In conclusion, *sustained neuronal reverberation during SW sleep, immediately followed by plasticity-related gene expression during REM sleep*, may be sufficient to explain the beneficial role of sleep on the consolidation of new memories." (Ribeiro et al., 2004, pp 134-135, italics added.)

These conclusions strongly support the theory that the creative replay of novel, numinous, and deeply meaningful life experiences is the essence of therapeutic suggestion and psychotherapy on the deepest levels of gene expression and brain plasticity (Rossi, 2002 and chapter 8 of this book). Ribeiro et al. (2004) document that this novelty-induced creative replay takes place between the hippocampus (that first encodes new memory and learning) and other areas of the brain (such as the cerebral cortex, putamen, and thalamus). It is this creative replay and growth of the brain during sleep that makes "*the morning wiser than the evening*". I hypothesize that, at their best, therapeutic hypnosis, psychotherapy, the creative arts, and sciences all utilize this natural process of gene expression and brain plasticity during sleep and dreaming. Perhaps, for example, this why so-called "post-hypnotic suggestion" and "hypnotically induced dreams and fantasies" can be so effective in facilitating behavior change and healing. Post-hypnotic suggestion reverberating during sleep may reinforce the novelty-induced recurrence of gene expression, new protein synthesis, and brain plasticity during slow wave sleep and REM state dreaming.

## 2:00 A.M To 4:00 A.M.: The Ultradian Cycle and Sleep Problems

That there is a natural ultradian 90–120 rhythm of gene expression, brain plasticity, creative adaptation that continues throughout sleep is well illustrated by the tendency for some people to awaken every few hours or so throughout the night. When this tendency is not understood it can have tragic consequences. Who has not seen the tragic pictures of swollen face and undue weight gain of the formerly beautiful Elvis Presley in the last years of his life when he was addicted to drugs? The story is told in his biography of how he required a handful of barbiturates to put him to sleep at night. Alas, after a while these drugs would wear off and he would awaken with a startle after an hour and a half or so of sleep. No one was there to assure him that he would soon drift back into a more natural sleep. In his alarm at awakening so soon and fearing that he would not get enough sleep to perform well the next day, Elvis would knock on his bed board to signal a member of his retinue to come in and give him another handful of "attack pills" to put him back to sleep. And thus it went throughout the night so that the normal complex adaptive psychobiological systems that regulated appetite, food intake, weight gain, growth, healing, memory, learning, creativity and many other delicately balanced life processes spiraled out of control until he finally died well before his time.

Contrast Elvis Presley's early demise with the advice of the 96-year old Nathaniel Kleitman (1963, 1969, 1970; Kleitman & Rossi, 1992) who described to me how it was entirely natural for some people to awaken every hour and a half or so throughout the night as a natural manifestation of the *Basic Rest Activity Cycle (BRAC).* He claimed that this was true especially as people got older. "So what?" he would exclaim. "When I awaken in the middle of the night I can enjoy it. I can read! I can think! I can even meditate if I feel like it. I can enjoy myself! I would never take drugs to go to sleep! What for, they cannot be good for someone my age."

## 4:00 TO 6:00 A.M.: Pain, Meditation, and Lucid Dreaming

Milton H. Erickson also had experiences of awakening out of sleep in the middle of the night, most often to deal with pain that arose as an after effect of the two bouts of polio he experienced, first when he was 18 and then again in his early fifties. Milton H. Erickson (1980) was an acknowledged master in dealing with pain problems through the use of his many innovative approaches to therapeutic hypnosis. He was very successful in using self-hypnosis for pain control in his own life—with one exception. When I spoke with him about it when he was in his seventies he assured me that he was always successful in his personal pain control when he was awake and he always was comfortable when he fell asleep at night. He also assured me that thorough the use of post-hypnotic suggestion most of his patients remained pain free through out the night. He did not know why, however, but for some reason or other, he might occasionally awaken with pain after the first few hours of sleep or sometime toward morning. Erickson had not ever heard of ultradian rhythms and the natural hour and a half or so Basic Rest Activity Cycle. To reassert his pain control he would have to awaken his wife, who then had to help him sit up in bed so that he could fully awaken, and then to listen to him as he described to her verbally his process of self-hypnosis. Erickson told me it was necessary for him sit up, and even use his wife as an audience, in order to awaken fully before he could proceed with pain control in the middle of the night. This is anecdotal evidence for the idea that arousal is important in many aspects of therapeutic hypnosis described in the previous chapter. So, how did Erickson do it? Here is the story of how he used self-hypnosis in the middle of the night as he told it to me when he was in his seventies. I can only quote his words from memory. I use ellipses (...) to indicate how Erickson would dramatically pause throughout his description to mark the transitional states of his self-hypnotic experience.

"Can you imagine how frightening it is for me in my condition to feel an intense, sharp, biting pain going right through my shoulder? I have to get fully awake even in the middle of the night, sit up, and fully experience that pain before I could do anything with it... After a few moments, or so, I realize it feels it like a hot wire going right through my shoulder... then I seems I forget myself for a while...and after a while I suddenly notice, what's this? ...It seems as if the hot wire is now resting on top of my shoulder!

[At this point Erickson glances at me sharply as if to check whether I understand the significance of this sensory-perceptual shift as the first indication of the spontaneous healing sifts characteristic of therapeutic hypnosis. At no time does Erickson suggest or tell himself what changes will take place during his self-hypnosis. He has no conscious expectancy of how the pain relief will take place. He appears to be simply observing and reporting how an implicate process evolved entirely on its own. The process continues with a life of its own somewhat as follows.]

"I wonder about that hot wire that is now resting on the *top* of my shoulder. Can you imagine what a great relief it is to me to realize its only on top of my shoulder not inside where it was really frightening to me...And I seem to forget myself for a little while again...Then, what is this!? ...Now it seems as if the whole top of my shoulder is hot...so hot that I can hardly stand it! But it is much better this way because I realize it is starting to spread... After a while I feel the heat spreading down my arm to my elbow... What a relief that is to me, I now realize that as it spreads its less hot. I sense it slowly moving down to my wrist and then my hand... by this time its only warm... pleasantly warm in fact.

"Pleasant warmth... umm... with the warmth I begin to feel a pleasant, comfortable fatigue!

[At this point Erickson again glances sharply at me with quizzically arched eyebrows. He wants to know if I am aware of the significance of this feeling of "comfortable fatigue." Of course I do not, so I silently shake my head slowly, no. Erickson proceeds to pedagogically explain.]

"From frightening pain in the center of my shoulder where the bones may be falling apart, to a hot wire feeling on top of my shoulder. From a hot wire to intense heat all over the shoulder. The heat gradually becomes warmth as it spreads down my arm. The warmth becomes comfortable and I feel fatigue! Warmth, comfort and fatigue...fatigue as in tired...tired as in *Sleep! I'm beginning to feel comfortable and sleepy!* I tell my wife I am sleepy and please shift me to a sleeping position...thank you and good night!"

Thus for Erickson, self-hypnosis for pain relief in the middle of the night required a sense of arousal and a trust that therapeutic transformations would happen all by themselves if he allowed himself to be in a receptive state. Exactly what the therapeutic transformations of his sensory-perceptual and psychobiological sifts would be, however, were left to an implicit or unconscious level of processing. This is very difficult to explain from a conventional theory of hypnosis, being a response to suggestion, but it could be explained as a natural result of expectancy. But, expectancy of what? I would purpose it is an expectancy that long experience teaches us that the natural ultradian shifts in our Basic Rest-Activity Cycle that will sooner or later allow us to slip into the natural psychobiological comfort of rest and healing that takes place every hour and a half or two. We know that changes in gene expression are associated with pain control (Ko et al., 2002). We now need to utilize DNA microarray technology to evaluate whether such changes in gene expression associated with pain control also take place naturally during the comfort of ultradian rest as well as during hypnotic suggestions for pain control.

Research has documented that the early morning period around 4 to 6 a.m. is the lowest point in our daily temperature (that normally varies between 36.3 degrees centigrade [97.88 Fahrenheit] around 4 a.m. to a peak of 36.9 degrees centigrade [98.42 Fahrenheit] at about 11 a.m.). It is no coincidence that the early morning low at about 4 a.m. is the period when there are more traffic accidents due to a lack of attention while the 11 a.m. peak period in temperature coincides with a peak in mental attention, as well as physical strength. It may seem to be a contradiction that most spiritual traditions regard this period from four to six a.m. as the clearest and most beneficial meditative period of the day. This apparent contradiction could be resolved when we realize that their circadian cycle may be shifted because these people tend to go to sleep earlier at sundown so that by 4 a.m. they have already had a full nights sleep. Lucid dreamers who are able to consciously recognize that they are dreaming while dreaming also report this early morning period as providing one of the best opportunities for communicating with their inner selves. If you are willing to interpret this "inner self" as your creative source then this is evidently one of the best times of the day for accessing it.

### 6:00 A.M. to 8:00 A.M.: Creative Awakening in the Morning

Awakening in the morning is a process of psychobiological arousal generated by *behavioral state-related gene expression*. We are able to cross the threshold between sleep and waking, in part, because of the activity of many genes associated with informational and energy processes at the cellular levels. The CYP17 gene, for example, codes for a protein that functions as an enzyme for converting cholesterol into cortisol, testosterone, and oestradiol. As we noted earlier in an earlier chapter, the threshold between sleep and waking in the morning is associated with cortisol and testosterone that achieve circadian peaks at this time. Ridley (1999) regards CYP17 as an example of what he calls a "social gene" because the hormones it codes for are associated with social interactions as well as awakening, performance, stress and emotional arousal. From our perspective we would regard the activity of CYP17 as an example of *behavioral state related gene expression* when it modulates the behavioral transition between the states of sleep and waking. Along with the families of immediate early genes such as per, jun and the clock genes, CYP17 modulates the circadian cycle of consciousness itself.

Awakening in the morning is "a call to adventure" from the mythological perspective. A major creative issue in the morning for most people, however, is their lack of awareness of how they changed during their night of sleep. They are usually not aware of how their consciousness, memory, learning, identity and behavior are continually changing on an implicit level during sleep and dreaming. Many creative workers, however, report that the seed of a new idea is often found in their first few minutes after awakening out of a dream. They are creative precisely because they have learned to enjoy the experience of their *consciousness as a novelty-seeking modality of life* (Rossi, 2002). The experience of novelty turns on gene expression and brain plasticity. The wondrous experience of creativity actually re-synthesizes the neural networks of their brain at deep molecular levels. Creative people take delight in exploring what is new, surprising, and numinous in their experience.

The philosopher Rene Descartes, for example, is said to have enjoyed the luxury of lingering in bed to dwell on these pristine early morning thoughts. This habit of accessing the inner mind-body source led to one of the most important discoveries in all of mathematics. One morning as he watched a fly crawl across his ceiling, he suddenly realized that it would require only two numbers to locate the fly precisely and thus discovered the famous Cartesian coordinates that became the basis of analytical geometry and higher mathematics.

As our energy and sexual hormones approach their maximum, we awaken. Even the consistency of our blood is a bit thicker as we get ready for daily activity. Yes, theoretically this would be the best time to have sex even though few of us do.

Apparently we are too distracted by what we imagine are all the other important things we are supposed to do this morning. But during honeymoons and vacations?

Anyway, within an hour and a half, or so, after we awaken we are already receiving mind-body cues (yawning, stretching and wanting another cup of coffee) for our first ultradian rest and recovery period. I think this is the time when Huxley sank into his easy chair in his study for his Deep Reflection to pull together the new that was synthesized within his dreams so that it could be expressed in creative writing.

### 8:00 A.M. TO 10:00 A.M.: Prime Creative Working Time for Larks

For larks this is usually the most productive part of the day for doing creative work. They are still close to the dynamics of gene expression, neurogenesis, and problem-solving that has been taking place in the final stages of sleep and dreaming. In fact, if the creative work that has been taking place on an implicit level during sleep and dream are to be realized they need to be reinforced as soon as possible after awakening. That is why it is a good idea to write your dreams upon awakening—even before you get out of bed. If you have asked yourself some important questions you would like your unconscious to answer during sleep and dream before you went to sleep, then this is the time to wonder if an answer is now more apparent to you in the first few moments of awakening. It is okay if one cannot remember a dream. Often the first few thoughts of the day are pregnant with the new that needs to be developed further.

Children are often discouraged with school learning. When first approaching new topics in reading, writing, and arithmetic, so much of it may seem strange that it initially seems impossible to grasp. One of the first bits of practical wisdom children need to learn is that *"the morning is wiser than the evening."* Things that seem so difficult in the evening when they are doing their homework will become clearer in the morning. There is something like a magical genie that helps us learn while we sleep and dream. Neuroscience now calls the magical genie "gene expression, synaptogenesis, and neurogenesis" (literally brain growth while we sleep) but for kids "magical genie" may be enough. Children can easily be taught to look forward to the new day so they can notice what the genie helped them learn while sleeping. They are usually delighted to realize how easy it now is to learn what seemed so difficult the night before.

### 10:00 A.M. TO NOON:
### Creative Work and/or Exercise

This is the time of the day when consciousness, creativity, and energy are at their daily peak for many people. This is naturally the period when it may be best for some people to exercise. Exercise can turn on and coordinate most of the hormonal messenger molecules and energy systems of mind and body to optimize performance in just about anything you do. The Ultradian Performance Rhythm tells the story of how there is a natural 90 to 120 ultradian rhythm that peaks for top performance when we are at our best in outer world tasks and enjoying life in general (Lloyd & Rossi, 1992, 1993; Rossi & Nimmons, 1991).

This is a good time to experience epiphanies. The poem "Vacillation" written by William Butler Yeats in 1934 (Finneran, 1997) describes it well.

My fiftieth year has come and gone,
I sat, a solitary man,
In a crowded London shop,
An open book and empty cup
On the marble table-top.

While on the shop and street I gazed
My body of a sudden blazed;
And twenty minutes more or less
It seemed, so great my happiness,
That I was blessed and could bless.

It is interesting that there is a cultural history of 20 minutes as a typical time period of peaks, as well as troughs, in our moods in everyday life (Deikman, 1980a & b; Pennebaker, 1997; Smith & Tart, 1998). Such 20 minute breaks at the high-phase of the hour and a half, or so, ultradian rhythm of Kleitman' Basic Rest Activity Cycle described by Yeats are balanced by the lows on the other end. After the 20-minute peak the mind-body then drifts down to a natural period of healing and rejuvenation for about 20 minutes, or so, that I call "The Ultradian Healing Response" when we are at our best for inner world tasks. When things are hot and you are deeply absorbed in work, as Huxley was in his morning of writing, you can easily override the rest ultradian rest phase. The ultradian rhythms easily adapt to help you maintain optimal performance when you need it for an extended period of work. When you habitually try to override your need for a healing break, however, you may be setting yourself up for The Ultradian Stress Syndrome that leads to malfunctioning and all the stress related problems of everyday life. We hypothesize that human clock genes are entrainable or modifiable in their expression to some degree by psychosocial cues but not by much—perhaps about 20 minutes, or so. It would be an interesting research project to determine if such entrainment windows occur during the 20-minute peaks and troughs of the Basic Rest Activity Cycle. Stress begins to build when we try to stretch the highs and lows too far beyond their natural limits.

### NOON TO 2:00 P.M.: Lunch and The Ultradian Diet

An ultradian approach to diet consists of six small meals of 200 to 300 well balanced calories spaced throughout the day to coincide with our natural 90–120 minute hunger rhythm. The cholesterol lowering, reduced need for insulin and generally health promoting aspects of such an ultradian eating pattern has been well documented by the Jenkins research group in *The New England Journal of Medicine* (1989). People can enjoy a light meal and then reward themselves with a 20 minute ultradian healing period during which the body will make an efficient beginning in digesting the food to the point where they feel completely satisfied.The popular saying that "you are what you eat" expresses a profound truth that we are only now coming to fully appreciate with the dramatic recent progress that is being made in research on how the nutrients in our food modulates gene expression for health and illness (Berdanier, 1996; Moustaïd-Moussa & Berdanier, 2001; Gstaiger, M. et al., 2003).

One dramatic discovery that now requires documentation in humans is the effect of the nutrient resveratrol in the typical Mediterranean diet found in Benevento, Italy. Resveratrol, which is found naturally in grapes and wine, extends the life span of yeast cells by 70 % by activating a longevity gene called "sirtuin" (SIR). This is the same gene that has been implicated in extending life span in mammals by caloric restriction. Research in the laboratory with animals now documents how resveratrol has potent effects against many age-related illnesses such as Alzheimer's (Jang, 2003), cancer (Ding et al, 2002; Lu et al., 1999; Narayanan et al., 2002), cardiovascular diseases as in heart attacks and stroke (Pace-Asciak, 1995; Sinha, K. et al., 2002), and inflammatory disorders (Yen, 2003).

The new DNA microarray technology, described in previous chapters, allows us to assess the expression of the entire human genome from a few drops of blood on a single gene chip. This means that we now have the ability to determine the effects of dietary factors such as resveratrol from a single meal on our general health and well being. International scientific congresses are currently exploring the feasibility of using gene chips to obtain an individual's gene expression profile in response to nutrients (Gstaiger, M. et al. 2003). This DNA innovation in gene chip technology is introducing the first truly scientific revolution in research on food and diet that can make a profound contribution to the health and well being of society.

These deep psychobiological changes at the genomic level in response to food explains why we all need an appropriate rest after eating. After lunch, in particular, we tend to experience important shifts in our mood, consciousness, stress levels, and behavior. Most people in most cultures would like to take a comfortable nap about this time. This brings up the interesting topic of naps and the very interesting research that has been done in this area recently.

### 2:00 P.M. to 4:00 P.M.: Creative Naps and the Ultradian Healing Response

Research on sleep and alertness by David Dinges and Roger Broughton (1989) has uncovered fascinating findings on the ultradian dynamics of naps. If we take a nap in the morning, we are more likely to experience a large proportion of the kind of Rapid Eye Movement sleep associated with dreaming. If we take an afternoon nap, it will contain a larger proportion of the kind of deeper Slow Wave Sleep associated with the release of growth hormone. This can have profound implications for senior citizens. It has recently been demonstrated that thrice weekly injections of growth hormone can help some seniors act and feel 20 years younger. The problem is these injections require a lot of medical attention to the tune of at least $12,0000 a year. So the question becomes, "Why not take an afternoon nap and see if you can get the same effect free the way nature probably intended. Research still needs to be done to clarify the possibilities in this area.

Seniors really have to do more than napping their way to well-being. As we age we tend to fall into a newly recognized condition called "Low Amplitude dysrhythmia." If we do not challenge ourselves a bit with exercise and effort everyday, we don't push ourselves to peaks of top performance as illustrated in The Ultradian Performance Rhythm illustrated in chapter two. This is entirely consistent with neuroscience research that documents how novelty, enrichment and physical exercise are needed for optimal gene expression and neurogenesis throughout the day. If one does not get sufficient exercise throughout the day, it has hidden but insidious consequences because your system then has less need to go down into deep slow wave sleep at night, so that you don't produce enough growth hormone for tissue repair and rejuvenation. You don't feel as if you have really had a "good nights sleep" and, in fact, you haven't. The next day you feel bad and tend to exercise less, so you again miss out on the top performance peaks of your ultradian performance rhythm and you even less need to go into deep sleep at night.

When this situation becomes chronic, it is easy to understand how your life can flatten out into the so-called "Low Amplitude Dysrhythmia." You do not reach high enough peaks of arousal in the daytime to really be fully awake and alert; you really do not get enough deeply satisfying slow wave sleep at night to release the growth hormones and immune system factors that lead to health and well-being. You feel poorly and depressed. Indeed, depression is another condition in which Low Amplitude Dysrhythmia is apparent.

Exercise is currently regarded as an important component for dealing with depression and some of the problems of the aging process. The kind of exercise you do now becomes an important issue. Is the sometimes recommended 20 minutes three times a week enough? Maybe it is enough for some sort of average existence but are you really satisfied with being average? If you are young, aerobic exercise for about an hour a day takes you a long way toward optimal being. But research is not yet consistent about what is best when you are older. A full appreciation of the implications of our ultradian 90–120 minute Basic Activity-Rest Rhythm, however, suggests that it may be better for older people (whatever that means) to enjoy fifteen to twenty minute exercise periods three or four times a day rather than an exhausting effort for an hour or more once a day that might overtax the body.

The relationship between exercise, neurogenesis, and new learning and memory remains controversial. As reviewed in previous chapters, most research documents how voluntary exercise such as running can increase the growth of new neurons (neurogenesis) from stem cells in the hippocampus throughout adult. Some research extends such neurogenesis to the cortex as well (Ahmadias et al., 2003) but this remains to be confirmed by further research. While many studies suggest that the neurogenesis in the hippocampus can enhance memory and learning, the exact mechanism remains unclear. Some research suggests that the benefits of exercise may be confined

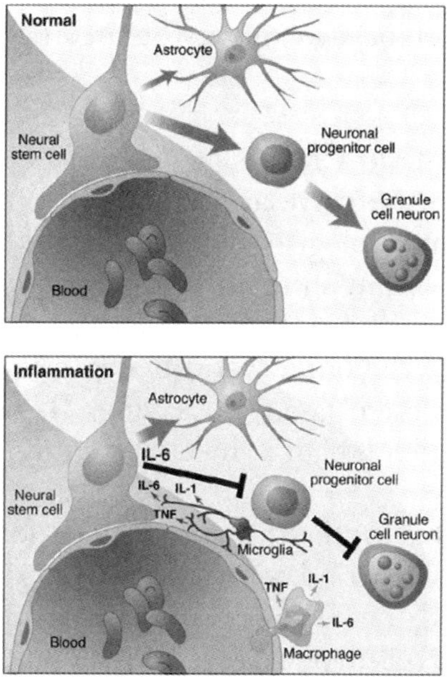


**Figure 10.1**. - The exercise-neurogenesis connection in the hippocampus of the brain is illustrated in the top image. Hippocampal neural stem cells have vascular "end-feet" in close proximity to blood vessels that supply nutrients, oxygen, and growth factors that can facilitate neurogenesis in the adult human brain. These neural stem cells are now thought to give rise to both astrocytes and neuronal progenitor cells. The neuronal progenitor cells have a high proliferative capacity, which then generate new granule cell neurons.

**During brain inflammation due to trauma, stress or infection, for example, neurogenesis is blocked (heavy black lines) by inflammatory cytokines such as Tumor Necrosis factor (TNF), IL-1, and Il-6 by activated Microglia and macrophages.**

to increasing motor coordination (Rhodes et al., 2003).

A recently published image of the close physical relationship between blood vessels in the brain and stem cells that are attached to them suggests how exercise may facilitate neurogenesis - the transition of stem cells into new neurons (Kempermann & Neuman, 2003). Exercise may literally pump nourishing blood cells close to the "vascular feet" of the neural stem cells that mature into new neurons. The top portion of figure 10.1 illustrates how under normal conditions the hippocampal neural stem cells, which are attached to blood vessels, can generate both neuronal progenitor cells and astrocytes (a type of brain cell that supports neurons). The bottom portion of figure 10.1 illustrates how inflammation in the brain due to infection or damage due to trauma or irradiation can block neurogenesis. Fortunately research now indicates that a common non-steroidal anti-inflammatory drug such as indomethacin can reverse this blockage and restore neurogenesis (Monje et al., 2003).

### 4:00 TO 6:00 P.M.: Breaking Point and/or Nurture and Healing

This the period that some biologists call the "breaking point." This is a time of important shifts in gene expression, neurogenesis and psychobiological experience on many levels. This is not the time to engage the outer world in challenging and difficult tasks. In India there is an old saying, "If you want to make an enemy of a friend, plan to meet at four in the afternoon."

Many people report that this can be the time for the deepest ultradian healing-restoration period of their day. Many of the psychotherapeutic methods of free association, inner dialogue, active imagination and self-hypnosis seem to be particularly profound at this time. If you have been skipping your normal ultradian rest-rejuvenation periods throughout the day, however, this is typically the time when our energy and emotions crash. The misinformed interpret this emergency state of physical fatigue as a psychological depression that means they are guilty of having done something wrong—they feel they are just not good enough. Self-esteem sinks to zero in spite of all political efforts to bolster it up. The glass always appears half empty instead of half full. This is not the time to make decisions and plans. This is a time for self-nurture and mutual family and friend support. Wait till morning when your mind-body is fully recovered to make self-assessments, plans, and important decisions.

In the precarious state of being stressed, overtired and worried toward the end of the workday, some people try to bolster themselves up artificially with drugs or too much food. From the psychobiological perspective this is the basis of our addictive society. The so-called "happy hour" in the cocktail lounge is anything but that. People need gentle social support and nurture not artificial chemical stimulation at this time. This may be one reason for the success of many of the twelve step programs such as alcoholics anonymous; they provide social support precisely at the time when addictive behavior is most tempting. This is the time to turn to family and friends for support and healing. It may also a good time to practice the final Ultradian Healing Response of the day.

### 6:00 P.M. To 8:00 P.M.: Family, Social Support and Healing

This period in the late afternoon, dinner, and early evening is the time of day when most families reassemble for better or worse. Returning to a happy, supportive family can be a welcome and healing period. It is an optimal time for nurturing others and ourselves. However, if we have ignored our ultradian calls for restoration and healing throughout the day and return home overwhelmed by stress and fatigue, this period can be a time of misunderstanding, arguments, and mental and physical abuse.

### 8:00 P.M. To 10:00 P.M.: Social Engagement and/or Self Nurture?

There is a wide range of options in this evening period ranging from family togetherness, diary writing, meditation, hobbies and the quiet nurture reading of Aldus Huxly to the social life of theater, movies, symphany, and parties. From the creative perspective all these activities from the most solitary to the most social can be best conceptualized as *implicit processing heuristics.* They are all sending messages and cues as well as provideing information that hopefully will be evokative in facilitating creative replays of gene expression and neurogenesis in our sleep and dreams. For Larks (people most awake in the morning), from the creative perspective, activities at this time can be best understood as setting ourselves up for the the rest and rejuvination that will hopefully take place in our sleep so we will be ready for the next mornings work. For owls (people most alert in the evening), however, this is the best period for creative work and/or play. Owls can be the life of the parties in the evening and night. Larks need the courage to recognize they need not try to compete with owls by overstimulating themselves at this time. This period can be one of the most important for providing cues to couples contemplating marriage. If one is an owl and the other is a lark this can bring up problems at this time of day. One wants to party while the other wants to rest. Wisdom suggests they work this out in the courtship period if they hope to enjoy a happy marriage.

### 10:00 P.M. To MIDNIGHT: Engaging the Creative Dream Cycle

During this time the mind-body prepares for sleep and rejuvenation at the deepest levels. There is a curious peak in cell division, for example, around ten p.m. that seems to set up natural periods of growth and healing as we approach sleep. Remember this is the period during which Huxley enjoyed his relaxed browsing in his home library. There can be something profound about comfort and quiet enjoyment during this final ultradian period of the day. Nature is telling us that it is okay to relax our grip on the outer world. She needs the time to do her inner work of healing and re-creation at the deepest levels of sleep and dream that we will soon enter. Clock genes and the natural process of behavioral state-related gene expression are initiating the dynamics of synthesizing the new proteins and hormones that will lead to rest, rejuvenation and healing during sleep

How well do you prepare yourself to enter those sacred portals? Do you write a bit in your diary to sum up the day and wonder about tomorrow? Do you write a question or two in your diary about what you hope your unconscious will work on during your dreams to help you move toward a resolution of the problems you would like to cope with tomorrow? Do you review healing attitudes of acceptance and/or reflectiveness so that you may prepare for your opportunities for a creative encounter with the gods within?

* * * *

While life is infinite in its natural variety we offered a daily outline for enjoying a creative day to optimize gene expression to optimize health, healing, and life extension in our style of everyday life. The rhythms of gene expression that regulate our health and well being can be facilitated on many levels ranging from our daily diet to our activity level of optimum performance and rest. The new DNA technology that enables us to assess the entire spectrum of human gene expression in relationship to our diet and life style is the current edge of scientific research on life extension as well as a variety of age related disorders ranging from Alzheimer's and cancer to the cardio-vascular and immune system diseases. Such research promises a quantum leap in improving the performance, health, creativity, and enjoyment of daily living.

APPENDIX 1

# HYPNOSIS AND ULTRADIAN CYCLES: A NEW STATE(S) THEORY OF HYPNOSIS? *

***Foreword***: This appendix is an exact copy of the 1982 paper that was Rossi's initial conceptualization of what Erickson called the "psycho-neuro-physiological" foundation of therapeutic hypnosis. By introducing the idea of ultradian cycles into hypnosis theory, this paper served as a transition between the earlier "psychophysioiogical" theories of hypnosis (that did not incorporate time as a significant variable) and our new and more comprehensive psychosocial and cultural genomics theory wherein time is crucial in helping us recognize the dynamic periodicities and pathways connecting mind and molecules. This paper is more than a historical curiosity, however, because it documents Rossi's original observations of the similarity between (1) Erickson's innovative approaches to therapeutic hypnosis and (2) the research literature of chronobiology, which supports the logic and generality of the psychosocial and cultural genomics theory of therapeutic hypnosis and psychotherapy, which now requires more direct assessment via research programs such as outlined in Appendix two.

The similarities between the behavioral characteristics of ultradian cycles (a multioscillatory system of psychophysiological processes involving many para-sympathetic and right-hemispheric functions which have a 90-mtnute periodicity throughout the 24-hour day) and those of the "common everyday trance" lead the author to propose a new state(s) theory of hypnosis. The background for this proposal developed over eight

* **Rossi, E. (1982).** *Hypnosis and ultradian cycles: A new state(s) theory of hypnosis?* The American Journal of Clinical Hypnosis, 25 (1), 21-32.

years of observing the clinical, hypnotherapeutic techniques of Milton H. Erickson, whose work appeared to utilize a similar 90-minute periodicity. The ultradian theory of hypnotherapeutic healing proposes that (1) the source of psychosomatic reactions is in stress-induced distortions'of the normal periodicity of ultradian cycles, and (2) the naturalistic approach to hypnotherapy facilitates healing by permitting a normalization of these ultradian processes. Research problems involved in the validation of these proposals are discussed, and the author's initial clinical approaches to utilizing ultradian cylces to facilitate hypnotherapy and posthypnotic suggestion are outlined.

A salient characteristic of Milton H. Erickson's (MHE) work with patients, as well as with hypnotherapeutic training groups, was that the sessions were lengthy and leisurely; they usually ranged between 90 to 120 minutes, or more. Even in his advanced years when his available energy was precarious, MHE's hypnotherapeutic sessions were still longer than the standard 50-minute hour. For many years the author wondered about this, and also wondered why MHE would choose a particular moment in the session to introduce trance. Sometime or other during those leisurely interviews with their traded memories and anecdotes, it would suddenly become apparent that the patient was quietly nodding his head in a slow and rhythmical fashion — with lids closing over faraway-looking eyes. A hypnotherapeutic trance was being induced in an indirect manner that once again had escaped the author's attention.

It soon became apparent that the author was missing the initial moments of hypnotic induction because he was focusing his attention on MHE rather than the patient. MHE was continuously observing the patient with a careful and highly focused attention: he would monitor the patient's heart rate (by observing pulsations in the face, throat, legs, or hands), the speed of reflexes (eyeblink, swallowing, respiration), and the degree of overall body movement. In his later years MHE would usually wait for the patient's physical and mental processes to "quiet down" before he induced trance, explaining that he was waiting to utilize those "natural" periods of quietness and receptivity. We soon began to call these quiet periods the ''common everyday trance" (Erickson & Rossi, 1976), because it seemed they were a part of natural everyday life. The housewife staring vacantly over a cup of coffee, the student with a faraway look in his eyes during the middle of a lecture, and the driver who automatically reaches his destination with no memory of the details of his route, are all varieties of the common everyday trance.

This author now conceptualizes the common everyday trance as a generalized movement toward parasympathetic and right-hemispheric dominance in all the psychophysiological and behavioral processes listed in Table. 1. In Erickson's work, as soon as a patient began to manifest an individual pattern of these behaviors in clinical interview, MHE would deepen it into a hypnotherapeutic trance through direct or indirect suggestions. MHE usually gave important suggestions only when he sensed that the patient had achieved "a state of *response attentive-ness:* that state of extreme attentiveness in responding to the nuances of communication presented by the therapist" (Erickson & Rossi, 1979, p. 2). In Table 1, response attentiveness is categorized as a form of social behavior in which the subject (patient, listener, client) is manifesting very close attention to the speaker (therapist, lecturer, friend) via (1) focused eye contact; (2) open and receptive body language; (3) an attitude of interest and expectation; (4) quietness of body (spontaneous catalepsy) except for occasional stereotypic agreement responses (head nodding yea, leaning closer to speaker, etc.); and (5) an acquiesence in accepting suggestions.

Most observers presumed that MHE evoked or induced the behaviors characteristic of the common everyday trance listed in Table 1 with his direct and indirect suggestions. Recently, however, the author had suggested that these behaviors might be a natural biological expression of the rest phase of the ultradian cycle that takes place physiologically every 90 to 120 minutes throughout the 24-hour day (Erickson & Rossi, 1981; Rossi, 1981). The author had proposed that the rest phase of the ultradian cycle and the common everyday trance were identical. Thus the essence of MHE's "naturalistic" approach

**Table 1.** - Behavioral processes changing during the common everyday trance, as summarized from Erickson and Rossi (1976-1980).

*Hypnotherapeutic Intervention Periodicity (90-120 Minutes)*

*Psychophysiological Processes*
- Heart rate Peripheral blood flow
  (blushing, blanching, coldness of extremeties, etc.)
- Eyeblink & motility
- Pupillary responses
- Swallowing
- Respiration
- Lack or exaggeration of startle response

*Motor Behavior*
- Response latency
- Economy of movement
- Decreased muscle tonicity (face "ironed out")
- Spontaneous catalepsy
- Altered vocal qualities

*Sensory-Perceptual Behavior*
- Relaxation and comfort
- Feeling ''distant"
- Spontaneous illusions and hallucinatory phenomena

*Cognitive Behavior*
- Amnesias
- Literalism
- Autonomous ideation and fantasy
- Time distortion and time lag in conceptual usage

*Social Behavior*
- "Take a break" periodicity (90 minutes)
- Response attentiveness
- eye contact
- open and receptive body language
- interest and expectation
- body quietness
- suggestibility

was that he utilized the ultradian cycle (the common everyday trance) by nudging it into a state of response attentiveness through his interesting stories, momentary surprises, fascinating facts, and so forth (Erickson, Rossi & Rossi, 1976), until a hypnotherapeutic trance was obviously manifest.

More recent research on the ultradian cycle, however, indicates that this proposal may have been too simple — research into ultradian cycles has become vastly more complex as it has proceeded. With this increasing complexity, though, the skeins of its possible relatedness to what we call "hypnosis" are becoming ever more intriguing. In the following section I shall outline some of the history of and current developments in ultradian cycle research, as well as demonstrate its relevance for hypnosis.

***Ultradian Cycles: A Multiosdilatory System of 90 to 120 Minute Variations in Psychophysiological Processes.*** Ultradian cycles are so named because they are of such short duration that many complete oscillations take place within the 24-hour circadian cycle. Research on ultradian cycles dates back at least a generation, when it was reported that stage REM occurred cyclically throughout the day, as well as at night in infants and during the day when adults slept. Dement and Kleitman (1957) concluded that the 90 to 120 minute cycle was a "Basic Rest-Activity Cycle" (the BRAC hypothesis), which ran continuously throughout the 24-hour day. Kleitman (1963) concluded that this basic rest-activity cycle (the BRAC model) was an endogenous oscillation of the central nervous system that had profound implications for physiology and behavior.

Tables 2 and 3 list the major behavioral processes identified as characteristic of the ultradian cycle, along with the researchers who have investigated them. Table 2 was arranged so that the correspondence between the behaviors of the ultradian cycle and those of the common everyday trance listed in Table 1 would be obvious. A general slowing of the reflexive processes of heart rate, pupillary response, eyeblink and .motility, respiration and swallowing, are characteristic of both the common everyday trance and of the rest phase in the ultradian cycle. However, Kleitman's original conception of the basic rest cycle has undergone significant modification: with more recent research, many of the diversified, cyclic, psychophysiological processes loosely conceptualized as ultradian in Tables 2 and 3 have been found to be out of phase with one another, and different neurophysiological sources have been found to account for them (Kripke, 1982a; Kripke et al., 1978). For example, there is no consistent phase relationship between the dominant REM-NONREM cycle and the very well documented gastric motility cycle (Lavie & Kripke, 1981). In general, however, the movement toward parasympathetic and right-hemispheric dominance that are characteristic of the common everyday trance can be found as psychophysiological polarities of the ultradian cycles presented in Tables 2 and 3.

Ultradian cycles are thus an incredibly complex, multioscillatory system of 90-to-100-minute variations in psycho-physiological and behavioral processes (Kripke, 1982a). They are no less complex, in fact, than the widely varied hypnotic phenomena which the author proposes to be associated with them. It is precisely this correspondence in complexity that lends credence to the view that ultradian cycles could be one of the important psycho-neuro-physiological foundations for hypnotic phenomena.

In a recent summary statement of ultradian cycle research, Kripke (1982a) indicated how many of these behavioral and psychological processes are associated with REM and endorine metabolism:

> The wealth of phenomenology which has been uncovered in pursuit of the BRAC hypothesis should console any investigator troubled that the hypothesis remains controversial. Dramatic behavioral cycles have been discovered, both in man and lower primates, and perhaps these cycles have important ethologic functions. Cycles in fantasy, hemispheric dominance, and perceptual processing have been described. The significance of these cycles in both normal and pathologic psychologic functioning deserves our attention. Episodic hormone secretion seems to be a fundamental property of endocrine metabolism, and it is somehow related

**Table 2.** - Behavioral processes that have been studied experimentally in ultradian cycle research and have been found in the common everyday trance and hypnosis.

| | |
|---|---|
| *Dream (REM) Steep Periodicity (90-120 minutes)* | Dement & Kleitman, 1957 |
| | Kleitman, 1963, 1969 |
| | Kripke, 1974, 1982a |
| *Hemispheric Late ratify* | Klein & Armitage, 1979 |
| | Kripke, 1982a |
| *Psychophysiological Processes* | |
| Heart rate | Lovett, 1980 |
| | Orr et al., 1974, 1976 |
| Peripheral blood flow. | Lovett, 1980 |
| | Romano & Gizdulich, 1980 |
| Eyeblink & motility | Krynicki, 1975 |
| | Ullner, 1974 |
| Pupillary response | Lavie & Schulz, 1978 |
| Swallowing | (See "Appetitive Behavior" in Table 3) |
| Respiration | Home & Whitehead, 1976 |
| *Motor Behavior* | |
| Response latency | Globus et al., 1970 |
| | Kripke, 1972 |
| | Kripke et al., 1978 |
| | Lovett & Podnieks, 1975 |
| | Meier-Kolletal., 1978 |
| | Orr et al., 1974 |
| | Podnieks & Lovett, 1975 |
| Muscle tonicity | Katz, 1980 |
| | Lovett et al., 1978 |
| | Rasmussen & Malven, 1981 |
| | Tierney et al., 1978 |
| *Sensory-Perceptual* | Lavie, 1976 |
| | Lovett, 1976 |
| | Lovett & Podnieks, 1975 |
| Rorschach alterations | Globus, 1966, 1968 |
| Visual illusions | Gopher & Lavie, 1980 |
| | Klein & Armitage, 1979 |
| | Lavie et al., 1974, 1975, 1977 |
| *Cognitive Behavior* | |
| Observing response | Kripke, 1972 |
| | Meier-Koll et al., 1978 |
| | Orr et al., 1974 |
| Fantasy | Kripke & Sonnenschein, 1978 |
| *Social Behavior* | Delgado et al., 1976 |
| | Lavie & Kripke, 1981 |
| | Maxim et al., 1976 |
| "Take a break" periodicity (90 minutes) napping time | Lavie & Scherson, 1981 |

**Table 3.** - Behavioral and psychophysiological processes that have been studied experimentally in ultradian cycle research that have not yet been associated with the common everyday trance and hypnosis.

| | |
|---|---|
| *Appetitive Behavior* | Friedman & Fisher, 1967 |
| | Hiatt & Kripke, 1975 |
| | Kripke, 1972 |
| | Lewis et al., 1977 |
| | Oswald et al., 1970 |
| | Reinberg et al., 1979 |
| | Weda, 1922 |
| *Hormonal Behavior* | Bykov & Katinas, 1979 |
| | Eriksson et al., 1980 |
| | Filicori et al., 1979 |
| | Friedman & Piopho, 1978 |
| | Levin et al., 1978 |
| | Millard et al., 1981 |
| | Quabbee et al., 1981 |
| | Shiotsuka et al., 1974 |
| | Simon & George, 1975 |
| | Steineret al., 1980 |
| | Tannenbaum & Martin, 1976 |
| | Tannenbaum et al., 1976 |
| | Ullner, 1974 |
| | Weitzman, 1974 |
| | Yenet al., 1974 |
| *Temperature Alterations* | Hunsaker et al., 1977 |
| *Urine Flow* | Lavie & Kripke, 1977 |
| | Luboshitzky et al., 1978 |
| *Psychoactive Drug Sensitivity* | Naber et al., 1980 |
| *Smoking* | Friedman & Fisher, 1967 |
| *Neuroendocrinal Functions* | Kripke, 1982a |
| *Penile Erections (in sleep)* | Ohlmeyer et al., 1944 |

to the REM cycle. The suggestion that the pituitary is only responsive to intermittent releasing hormone stimulation is particularly exciting, for it suggests one way in which ultradian cycles may be a functional necessity. The early stage of search for cycles must now give way to more analytic studies which explore the mechanisms and functional implications of each ultradian cyclic process, (pp. 30-31 in unpublished paper; 1982a in the forthcoming publication)

If our proposed identification and/or association of ultradian cycles with hypnosis is correct, the above quotation suggests incredible new vistas for the future development of therapeutic hypnosis.

***The Ultradian Cycle, Stress, and Psychosomatic Reactions.*** Orr, Hoffman, and Hegge (1974) reported that almost all their human subjects showed a stable ultradian cycle when under quiet, bedrest conditions. When subjects were stressed with extended performance tasks (observing responses monitoring three panel meters hidden behind front-surfaced mirrors), however, they experienced major disruptions in the amplitude and patterning of their ultradian cycles. In association with Stroebel's earlier work with rhesus monkeys, the investigators concluded that "psychosomatic-like[1]' responses (heart rate alterations, ulcers, gastritis, asthma, and skin rashes) resulted from stressors' continual disruption of the ultradian cycle. The highly individualized dynamics of this disruption process, which is also characteristic of psychosomatic (and hypnotic) re-sponsivity, was emphasized in their conclusions:

> .... (a) The same stressor can produce quite different physiological and behavioral response patterns in different subjects; (b) behavioral stress can produce disruptions which are clearly manifest in terms of altered biological rhythms; and (c) there appears to be no simple relationship between a physiological response and specific behavioral responses, (p. 1000)

The implications of this association between disruptions of the ultradian cycle by stress and psychosomatic illness are profound. If the major proposal of this section is correct — that therapeutic hypnosis involving physiological processes is actually a utilization of ultradian cycles — then we can finally understand in psychophysiolo-gical terms why hypnosis traditionally has been found to be an effective therapeutic approach to psychosomatic problems: *Individuals who override and disrupt their own ultradian cycles (by ignoring their natural periodic needs for rest in any extended performance situation, for example) are thereby setting in motion the basic physiological mechanisms of psychosomatic illness.* Most of this self-induced stress could be conceptualized as left-hemispheric processes overriding their ideal balance with right-hemispheric processes and associated parasympathetic functions. *Naturalistic therapeutic hypnosis provides a comfortable state wherein these ultradian cycles can simply normalize themselves and thus undercut the processes of psychosomatic illnesses at their psychophysiological source.*

A more exact understanding of how an individual's ultradian cycles manifest themselves could lead to the development of more deeply involving hypnotic states wherein the ultradian characteristics of therapeutic trance are utilized for psychosomatic healing. In the following section, I shall examine how clinical research in this area could proceed.

### *Clinical Research on the Ultradian Theory of Hypnosis*

It would be tempting to immediately propose the obvious research approach of determining if there is a correlation between scores on hypnotic suggestibility scales and indices of the ultradian cycles. There are two sets of problems with such an approach, however: problems from the side of ultradian research, and problems from the side of the "kinds of hypnosis" involved in the usual laboratory or clinical use of hypnotic suggestibility scales.

From the ultradian side, Kripke — who is perhaps the most prominent researcher in this area — recently commented to the author, "The more we study the ultradian cycles, the more elusive they seem" (1982b). As was noted in the previous section, it is the very irregularity and variability of these cycles that recommends a possible commonality with the similar variability of hypnosis and hypnotic phenomena. Because of this wide variability together with the many sources of large but indeterminant errors of measurement, however, we would not expect to find many statistically significant correlations. With so many imponderables of experimental design involved in investigating these elusive phenomena, any lack of statistical correlation would simply not be convincing evidence against the proposed relationship between ultradian cycles and hypnotic phenomena. As in high energy particle physics, the elusiveness of the phenomena under investigation, compounded by the limitations of our methods of measurement, leads to an uncertainty principle. Perhaps because of this factor, Kripke demurred when the author proposed the familiar type of correlational investigation — and, to date, no such work has been done.

From the side of the "kinds of hypnosis" used with standardized hypnotic suggestibility scales, there are difficulties related to the above problems of measurement (Bowers, 1982; Hilgard, 1982a & b). Further, a standardized administration of scales like the Stanford and Spiegel's Hypnotic Induction Profile does not take into account that quality of *response attentiveness* that Erickson felt was of essence in his form of therapeutic hypnosis. A high degree of clinical skill is required to recognize, facilitate, and utilize response attentiveness in each unique clinical situation, so it remains an open question as to how it could be standardized and scored in an objective manner.

Until these experimental design problems are solved in a convincing manner, we must make do with more traditional clinical approaches to explore the proposed relationship among hypnotic phenomena, ultradian cycles, and psychosomatic illness and healing. In an effort to develop a systematic clinical approach to this problem, the author has begun to ask himself a basic question: *How could hypnotic suggestions be associated with ultradian cyclic phenomena to produce specific therapeutic effects or new hypnotic phenomena that could not be produced in any other way?* For example, there are a myriad of possible associations between hypnotic suggestion and the neuroendocrinal and hormonal factors in ultradian cycles listed in Table 3 that could be explored to produce fascinating and entirely new effects. A renaissance of exploration in therapeutic hypnosis will be initiated by those workers with the requisit clinical skills and psychophysiological knowledge to investigate these new vistas. To date, however, the author has been limited in his private practice to investigating the use of posthypnotic suggestion with ultradian cycles, as outlined in the following section.

### *Utilizing Ultradian Cycles in Posthypnotic Suggestion*

The basic principle utilized in this section is Erickson's view that the most effective means of facilitating posthypnotic suggestions is to associate them with behavioral inevitabilities. We know that it is inevitable that a patient will walk out of the office, go home, eat, sleep, dream, go to the bathroom, work, play, etc. Erickson usually associated posthypnotic suggestions with these kinds of inevitable daily behaviors (Erickson & Rossi, 1979).

The experienced clinician could immediately think of many ways of utilizing the ultradian cycles as psychophysiological and behavioral inevitabilities that would give posthypnotic suggestions associated with them continual reinforcement throughout the 24-hour day. This report can only outline a few approaches the author has been exploring with some success in clinical practice during the past three years. Obviously the needs of the individual patient determine which approach will be used. (These approaches are not mutually exclusive, and it is often more effective to blend two or more together.)

1. *Associating a Heightened Sensitivity to Ultradian Behavior with Hypnotic Responsiveness and Posthypnotic Suggestion*

In clinical interview, when the patient's behavior indicates that a quiet period of the ultradian cycle is being experienced (each patient has his own pattern of manifesting the behaviors of Tables 1 and 2, which the observant therapist must come to recognize), the author will utilize it to initiate a

hypnotherapeutic trance. He uses a variable pattern of double binding questions and statements that comment upon whatever ultradian (common everyday trance) behavior the patient is actually manifesting at the time. For one patient, it might run somewhat as follows:

> "I notice your body seems to be getting quieter in these last few moments, and you're not saying much as you look out the window that way. I wonder if that means your unconscious is ready to enter a comfortable period of theratpeutic healing: And will it continue to develop as I continue talking to you? And how will you know? Can you feel your body getting even quieter? Are some parts of your body now more comfortable than others? That's right, you can adjust your arm just like that, so it can be easy and comfortable.
>
> "And are you letting yourself enjoy that comfort even more as it continues? Do you notice how your breathing and heart rate are quietly changing? Do your eyelids feel like they may become heavy and want to close, sort of by them-

> selves? Will you be aware of the exact moment they close, or will you already be daydreaming by then — just wondering how healing happens. And will it happen mostly by itself, so you can just quietly enjoy it, or will you know there is something you must do to help it?"

This associative network focuses attention inward on the many behaviors that are compatible with the parasympathetic and right-hemispheric dominant portions of the ultradian cycle. The behaviors more characteristic of traditional clinical hypnosis — such as catalepsy, arm levitation, ideomo-tor head and finger signalling — can then be explored to assess the patient's readiness for hypnotherapeutic work. If it appears that deepening is required, the patient may be awakened so that the therapist can receive a report on just which of the ultradian cycle behaviors the patient was most aware of during this initial phase of the trance induction. Trance is then reinduced by the therapist, who now focuses his attention and pattern of suggestions and statements on those ultradian behaviors which the patient is most aware of.

After the body of hypnotherapeutic work is accomplished, posthypnotic suggestion can be greatly facilitated by associating it with the inevitable behaviors of the ultra-dian cycle throughout the 24-hour day. The patient is given suggestions that the hypnotherapeutic work can reinforce itself automatically during dreaming at night, and during those periods throughout the day when the patient just feels like "taking a break" — during which time he can rest while his unconscious continues the healing process.

2. *An Intellectual Presentation of the Ultradian Cycle with Posthypnotic Suggestion: Hypnosis without Awareness of It*

In this approach the therapist also begins by noting when the patient is experiencing some quiet portions of the ultradian cycle, but he does not tell this to the patient. Instead the therapist simply introduces an intellectual discussion of the "fascinating research that's being done on the ultradian cycle." The therapist indirectly induces trance by firmly focusing the patient's attention and facilitating *response atten-tiveness* with a series of questions and statements that utilize the patient's own special areas of interest — particularly the patient's interest in effective therapeutic work being accomplished that day. It could run somewhat as follows:

"You know, Mary since you are an engineer, I'm *wondering* if you are *aware* of the *fascinating* research that's *being done* on the ultradian cycle?"

The author never had a patient who was not interested, so he goes on to explain in impressive terms (adjusted to the patient's own vocabulary level and frames of reference) just what the ultradian cycle is. When the author comes to the part about the characteristic behaviors of the quiet (paras ympathetic or right-hemispheric) portions of the ultradian cycle, he tends to slow down and emphasize with a subtle tone of inquiry those behaviors that the patient is currently manifesting. He then tells the patient that while it is only a "clinical hunch," there may be a relation between the ultradian behaviors and the patient's psychosomatic symptom (habit problem, or whatever). The patient is then enjoined to begin to tune into her ultradian behaviors throughout the day and notice if there is any relationship. Does the symptom tend to flare up when the patient does not allow herself to take a break during those portions of the ultradian cycle when the symptoms are manifest? Posthypnotic suggestions are then introduced by having the patient remind herself of the therapeutic work that needs to proceed "all by itself during those "take-a-break" periods.

In this approach, then, a period of therapeutic trance is introduced by fixing and focusing the patient's attention and response attentiveness with an appropriately fascinating discussion that associates therapeutic suggestions with ultradian behaviors actually being manifested by the patient during this moment of the therapeutic encounter. The patient is not aware that associative connections are being formed between these ultradian behaviors currently being experienced and the apparent "intellectual discussion" that is actually loaded with therapeutic suggestions. When the patient later experiences these inevitable ultradian behaviors throughout the 24-hour day, the therapeutic associations will be evoked "covertly," and reinforced automatically within themselves, usually on an unconscious level. When the ultradian behaviors are particularly evident to the patient, she may become consciously aware of them and then reinforce some of the "suggestions" consciously, as directed by the therapist. The patient is consciously aware of the suggestions the therapist told her to associate with the ultradian behaviors, but she tends to be unconscious of other suggestions the therapist was associating with her ultradian behaviors under the guise of an intellectual discussion, interesting story, metaphor, etc. Thus both conscious and unconscious posthypnotic suggestions can be associated with ultradian behaviors so that they wilt be continuously reinforced on conscious and unconscious levels. The patient has no conscious awareness that a "hypnotic process" was used. Indeed, it is perhaps a matter of semantics. Is the process of associating conscious and unconscious processes with behavioral moods or ultradian cyclic behaviors a form of hypnosis? It is if one identifies ultradian behaviors with the Erickson-Rossi concept of the common everyday trance. Otherwise, the approach could be conceptualized as a new form of state dependent learning (Fischer, 1971).

* * * *

The ultradian theory of hypnotic experience would view MHE as a brilliant innovator of many therapeutic approaches to facilitating naturally occurring oscillations of psychophysiological behavior characteristic of the ultradian cycle. Ultradian cycle research indicates that normal mental and physical health depend upon numerous naturally occurring rhythms, wherein para-sympathetic and right-hemispheric processes have varying degrees of dominance every 90 to 100 minutes throughout the 24-hour period. Many, perhaps most, mental and psychosomatic problems could be understood as distortions of these naturally occurring "rest-activity cycles" induced by stress. Mental and physical stress distorts the normal periodicity of ultradian cycles and disrupts a variety of psychophysiological balances necessary for optimal functioning. Trance readiness, or the common everyday trance, may be understood as highly individual and variable but behaviorally recognizable portions of the ultradian cycle. Hypnotherapists can learn to recognize the behavioral characteristics of this naturally occurring cycle and facilitate hypnotherapeutic trance at this time by utilizing Erickson's naturalistic approaches. Hypnotherapy may be conceptualized as a facilitation of these naturally occurring ultradian cycles, during which parasympathetic and right-hemispheric processes can be maximized to facilitate healing.

APPENDIX 2

# A NEW DISCOURSE WITH OUR GENES: 64 MORE RESEARCH PROJECTS IN SEARCH OF A GRADUATE STUDENT

*This New Discourse with Our Genes* made possible by the Human Genome Project and current neuroscience opens a vast universe of understanding and research that will greatly extend the realization of our human potentials. When I published my original set of *64 Research Projects in Search of a Graduate Student* (Rossi and Cheek, 1988) I had no idea that the gratifying response would open an entirely new worldview of mind, brain, gene, culture, and spirit that I now call *Psychosocial and Cultural Genomics* (Lloyd & Rossi, 1992; Rossi, 1985, 1986c/1993, 1990, 1994, 2002).

Courage and persistence will be required for a new generation of students and researchers to integrate the many previously artificially separated areas of scientific studies and humanistic scholarship that now need to be integrated to advance the human condition. In this brief overview of *64 More Research Projects in Search of a Graduate Student* I begin with the story of how my early studies of Milton H. Erickson's neuro-psycho-physiological concept of therapeutic hypnosis led to my chronobiological theory of therapeutic suggestion originally published as "Hypnosis and Ultradian Cycles – A New State(s) Theory of Hypnosis" (Rossi, 1982, reprinted Appendix 1). I will then describe how this early chronobiological theory was updated in my current formulation of *Psychosocial and Cultural Genomics*: how creative and deeply meaningful psychological, social, cultural experiences can modulate gene expression and brain plasticity in health or illness. Therapists and clinicians will find in this new set of "64 More Research Projects in Search of a Graduate Student" many visionary proposals ranging from the development of new scales of mind-molecular hypnotic skills to a futuristic mind-gene biofeedback device that may enable us to establish more direct channels of mind-molecular communication to facilitate healing via gene expression and brain plasticity.

### *The Chronobiological Hypothesis of Therapeutic Hypnosis and Psychotherapy*

Scientists often stumble upon important themes in their research in a most haphazard fashion. The significance of the chance encounters, connections, and accidental discoveries that lead to an original theory only appear much later in retrospect. This certainly was the case with my first insight into the possible connections between chronobiology, the natural biology of time, and what Milton H. Erickson (1958/1980, 1959/1980) called his "naturalistic" or "utilization" approache" to therapeutic hypnosis, psychosomatic medicine, and problem solving.

When I pointed out to Erickson the similarity between our published list of his natural, observable, behavioral "indicators of trance development" (Erickson et al., 1976, p. 306) and the biological indicators of the bodies natural ultradian rhythms (Kripke, 1974; Kripke & Sonnenschein, 1978) there seemed to be a remarkable fit. When I remarked that the typical length of the most prominent ultradian rhythm, Kleitman's Basic Rest-Activity Cycle, was 90 to 120 minutes - the typical time of Erickson's hypnotherapeutic sessions – we both became misty eyed about the profound implications. *Perhaps Erickson's great success with treating many types of psychosomatic disorders was due to the fact that he was intuitively observing and facilitating the patient's natural behavioral manifestations of their ultradian mind-body rhythms of activity, rest, and healing.* Perhaps this was the reason why his naturalistic approach to mind-body problems was so clinically effective – Erickson was utilizing his highly developed skill in observing the natural behavioral states of his patients and facilitating nature's own psychobiological time table for arousal, rest, creative problem solving, and healing with therapeutic suggestions

In the last 8 years of his life when I studied with him,

Erickson used 2 major approaches to hypnotic induction depending upon the patient's observable behavior and activity level during the therapy session. If the patient was in a rather quiet or passive mode, Erickson would use a simple eye focusing approach with what we called his "Early Learning Set Induction" (Erickson et al., 1976) wherein Erickson's quiet voice would evoke early memories of learning as a child in kindergarten. This pleasant recall of early learning created a creative context with the therapeutic implication that the patient would learn how to solve problems as simply and easily as a child again.

If, on the other hand, the patient entered the therapy session with high activity and arousal Erickson tended to use his "Hand Levitation Approach" to utilize the patient's energy by channeling it appropriately with a variety of therapeutic suggestions. *This observation became the basis of my original chronobiological hypothesis of therapeutic hypnosis that gradually evolved into my current psychosocial and cultural genomics theory of therapeutic suggestion.* This eventually led to a new state(s) theory of therapeutic hypnosis (Rossi, 1981, 1982, 1986a, b, 1986c/1993). *This new chronobiological states theory proposed that the neuro-psycho-physiological dynamics of Erickson's clinical effectiveness were mediated, in great part, by utilizing the patient's natural ultradian rhythms of activity, rest, and healing that Kleitman called "The Basic Rest-Activity Cycle (BRAC)."* How to assess this chronobiological theory of therapeutic suggestion was pursued experimentally by a number of investigators in the 1980's and 1990's with varying levels of success as reviewed in the 10 visionary hypotheses of chapter 3 of this book.

### *10 Visionary Hypotheses for a Deep Psychobiology of Therapeutic Hypnosis*

**Research Projects # 1 to # 10** Most of the 10 visionary hypotheses for a deep psychobiology of therapeutic hypnosis in the new millennium that inspired chapter 3 of this book were assessed with pilot studies that now need to be replicated with larger numbers of subjects, appropriate controls, and attention to their implications for practical clinical practice. Replicate these pilot studies with the updated approaches of bioinformatics (chapter 5) and systems biology (chapter 7).

# 11 Rossi's early pilot studies of the monitoring of DC body potentials during hypnotic catalepsy induced by Erickson (Erickson & Rossi, 1981) were recently replicated with a more generalized experimental paradigm relating the electronic monitoring of ultradian rhythms, hypnosis, and hypnotic phenomena by the students of Eduardo Muñoz, Professor of Psychobiology, at the Universidad Diego Portales in Santiago, Chile duing 2003 (eduardo.munoz@psicooncologia.cl). Coordination and further integration of these research paradigms with DNA microarray technology and brain imaging would extend the main line of our visionary hypotheses of the deep psychobiology of therapeutic suggestion in the new millennium.

### *Do We Have a Fruitful Theory of Therapeutic Hypnosis Today?*

As we reviewed earlier in chapter 8, Weitzenhoffer (2001), in his scholarly, authorative text, *The Practice of Hypnotism,* concluded:

> My position today (Weitzenhoffer, 2000) is that any theorizing regarding hypnosis has been and continues to be premature. There is still much groundwork to be done before anything fruitful of the sort can be accomplished. Today I have little in the line of a theory – just a few hypotheses which are insufficient to account for all the facts that have been satisfactorily established (2001, p. 157).

# 12 Do our 10 visionary hypotheses of the deep psychobiology of therapeutic hypnosis for a new millennium constitute adequate "groundwork" that meets Weitzenhoffer's criteria for a "fruitful" theory of hypnosis today? Write and publish a conceptual review of the currently contending theories of hypnosis. Are there any other theories that cover the entire

range of hypnotic phenomena (Rossi, 1982, 1986a) as completely as our 10 visionary hypothesis of the deep psychobiology of therapeutic hypnosis? Are there any other theories that are as comprehensive and amenable to the research methods of neuroscience, functional genomics, and humanistic studies as our psychosocial and cultural genomics theory?

# **13** Carefully review and update tables 1, 2, & 3 of Appendix 1 to determine how you could devise new experimental paradigms to illustrate the chronobiological hypothesis of therapeutic suggestion. Note that table 2, which documents the research data supporting the chronobiological theory of therapeutic hypnosis, does not mention the genomic level even in my updates of this table (Rossi, 1986, Rossi & Lippencott, 1992). Publish a conceptual review illustrating how the addition of the genomic level to table 2 marks the critical transition from my original chronobiological theory to the more complete psychosocial and cultural genomics theory of therapeutic suggestion that is entirely consistent with modern systems biology (Rossi, 2002).

# **14** Carefully note that chapter 7 on "Systems biology and the Psychosocial Genomics of Therapeutic Hypnosis" illustrates in figure 1 *only the hypothesis* of the possible psychosocial genomic relationships between the (1) cognitive, emotional, and behavioral levels of hypnosis, (2) the physiological level of core body temperature, and (3) the expression of the *Thra* gene associated with rate of metabolism and body temperature during creative hypnotherapeutic work. Devise a direct experimental test of this psychogenomic hypothesis of therapeutic suggestion with groups of high and low hypnotic susceptibility subjects assessed via cDNA microarray technology.

# **15** Explore mathematical models of the data relating (1) measurements of creative hypnotherapeutic work on the phenotypic or behavioral level to (2) some measurement on the physiological level and (3) the gene expression levels via

cDNA technology in the previous three research projects. Do these relationships between these three levels tend to be linear or non-linear? Do they tell you anything about the possibilities of mind-body communication and as well as the limits of conscious information processing (Rossi, 1998)? Are they modeled better by rhythmic, oscillating differential equations (Rossi, 1996, 2000) or the stochastic dynamics of the master equation (Sjöberg, 2002 )?

**# 16** Write and publish an updated conceptual review of the practical clinical applications of the shift from the original chronobiological theory of therapeutic suggestion as presented by Rossi, Lippincott, & Bessette (1994, 1995) to the more complete psychosocial and cultural genomics theory.

### *The Bioinformatics of Psychosocial and Cultural Genomics from Mind to Molecule*

What is the bioinformatics of *Psychosocial and Cultural Genomics* and how do we study it? Classical genetics, behavioral genetics, and evolutionary psychology focus primarily on how genes modulate behavior; it is the "bottoms up" - from "molecules to mind" perspective that has been so successful in science. Our interest is the opposite, however. *Psychosocial and cultural genomics* focuses on how psychological, social, and cultural processes modulate gene expression, brain plasticity, mind-body healing. *Psychosocial and Cultural Genomics* is a "tops down" - from "mind to molecules" perspective.

Textbooks of classical genetics usually acknowledge that gene expression is actually an interaction between the gene and the environment that is sometimes called "*epigenesis*." Most elementary textbooks, however, focus on how genes, as the agents or units of heredity, determine observable traits from one generation to the next. These laws of heredity are usually illustrated by tree-like branching diagrams of how dominant and recessive genes are distributed from one generation to the next with little or no regard for environmental factors. It is only recently that studies of *complex behavior* such as memory, learning, and mental illnesses have begun to clarify how trauma, stress, psychosocial, and cultural factors modulate gene expression and brain plasticity. *Biology is not destiny; biology cooperates with culture in creating individual human experience and personal destiny.* The bioinformatics research projects outlined in this section are oriented to *assess our core theory of psychosocial and cultural genomics* to better facilitate our perpetually unfolding and creative personal life.

**# 17** Do a literature review of the basic hypothesis of psychosocial and cultural genomics by entering searches on your computer browser (e.g. Google) relating with key words such as *"gene expression, brain plasticity, and healing"* and your special areas of interest such as: "behavior, memory, learning, personality, aggression, shyness, drug addiction, cocaine, drugs, stress, trauma, neurotoxicity, excitability, psychology, psychobiological arousal, diet, longevity, rehabilitation, psychosocial, culture, dreams, meditation, social factors, prayer, spirit, ritual, psychotherapy, hypnosis, art, drama, myth, poetry, music, dance, exercise, numinous, creative, 4-stage creative process, novelty etc. By simply counting all the relevant connections between these key words, do you note any bias between "bottoms up" versus "top down" perspectives? How do you account for this bias?

**# 18** For an interesting contrast run the same key word searches on my home page: http://ErnestRossi,com. How would you integrate you findings from these 2 key word searches from the "bottoms up" and the "top down" perspectives? How many new and different research areas have you uncovered in these searches, which require further investigation?

**# 19** What new ideas occur to you from this "conceptual review" contrasting the bottoms up and the top down perspectives? What recommendations would you make for further research? What peer reviewed scientific and professional journals will publish your conceptual review in? Which of your research recommendations could you carry out utilizing your special areas of clinical expertise? Pain researchers are finding that there are special classes of genes associated with pain, for example (Ko et al., 2002). Do your clinically effective approaches to modulating pain with therapeutic suggestions evoke such pain associated gene expression? Replicate the classical research of Rainville et al. (2002) to determine whether gene expression is evoked in the anterior cingulated cortex (ACC) by hypnotherapeutic suggestion as would be predicted by the bioinformatics of psychosocial and cultural genomics of pain relief.

**# 20** The core concept of psychosocial and cultural genomics is that the (1) the ultradian periodicity of gene expression/protein synthesis cycle gives rise to (2) the basic psychophysiology of mind-body rhythms (Kleitman's Basic-Rest Activity cycle), which are (3) manifest in the 4-stage creative cycle that is evident in the numinous psychological experiences of art, science, and psychotherapy, which can facilitate another gene expression/protein synthesis cycle. The essence of creativity is this continually iterating cycle of mind-gene communication illustrated as the *"adaptive bioinformatic cycle of sequence, structure, function, and experience"* in figure 1 of Chapter five. The novel and distinguishing concept of psychosocial and cultural genomics is the missing link I propose between *human experience* and gene *sequence: novel, numinous, and enrichening psychological experiences can turn on the gene expression/protein synthesis cycle. This is the missing link that is needed to update classical bioinformatics into the psycho-bioinformatics of psychosocial and cultural genomics. There has never been a direct test of this missing link of psycho-bioinformatics that is the core of psychosocial and cultural genomics.* This missing link is *implied* by the research of Kandel, Schanberg, Soreq, Levsky and many others as reviewed in most chapters of this book. Replicate and extend the pioneering studies of these biological researchers by introducing psychological variables into their research paradigms. How, for example, could you directly test the *missing link (numinous psychological experience in the 4-stage cre-*

*ative cycle can turn on the gene expression/protein cycle)* as illustrated in figure 4 of chapter eight?

### *Epigenetics as a Core Discipline of Psychosocial and Cultural Genomics*

Epigenetics is a currently emerging biological discipline that studies how genes are turned on or off in response to signals from the environment. Epigenetic signals modulate gene expression to generate emergent and sometimes surprising behavior rather than the fixed, hardwired traits and apparently lawful characteristics (phenotypes) we usually expect from genes. It is not yet known to what degree psychosocial and cultural processes can function as environmental factors that modulate the epigenetics of normal human development as well as stress related dysfunctions. It is clear, however that the current formation of the Human Epigenetic Project (HEP) will provide important data to answer these questions. Here is one recent effort to explain *epigenomics* (Bradbury, 2003, p. 316) that helps us understand how it may become a core discipline of psychosocial and cultural genomics.

> "Most people have a fair idea of what is meant by genetics. They know that characteristics such as eye colour are specified by the DNA sequence within their genome. But not everything is that simple. For example, genetically identical twins can be very different. 'One might be normal, while the other is autistic', explains chromatin researcher David Allis (Rockefeller University, New York, United States). 'We can't explain that on the basis of pure genetics because the DNA is identical. Something else must be at play'.
>
> "That 'something else' is chemical modifications of genes that are heritable from one cell generation to the next [within an individual] and that affect gene expression but do not alter the DNA sequence. Epigenetic modifications... [have been described] *as 'red and green traffic lights that are superimposed on top of the genome to tell the genes whether they should be active or inactive'. For individual cells, this code on top of a code helps to determine whether a cell is a blood cell, a fat cell, or something else.*
>
> *"The study of these modifications—what they are, how they are laid down, and the processes that they control—is the field of*

> *research known as epigenetics. An epigenome is the description of these modifications across the whole genome, but unlike the genome DNA sequence, each organism has multiple epigenomes — for example, in different cell types—that may change during its lifetime in response to environmental cues."* [Italics added here]

**# 21** Monitor the following epigenetic web sites for emerging data that could be used to design new experimental paradigms to assess whether psychosocial stress, therapeutic suggestion, and/or psychotherapy modulate the molecular signals that act as traffic lights to turn gene expression on and off.
http://www.epigenome.org/index.php
http://www.epigenome.org/index.php?page=project
http://www.sanger.ac.uk/PostGenomics/epigenome/

**# 22** Epigenomics is one of the key areas that focuses on the dynamics of gene expression in response to environmental signals to explore the "plasticity" (transformability) of our genome. For students familiar with the colorful but often erroneous history of therapeutic hypnosis as a "universal panacea," this idea of the plasticity of our genome in response to psychosocial and cultural signals may seem to be too good to be true – it may even seem to have a bit of a mystical aura about it. Many hard-nosed scientists would rather eschew psychosocial and cultural genomics least it lull their critical faculties to the point where they fear they would lapse back to the futile, intuitive, and metaphorical byways of past centuries. Fear not! We have a bracing cold shower of scientific research and ultra-rationality to affirm the critical scientific point of view. Check out the Sanger web site and learn to use its Ensemble Genome Browser to explore the hard data of epigenetics to develop new experimental paradigms of psychosocial and cultural genomics, therapeutic suggestion, and psychotherapy at: http://www.sanger.ac.uk/PostGenomics/epigenome/instructions.shtml?num=2 .

**# 23** In chapter 6 we reviewed how the psychosocial stress of traumatic war experiences could lead to alternative splicing of the acetylcholinesterase gene. Explore the epigenetic web site to determine what other families of genes are modulated by psychosocial stress and the molecular pathways by which these genes generate the proteins that lead to the synthesis of the neurotransmitters, hormones, growth factors, cytokines etc. that result in posttraumatic and psychosomatic disorders.

# 24 Design new experimental paradigms of therapeutic hypnosis and/or psychotherapy to assess the efficacy of *implicit processing heuristics* (Rossi, 2000) in correcting the dysfunctional molecular pathways of stress induced alternative gene splicing. *Hint.* Utilize creative replay as the essence of these new psychotherapy paradigms (outlined in chapter 8). Utilize the 4-stage creative cycle described in chapters 6 and 9 as an activity-dependent process to evoke, replay, and resynthesize memory, learning and behavioral systems via gene expression and the dynamics of brain plasticity that encode these psychotherapeutic experiences as new neural networks in the brain that can be assessed with real time brain imaging technology.

### *Individuality and the Dynamics of Psychosocial and Cultural Genomics*

Currently emerging research on gene expression has localized the deep psychobiological sources of our individuality in the unique patterns of gene expression found in all the organs, tissues, and cells of the brain and body. In a model experiment worthy of emulation, Whitney et al. (2003), for example, found that a careful characterization of individual variation in gene expression patterns in the blood cells of healthy human volunteers was an essential foundation for recognizing and interpreting the significance of fundamental physiological changes related to illness such as cancer. Our basic hypothesis of psychosocial and cultural genomics leads us to expect important changes in gene expression in the blood due to trauma and psychological stress. In chapter 6, for example, we reviewed research on stress-induced alternative expression of the acetylcholinesterase gene and its relation to creativity as well as mind, memory, mood disorders, and CNS pathology.

**# 25** Replicate the Whitney et al. (2003) model study using cDNA microarrays to explore individuality and variation in gene expression patterns in healthy human volunteers. Whitney found that gender, age, and time of day all were associated with individual variation in gene expression patterns in subsets of the peripheral blood (lymphocytes etc.). They also found that circulating lymphocytes could be viewed as scouts maintaining a continuous surveillance of the body for signs of infection or other threats. They found that the variation in gene expression patterns in disease states (bacterial infection, cancer), for example, was strikingly greater than that in healthy volunteers. *We now need a broad research program to systematically explore individuality and variation in gene expression patterns during therapeutic hypnosis. Does therapeutic hypnosis increase variation in gene expression?*

**# 26** Determine whether therapeutic hypnosis modulates *immediate-early, activity-dependent, behavioral state-related, or alternative* gene splicing in blood as assessed by DNA microarrays during waking, sleeping, and hypnosis.

**# 27** The subjective experience of hypnosis has been compared with dreaming, somnambulism, and sleep-walking. Are such comparisons mere metaphors or are there special classes of genes expressed in common by all these states? Is the zif-268 gene that is turned on during REM dream sleep after daytime experiences of novelty (Ribeiro et al. 1999), for example, also turned on after novel, activity-dependent, creative experiences of therapeutic hypnosis?

**# 28** It is this creative replay and growth of the brain during sleep that makes "the morning wiser than the evening." I hypothesize that, at their best, therapeutic hypnosis, psychotherapy, the creative arts, and sciences all utilize this natural process of gene expression and brain plasticity during sleep and dreaming. Design an experiment to test the hypothesis that it is this creative replay and growth of the brain during sleep that makes "the morning wiser than the evening." How would you asses the hypothesis that therapeutic hypnosis and psychotherapy utilize this natural process of zif-268 gene expression and brain plasticity during sleep and dreaming?

**# 29** Explore the hypothesis that post-hypnotic suggestion reverberating during sleep may facilitate novelty-induced of gene expression, new protein synthesis, and brain plasticity during sleep and dreaming. Design an experimental study to evaluate the hypothesis that "post-hypnotic suggestion" and "hypnotically induced dreams and fantasies" facilitate behavior change and healing by facilitating gene expression and brain plasticity during sleep. How would you test the idea that post-hypnotic suggestions reverberating during slow wave and REM sleep may reinforce the novelty-induced recurrence of zif-268 gene expression, new protein synthesis, and brain plasticity?

**# 30** Are the classes of genes associated with therapeutic hypnosis more similar to the waking, sleeping or dreaming (REM) states as illustrated in the phase histogram of circadian gene expression by Storch et al. (2002) in chapter seven?

**# 31** Replicate the Sanders and Mann (1995) study of the effects of light, temperature, hypnotic trance length, and time of day on divergent circadian gene expression in various tissues of the body with DNA microarrays as illustrated by Storch et al. (2002).

**# 32** Replicate the Lippincott (1992, 1993) studies by adding DNA microarrays to further explore relationships between hypnotizability, circadian rhythms, owl and lark chronobiology and gene expression.

**# 33** Design a study to determine whether the various techniques of inducing therapeutic hypnosis such as the *sleep and relaxation suggestions* of the standardized hypnotic susceptibility scales versus the *alert hypnosis* (Barabasz & Barabasz, 1996). Do they modulate different patterns of gene expression?

# 34 Do the various techniques of therapeutic suggestion (direct suggestion, metaphors, implicit processing heuristics, etc.) actually modulate the gene expression patterns characteristic of disease and dysfunctional states (cancer, arthritis, hypertension, depression) that hypnosis is purported to ameliorate?

# 35 Do the various techniques of therapeutic suggestion (direct suggestion, metaphors, implicit processing heuristics, etc.) facilitate the gene expression patterns characteristic of exercise and optimal performance in sports?

### *Brain Plasticity in Therapeutic Hypnosis and Psychotherapy*

How would you determine whether hypnotic suggestion can facilitate gene expression and cortical brain plasticity in human subjects? Dragansky et al. (2004) have developed an ingenious non-invasive method of assessing *activity or experience-induced brain plasticity* in the gray matter of the human brain using voxel-based magnetic resonance imaging. Their research documents that experience-induced exercises such as developing new juggling skills induces anatomical changes in the macroscopic structure of the human brain cortex (cortical plasticity). They report that "juggling, and consequently the perception and spatial anticipation of moving objects, is a stronger stimulus for structural plasticity in the visual areas (used for he retention of visual-motion information) than in the motor execution of coordinate motion – that is, the supplementary motor area and/or the motor cortex, cerebellum and basal ganglia)." (pp. 311) Utilize their experimental paradigm to determine whether high and low susceptible hypnotic subjects experience greater and/or lower brain plasticity.

That is, evaluate whether hypnotic suggestion can facilitate cortical brain plasticity as well as changes in gene expression in human subjects (Ribeiro et al., 1999, 2002, 2004).

**# 36** Design an experimental study to determine whether hypnotic suggestion can facilitate creative insight and *experience-induced brain plasticity* with the voxel-based magnetic resonance imaging method developed by Dragansky et al. (2004). Utilize DNA microarray technology to determine whether changes in gene expression in human blood accompany these changes in brain plasticity (Whitney et al. 2003).

**# 37** How would you determine that hypnotic and post-hypnotic suggestion can facilitate gene expression and cortical brain plasticity in humans experiencing a creative moment of psychological insight? Maquet and Ruby (2004) summarize the experimental research of Wagner et al. (2004) that supports the idea that different stages of sleep and dreaming can facilitate creative thinking to discover a "hidden rule." They ask the following questions that are now ripe for further experimental assessment. "During wakefulness, learning of the hidden rule is known to be related to the activation of two parts of the brain in particular – the perirhinal cortex and the superior parietal lobe. Do these areas participate in the off-line processing during sleep? What are the neuronal interactions, reinforced during sleep, that underpin the emergence of the conscious knowledge?" (p. 304). Replicate the Wagner et al. (2004) study by adding a hypnotic condition to the sleep and waking conditions at different times of day that they used. Does the hypnotic state facilitate insight as well or better than the sleep condition?

## Psycho-Bioinformatic Scales of Human Hypnotic Skills

What is the future of therapeutic hypnosis and psychotherapy to be? Shall psychology continue to treat the brain and body as a black box by measuring only cognitive, sensory, and behavioral inputs and output responses on the phenotype level? Or will psychology join neuroscience by exploring how we can assess and facilitate exactly how mind can modulate molecules? With the current revolutions in neuroscience, systems biology, and bioinformatics it will be a daunting but not impossible task to develop new *Psycho-Bioinformatic Scales of Human Hypnotic Skills* to update and replace the standardized scales of hypnotic susceptibility that are now almost two generations old (Hilgard, 1965).

**# 38 to # 43** Developing Psycho-Bioinformatic Scales of Human Hypnotic Skills will be an industrial scale operation for psychology as was the human genome project for biology. We might begin by using our current measures (e.g. Harvard Group Scale of Hypnotic Susceptibility) for screening large populations to select groups of high and low subjects. We could then use these subjects with high and low hypnotic susceptibility to evaluate the range of psycho-bioinformatic variables we are initially interested in. We would test therapeutic suggestions in the form of *implicit processing heuristics* (Rossi, 2002) to facilitate psycho-bioinformatic variables on all levels of systems biology. These levels would include but not be limited to measures of gene expression, alternative gene splicing, proteomics, neurotransmitters, growth factors, brain plasticity (synaptogenesis, neurogenesis, astrocytes, microglia, etc), psychoneuroimmune system, motor system, insight system, novelty sensitivity, new memory and learning, abstractness, music, color, etc. Verbal therapeutic suggestions in the form of *implicit processing heuristics* that are most effective in facilitating molecular responses on all levels of systems biology would then become the items on our new Psycho-Bioinformatic Scales of Human Hypnotic Skills.

### *Creative Replay of Numinous Experience is the Essence of Therapeutic Hypnosis, Psychotherapy, and Humanistic Arts*

The mission of psychosocial and cultural genomics is to access and turn on the gene expression/protein synthesis cycle to facilitate brain plasticity, creative adaptation, and the successful resolution of life problems on all levels from mind to molecules. *This is a new way of describing the essential mission of therapeutic hypnosis, psychotherapy, and all the humanistic arts in general through out all of history.*

**# 44** Optimize your genius by drawing pictures and diagrams that summarize in one image the creative replay of psychosocial and cultural genomics from mind to molecule from the personal, artistic, scientific, and professional perspectives that are most interesting to you (Gelb, 1998). Many chapters of this book (e.g. Chapters 2, 3, 4, 8, and 9) as well as Rossi (2002) attempt such Gaia-mind-gene illustrations, usually in 4 stages for simplicity sake. What do your pictures suggest is the next step you need to take in focusing your personal and professional activities for greater effectiveness? What new research paradigms do your images suggest?

**# 45** Some of the most dramatic scientific data that implies the idea that the *creative replay of novel and numinous psychological experience from mind to gene is the essence of all the psychotherapies and cultural arts* comes from the research of the research team of Sidarta Ribeiro (Ribeiro et al., 1999, 2002, 2004). The Ribeiro research uses primarily animals as their subjects, however. This means that any implications we draw from their research about human psychology needs to be assess by replication with human subjects. Since this is not always ethically possible a great deal of ingenuity is now needed to assess the role of creative replay as the essence of all the psychotherapies.

**# 46** One of the most controversial issues today in the use of therapeutic hypnosis and psychotherapy today is the reactivation of traumatic and stressful memories in order to resolve them. In chapter 8 we summarized the research and logic of this approach with these words. "The *activity-dependent process of reactivating an old traumatic memory during the creative process in order to re-construct it on the level of gene* *expression, protein synthesis, and brain plasticity is the essence of creative replay in the humanistic and cultural arts as well as psychotherapy.* Research on the psychosocial genomics of creative replay would be an important step toward a general neuroscience theory of the efficacy of therapeutic suggestion on all levels from the cognitive-behavioral to the genomic." Replicate the experimental paradigm of Wagner et al. (2004) with PTSD patients who suffer from dreams that replay their stress and trauma with no apparent resolution (that is, their dream drama becomes stuck in stage 2 of the creative process). Add a hypnotic condition to the sleep and waking conditions used Wagner wherein patients are given a posthypnotic suggestion to creatively replay their dream drama to a successful conclusion. Do such posthypnotic suggestions facilitate insight and a creative resolution of their problematic dreams?

**# 47** Replicate the classical research paradigms of Barabasz et al. (1999) to determine whether electroencephalographic cortical P300 event-related potentials (ERPs) are evoked during creative replays of the 4-stage creative replays of therapeutic hypnosis.

**# 48** Expand on the creative replay as the essence of psychotherapy theme of chapter 8 by publishing a conceptual review of all the typical methods of psychosocial and cultural genomic *creative replay* that have evolved in different societies throughout human history. What research projects does this conceptual review suggest?

**# 49** Develop new experimental designs to assess the degree to which social and cultural rituals are creative replays that evoke gene expression, synaptogenesis, and brain plasticity in general. To what degree do prayer, political caucuses, national elections, business meetings, educational classes, art, drama, novels, relationships & anniversaries, birthdays, national holidays, mourning, weddings, rituals of all sorts (religious, educational, personal, cultural, diets exercise, sporting events, creative work etc. evoke *creative replays* of characteristic patterns of gene expression?

**# 50** Are there characteristic patterns of gene expression involved in all cultural replays or do certain rituals have a special efficacy in evoking specific patterns of activity-dependent and behavior state-related gene expression? To what extent can the entire field of chronobiology, circadian, and ultradian rhythms be considered as *creative replays* of archetypal life patterns of gene expression?

**# 51** Compare and contrast our psychosocial and cultural genomics perspective with those of the various schools of anthropology, sociology, psychology, and psychotherapy (Freud, Jung, Adler, etc.) on the major institutions of our culture and their theories of the role and functions of art, drama, folk tales, songs story telling, religion, war, sin, guilt, the concept of the "human spirit" etc.

**# 52** Comment on Freud's questions regarding religion as the future of an illusion. Is religion just an illusion or does religious belief and the practice of religious (e.g. prayer, confession, meditation, fasting) have some special efficacy in evoking unique patterns of gene expression, brain plasticity, problem solving, and mind-body healing?

**# 53** Does attendance at a great drama, reading an absorbing novel, or falling in love evoke different patterns of gene expression, brain plasticity, hormones, neurotransmitters, and mind-body healing?

**# 54** Update and expand the new theory of aesthetics proposed in chapter 1, which "reifies Keats' poetic and philosophical conundrum, 'Beauty is truth, truth [is] beauty, - that is all ye know on earth, and all ye need to know.' If beauty and truth are both numinous experiences that activate gene expression and brain plasticity, then beauty could actually reconstruct our brain to generate new experiences of truth and visa versa." (Rossi, 2004) Likewise, is "falling in love" a highly numinous experience that could activate gene expression and brain plasticity via the *novelty-numi-*

*nosum-neurogenesis effect*? Design experimental assessments of such beauty/truth and love hypotheses of gene expression and brain plasticity.

### Society, War, Peace, Gene Expression and Brain Plasticity

**# 55** To what degree are the phenomena of sociology, war and peace, and group dynamics in general all phenotypic manifestations of behavior state-related and activity-dependent gene expression? Design experimental paradigms to assess this question with DNA microarrays and brain imaging.

**# 56** Precisely what methods and techniques of social influence are most efficacious in evoking gene expression and the deep psychobiological dynamics of brain plasticity? What combination of DNA microarray and brain imaging research tools will be needed to assess this question?

### The 4-Stage Creative Process in Art, Culture, Life, and Science

**# 57** There have been many versions and theories of the creative process in the arts and sciences. Write a conceptual review to compare, contrast, and update descriptions of the creative process presented in many parts of this book (particularly chapters 2, 3, 6, 8, 9, & 10) with others you may find from your special areas of interest. Gelb (1998), for example, has outlined 7 principles of the creative process as they apparently were experienced and described by Leonardo da Vinci (*Curiosità, Dimonstrazione, Sensazione, Sfumato, Arte/Scienza, Corporalita, Connessione*). I find that his concept of *Sfumato* (literally "going up in smoke") is a strikingly apt description of how the negative emotional conflicts of stage 2 of the creative process in psychotherapy are transformed into the positive new insights of the heuristic breakouts of stage 3, which I imagine Leonardo da Vinci would recognize as the essence of *Arte/Scienze.* How would you map Leonardo's other principles on to our 4 stage creative cycle?

**# 58** Write a conceptual review on the theme that *psychological conflict (i.e. stage 2 of the creative process) is ultimately and fundamentally a clash of opponent families of behavior state-related gene expression.* Develop new experimental paradigms for assessing alternative gene expression during conflict, stress, problem solving, creative work, and psychotherapy as reviewed in chapter six.

**# 59** Does psychological *conflict within individuals* evoke the same or different families of gene expression and brain plasticity as *conflicts between groups*? A combination of DNA microarray and brain imaging will be required to answer this question.

**# 60** The transition between stage 2 (emotional conflict, the dark night of the soul) and stage 3 (the creative breakout, positive new insights) is the most critical phase of the creative cycle where most people get stuck and often need professional help. I suspect this the place where most people feel overwhelmed and helpless in dreams, nightmares, fantasies, obsessions and the ordinary stresses of everyday life when they cannot solve a problem or move out of a difficult situation. Design new experimental paradigms to investigate how this difficult phase transition of the creative process manifest on the levels of gene expression, enzymes, neurotransmitters, synaptogenesis, neurogenesis, and neuroanatomical pathways.

**# 61** In chapter 6 it was hypothesized that the transition from a highly stressful stage 2 of the creative process to its resolution in stage 3 could be mediated by the alternative splicing of the acetylcholinesterase gene from its AChE-S form (associated with high stress) to its AChE-R form, which acts as a stress modulator in mammalian brain. Devise an experimental paradigm to assess this hypothesis utilizing the molecular biological techniques developed in Soreq lab.

**# 62** An insight into the exploration of this difficult phase transition of the creative process that I call "Period of Private Inner Work & Creative Replay" in the 4-stage creative cycle (illustrated in figure 4 of chapter six) may be related to the neurobiology expectancy and surprise reviewed in chapter five.

### The Psychosocial Genomics of the Numinous, Surprise, Uncertainty, Gambling, & Reward

Chapter 5 reviews neuroscience research that indicates how Milton H. Erickson's neuro-psycho-physiological approaches to therapeutic hypnosis may engage the complementary dynamics of expectancy and surprise on many levels of human experience from mind and brain plasticity to gene expression. I speculated that *negative experiences* of trauma and stress as well as *positive experiences* of wonder, surprise and the numinosum that evoke psychobiological arousal may evoke gene expression and protein synthesis to initiate cascades of molecular signaling and growth to facilitate brain plasticity, creative problem solving, and healing. Psychosocial experiences of *expectancy*, by contrast, facilitated more predictable hypnotic responses by modulating neurotransmitter release at the synaptic connection between neurons. It now will require ingenious extensions of current neuroscience theory and methodology to measure more precisely how hypnosis mediates the psychosocial genomics of surprise and expectancy at the cellular-genomic and synaptic levels respectively.

Recent developments in neuroscience indicate that single dopamine neurons are involved in the dynamics of surprise, uncertainty, and reward systems of the brain described by Powell (2003, p. 313) as follows.

> "One laboratory has shown that neurons can indeed 'code' for some of the variables weighed during simple decisions or choices... What they found were two distinct ways in which the same dopamine neurons code for probability and

> uncertainty. With decreasing probability of reward, the monkey's dopamine neurons fired stronger bursts at the time of the reward delivery. At the same time, with greater uncertainty of reward, a sustained increase in activity occurred between the flashing of the visual cue and the reward delivery... "To make decisions about rewards or money, a person has to make predictions about the future, and in any prediction there is some uncertainty that is critical," says Christopher Fiorillo, a neurophysiologist in Schultz' lab who led the study. "This is the first demonstration of a single neuron coding uncertainty." Fiorillo says there are probably many other types of neurons in the brain that can code uncertainty, but the fact that dopamine neurons do it adds another intriguing layer to decision-making behavior.
>
> "Dopamine neurons, he explains, have been shown to have a reinforcing effect, so that an animal will seek out stimuli or actions that are followed by a release of dopamine. So, Fiorillo says, he was surprised to see the activity of dopamine neurons increased by uncertainty about a reward, as if uncertainty itself were rewarding in some way. The finding might help explain why people are drawn to gambling even though they tend to lose money on average. Fiorillo and his colleagues speculate that outside the artificial conditions of a laboratory or a casino, an uncertain situation presents a learning opportunity that may help the decision-makers "beat the odds" the next time they face it. And so *evolution would favor paying attention to highly uncertain scenarios."* [Italics added here]

# 63 Write a conceptual review of neuroscience research contrasting how (1) *prediction and expectancy* versus (2) *uncertainty, surprise, and gambling* are associated with *neurotransmission* and *gene expression* respectively in dopamine neurons in the brain that are associated with reward. Is there any evidence to support the philosophical view that the significance of the *numinosum (fascination, mysterious & tremendous experiences)* in the spiritual and humanistic studies is "explained" in any sense by neuroscience research on *novelty, uncertainty, and surprise* in turning on the gene expression/brain plasticity cascade in the 4-stage creative process in the arts as well as psychotherapy? Could you find evidence to support the hypoth-

esis that the numinosum is a special sense (experience) *"that evolution would favor paying attention to highly uncertain scenarios"* as described above by Fiorillo and his colleagues?

### # 64 Is a Mind-Gene Biofeedback Device Possible?

Will it be possible to develop a *mind-gene biofeedback device* in the future that would allow us to modulate gene expression and brain plasticity just as we now use inexpensive biofeedback devices to modulate muscle relaxation? This would be the ultimate kind of mind-body biofeedback that theoretically could facilitate to any type of psychophysiological healing at the molecular-genomic level.

To make a mind-gene biofeedback device we need a *mind-gene transducer*. That is, we need to invent a transducer or "transformer" that converts a subjective psychological experience (thought or neural energy) into some kind of molecular signal that would turn on gene expression and brain plasticity. Recent research in nano-technology suggests how this may be possible.

"Nano" means billionths of a meter, or a hundred-millionths of an inch – much smaller than the usual meaning of small. Individual atoms are just a few nanometers in diameter. This is the scale at which atoms are assembled to form the molecules that create the basic building blocks of life such as DNA, genes, and proteins. With but few exceptions every cell in the body contains the entire genetic code consisting of about 3 billion "base pairs" of nucleotides for making the body's 10 trillion cells. Since these cells replace themselves every few years, the human body makes new DNA at the estimated rate of 10,000 miles an hour! Nature, of course, makes all this DNA at an implicit or unconscious level. Our bold proposal for a mind-gene biofeedback machine would be a way of introducing consciousness more directly into the process of DNA production, gene expression, brain plasticity, and healing. This may be possible via the new "nanolab chips" currently under development.

Researchers at the California Institute of Technology,

UCLA, IBM, and the Institute for Systems Biology in Seattle are creating nanolab chips that greatly improve the speed and efficiency of our current DNA microarray (gene chip) technology. Nanolab chips will be able to analyze what is happening within individual cells in *a few seconds or minutes* what it takes current microarray technology hours to accomplish with much larger tissue samples. These nanolab chips contain nano-mechanical sensors to detect protein and gene interactions (i.e. gene transcription) that could become part of a mind-gene transducer that would be the essence of our proposed mind-gene biofeedback device. The heart of these nano-mechanical mind-gene transducers could be a nanowire sensor that produces an electric signal when it binds to a gene and/or protein as well as a wide range of other biological molecules (Hahm & Lieber, 2004). This electric signal can then be amplified to produce an image on a computer screen that would enable *human subjects to use their consciousness to detect when they are in contact with gene expression and its contribution to brain plasticity – this is what would make direct mind-gene communication possible in real time for a practical mind-gene biofeedback device.* It already has been proposed that such direct detection of gene expression could enable researchers use a single drop of blood to screen patients for all the known human genetic disorders (Goho, 2004).

Before nano-technology can create such a mind-gene biofeedback device, however, many fundamental issues must be resolved. We need to know much more precisely how we can use our verbal "implicit processing heuristics" to modulate gene expression and brain plasticity. Utilize internet nano-technology web sites such as the following to stay informed of breakthroughs that will make a mind-gene biofeedback device a practical tool for facilitating mind-body healing in the future.

# References

## CHAPTER 1


Erickson, M. (1958/1980). Naturalistic techniques of hypnosis. In Rossi, E. (Ed.), *The Collected Papers of Milton H. Erickson on Hypnosis. Vol. I. The Nature of Hypnosis and Suggestion* (pp. 168-176). New York: Irvington.

Erickson, M. (1959/1980). Further clinical techniques of hypnosis: Utilization techniques. In Rossi, E. (Ed.), *The Collected Papers of Milton H. Erickson on Hypnosis. Vol. I. The Nature of Hypnosis and Suggestion* (pp. 177-205). New York: Irvington.

Jung, C. (1916/1960). The Transcendent Function. Bollingen Series XX, Volume 8. *The Collected Works of C. G. Jung*. (R. F. C. Hull, Trans.). New York: Pantheon Books.

Jung, C. (1959). *Aion: Researches into The Phenomenology of the Self.* Bollingen Series XX, Volume 9, 2. *The Collected Works of C. G. Jung*. (R. F. C. Hull, Trans.). New York: Pantheon Books.

Jung, C. (1966). The Synthetic or Constructive Method. In *Two Essays on Analytical Psychology. Vol. 7. The Collected Works of C. G. Jung.* (R. F. C. Hull, Trans.). Bollingen Series XX. Princeton, New Jersey: Princeton University Press, 80-89.

Keats, J. (1990). *The Poetic Works of John Keats.* Garrod, H. (Ed.), New York: Oxford University Press.

Kempermann, G. & Gage, F. (1999). New nerve cells for the adult brain. *Scientific American,* 280, 48-53.

Kempermann, G., Kuhn, G. & Gage, F. (1997). More hippocampal neurons in adult mice living in an enriched environment. *Nature,* 386, 493-495.

Otto, R. (1923/1950). *The Idea of the Holy.* NY: Oxford University Press.

Ribeiro, S., Goyal, V., Mello, C. & Pavlides, C. (1999). Brain gene expression during REM sleep depends on prior waking experience. *Learning & Memory, 6:* 500-508.

Ribeiro, S., Gervasoni, D., Soares, E., Zhou, Y., Lin, S., Pantoja, J., Lavine, M., & Nicolelis, M. (2004). Long-lasting novelty-induced neuronal reverberation during slow-wave sleep in multiple forebrain areas. *Public Library of Science, Biology. (PLoS),* 2 (1), 126-137.

Rossi, E. (1996). *The Symptom Path to Enlightenment: The New Dynamics of Self-Organization in Hypnotherapy.* N.Y.: Zeig,Tucker, Theisen, Inc.

Rossi, E., (2000a). *Dreams, Consciousness, Spirit: The Quantum Experience of Self-Reflection and Co-Creation.* (Third edition of *Dreams and the Growth of Personality,* 1972/1985/2000). Phoenix: Zeig, Tucker, Theisen, Inc.

Rossi, E. (2000b). Psyche, Soul, and Gene Expression. *Psychological Perspectives,* 42, 80-88.

Rossi, E. (2002). *The Psychobiology of Gene Expression: Neuroscience and Neurogenesis in Therapeutic Hypnosis and the Healing Arts.* NY: W. W. Norton Professional Books.

Van Praag, H., Schinder, A., Christie, B., Toni, N., Palmer, T., & Gage, F. (2002). Functional neurogenesis in the adult hippocampus. *Nature,* 415, 1030-1034.

Whitney, A., Diehn, M., Popper, S., Alizadeh, A., Boldrick, J., Relman, D. & Brown, P. (2003) Individuality and variation in gene expression patterns in human

blood. *Proceedings of the National Academy of Sciences, PNAS*, 100, 1896-1901.
Yeats, W. (1934/1997). *Vacillation*. In Finneran, R. (Ed.) *The Poems (2nd Ed.), Collected Works of W. B. Yeats, Vol 1.* New York: Schribner.

**CHAPTER 2**

Born, J., Hansen, K., Marshall, L., Molle, M. & Fehm, H. (1999) Timing the end of nocturnal sleep. Nature, 397, 29-30.

Erickson, M. (1948) Hypnotic Psychotherapy. In Rossi, E. (Ed.) The Collected Papers of Milton H. Erickson, Vol. 4,35-48.

Eriksson, P., Perfilieva, E., Björk-Ericksson, T., Alborn, A-M. Nordborg, C., Peterson, D., & Gage, F. (1998). Neurogenesis in the adult human hippocampus. *Nature Medicine, 4*, 1313-1317.

Glaser, R., Kennedy, S., Lafuse, W., Bonneau, R., Speicher, C., Hillhouse, J. & Kiecolt-Glaser, J, (1990), Psychological stress-induced modulation of interleukin 2 receptor gene expression and interleukin 2 production in peripheral blood leukocytes. *Archives of General Psychiatry*, 47, 707-712.

Glaser, R., Lafuse, W., Bonneau, R., Atkinson, C., & Kiecolt-Glaser, J, (1993) Stress-associated modulation of proto-oncogene expression in human peripheral blood leukocytes. *Behavioral Neuroscience*, 107, 525-529.

Jung, C. (1960) The Transcendent Function in The Structure and Dynamics of the Psyche. Vol. VIII. The Collected Works of C. G. Jung. (R. F. C. Hull, Trans.). Bollingen Series XX. Princeton, New Jersey: Princeton University Press, p 67-91.

Lloyd, D. & Rossi, E. (1992) *Ultradian Rhythms in Life Processes: A Fundamental Inquiry into Chronobiology and Psychobiology.* New York: Springer-Verlag.

Lydic, R. (Ed.) (1998) *Molecular Regulation of Arousal States.* New York: CRC Press.

Merchant, K.(1996)*Pharmacological Regulation of Gene Expression in the CNS,* Boca Raton: CRC Press.

Rosenberg, S. & Barry, J, (1992). *The Transformed Cell: Unlocking the Mysteries of Cancer.* New York: Putnam/Chapmans.

Rossi, E., (1972/1985/2000). *Dreams, Consciousness & The Human Spirit: The New Dynamics of Self-Reflection and Co-Creation.* Third Edition. Palisades Gateway Publishing: 23708 Harbor Vista Drive, Malibu, CA.

Rossi, E.(1986/1993) *The Psychobiology of Mind-Body Healing,* Revised Edition. New York: Norton.

Rossi, E. (1996) *The Symptom Path to Enlightenment: The New Dynamics of Hypnotherapy.* Palisades Gateway Publishing: 23708 Harbor Vista Drive, Malibu, California.

Rossi, E. & Cheek, D.(1988) *Mind-Body Therapy: Ideodynamic Healing in Hypnosis.* New York: W. W. Norton.

Rossi, E. & Nimmons, D. (1991) *The Twenty Minute Break: Using The New Science of Ultradian Rhythms.* Palisades Gateway Publishing: 23708 Harbor Vista Drive, Malibu, CA.

Stonier, T. (1990) *Information and the Internal Structure of the Universe. New York: Springer-Verlag.*

Tölle, T., Schadrack, J., & Zieglgansberger (1995) Immediate-early genes in the CNS. New York: Springer-Verlag.

Wilczek, F.(1999) Getting its from bits. Nature, 397, 303-306.

## CHAPTER 3

Aldrich, K., & Bernstein, D. (1987). The effect of time of day on hypnotizability. International Journal of Clinical & Experimental Hypnosis, 35(3), 141-145.

Amigo, S. (1994). Self-regulation therapy and the voluntary reproduction of stimulant effects of epinephrine: possible therapeutic applications. Contemporary Hypnosis, 11, 3, 108-120.

Barabasz, A. (2002). The reality of trance. Paper presented at The Society for Clinical and Experimental Hypnosis: 53rd Annual Scientific Meeting November 9-10, Boston, Massachusetts.

Bailey, C., Bartsch, D. & Kandel, E. (1996). Toward a molecular definition of long-term memory storage. Proc. Natl. Acad. Sci. USA, 93:13445-13452.

Barabasz, A. & Barabasz, M. (1996). Neurotherapy and alert hypnosis in the treatment of attention deficit hyperactivity disorder. In Lynn, S., Kirsch, J. & Rhue, J. Clinical Hypnosis Handbook, APA 271-291.

Barabasz, A., & Lonsdale, C. (1983). Effects of hypnosis on P300 olfactory-evoked potential amplitudes. Journal of Abnormal Psychology. 92, 520-523.

Barber, J. (1990). Miracle cures? Therapeutic consequences of clinical demonstrations. In Zeig, J. and Gilligan, S., (Eds.), Brief Therapy: Myths, Methods, and Metaphors. New York: Brunner/Mazel, pp. 437-442.

Barber, T. (1984). Changing unchangeable bodily processes by (hypnotic) suggestions: A new look at hypnosis cognitions, imagining, and the mindbody problem. Advances, 1, 7-40.

Bentivoglio, M., & Grassi-Zucconi, G. (1999) Immediate early gene expression in sleep and wakefulness. In Lydic, R. & Baghdoyan, H. (Eds.) 1999. Handbook of Behavioral State Control: Cellular and Molecular Mechanisms. New York: CRC Press.

Bernheim, H. (1886/1957). Suggestive Therapeutics: A Treatise on the Nature and Uses of Hypnotism. Westport: Associated Booksellers.

Bongartz, E. (Ed.), (1992) Hypnosis: 175 Years after Mesmer: Recent Developments in Theory and Application. Konstanz: Universitatsvergag.

Born, J., Hansen, K., Marshall, L., Mölle, M. & Fehm, H. (1999). Timing the end of nocturnal sleep. Nature, 397, 29-30.

Borrelli, R. and Coleman, C. (1998). Differential Equations: A Modeling Perspective. New York: Wiley.

Braid, J. (1855/1970). The Physiology of Fascination and the Critics Criticized. In Tinterow, M. (1970). Foundations of Hypnosis. Springfield, Ill: C. C. Thomas.

Brey, D. (1995). Protein molecules as computational elements in living cells. Nature, 376, 307-312.

Brown, T. (1999). Genomes. New York: Wiley-Liss.

Brown, P. (1991a). Ultradian rhythms of cerebral function and hypnosis. Contemporary Hypnosis, 8, 17-24.

Brown, P. (1991b). The Hypnotic Brain: Hypnotherapy and Social communication. New Haven: Yale University Press.

Brush, F. R. & Levine, S. (1989). Psychoendocrinology. San Diego: Academic Press.

Cahill, L., Prins, B., Weber, M., & McGaugh, J. (1994). B-Adrenergic activation and memory for emotional events. Nature, 371, 20, 702-704.

Castes, M., Hagel, I., Palenque, M., Canelones, P., Corano, A., & Lynch, N. (1999). Immunological changes associated with clinical improvement of asthmatic children subjected to psychosocial intervention. Brain & Behavioral Immunology, 13(1), 1-13.

Cirelli, C., Pompeiano, M. & Tononi, G. (1998). Immediate early genes as a tool to understand the regulation of the sleep-wake cycle: In situ hybridization, and antisense approaches. In Lydic, R. (Ed.), Molecular Regulation of Arousal States. New York: CRC Press.

Crasilneck, H. (1997). Advanced Workshop in Pain: The case of Janelle. 14th International Congress of Hypnosis, San Diego, California.

Davidson, E. et al. (2002). A Genomic regulatory network for development. Science, 295, 1669-1678.

DeBenedittis, G., Cigada, M., Bianchi, A., Signorini, M., & Cerutti, S., (1994). Autonomic changes during hypnosis: A heart rate variability power spectrum analysis as a marker of sympatho-vegal balance. International J. Clinical and Experimental Hypnosis, 42, 2, 140-152.

Edmonston, W. (1986). The Induction of Hypnosis. New York: Wiley.

Erickson, M. (1948) Hypnotic Psychotherapy. In Rossi, E. (Ed.) The Collected Papers of Milton H. Erickson, Vol. 4,35-48.

Erickson, M. (1958/1980). Naturalistic techniques of hypnosis. In E. Rossi (Eds.), The Collected Papers of Milton H. Erickson on Hypnosis. Vol. I. The Nature of Hypnosis and Suggestion (pp. 168-176). New York: Irvington.

Erickson, M. (1959/1980). Further clinical techniques of hypnosis: Utilization techniques. In E. Rossi (Eds.), The Collected Papers of Milton H. Erickson on Hypnosis. Vol. I. The Nature of Hypnosis and Suggestion (pp. 177-205). New York: Irvington.

Erickson, M., Rossi, E.,& Rossi, S. (1976). Hypnotic Realities. New York: Irvington.

Erickson, M. & Rossi, E. (1979). Hypnotherapy: An Exploratory Casebook. New York: Irvington.

Eriksson, P., Perfilieva, E., Björk-Ericksson, T., Alborn, A-M., Nordborg, C., Peterson, D., & Gage, F. (1998). Neurogenesis in the adult human hippocampus. Nature Medicine, 4, 1313-1317.

Escera, C., Cilveti, R. & Grau, C., (1992). Ultradian rhythms in cognitive operations: Evidence from the P300 component of the event-related potentials. Medical-Science-Research, 20(4), 137-138.

Ewin, D., (1986). Emergency room hypnosis for the burned patient. American Journal of Clinical Hypnosis, 29, 7-12.

Eysenck, H. (1991). Is Suggestibility? In Schumaker, J. (Ed.) Human Suggestibility: Advances in Theory, Research and Application. New York: Routledge.

Felker, B. & Hubbard, J. (1998). Influence of mental stress on the endocrine system. In Hubbard, J. and Workman, E. (Eds.) Handbook of Stress Medicine: An Organ System Approach. Boca Raton: CRC Press.

Glaser, R., Kennedy, S., Lafuse, W., Bonneau, R., Speicher, C., Hillhouse, J. & Kiecolt-Glaser, J. (1990). Psychological stress-induced modulation of interleukin 2 receptor gene expression and interleukin 2 production in peripheral blood leukocytes. Arch. of Gen. Psychiatry, 47: 707-712.

Glaser, R., Lafuse, W., Bonneau, R. Atkinson, C., & Kiecolt-Glaser, J. (1993). Stress-associated modulation of proto-oncogene expression in human peripheral blood leukocytes. Beh. Neuroscience, 107: 525-529.

Gorton, B. (1957). The physiology of hypnosis, I. Journal of the Society of Psychosomatic Dentistry, 4(3), 86-103.

Gorton, B. (1958). The physiology of hypnosis: Vasomotor activity in hypnosis. Journal of the American Society of Psychosomatic Dentistry, 5(1), 20-28.

Gould, E., Tanapat, P., McEwen, B., Flügge, G. & Eberhard, F. (1998). Proliferation of granule cell precursors in the dentate gyrus of adult monkeys is diminished by stress. Proceeding of the National Academy of Sciences, 95, 3168-3171.

Gould E., Tanapat, P., Reeves, A., & Shors, T. (1999). Learning enhances adult neurogenesis in the hippocampal formation. Nature Neuroscience, 2(3), 260-265.

Green, R. & Green, M. (1987) Relaxation increases salivary immunoglobulin A. Psychological Reports, 61, 623-629.

Gruzelier, J., Smith, F., Nagy, A., & Henderson, D. (2001). Cellular and humoral immunity, mood and exam stress: The influences of self-hypnosis and personality predictors. International Journal of Psychophysiology, 42(1), 55-71.

Guastello, S. (1995). Chaos, Catastrophy, and Human Affairs. Hillsdale, NJ: Lawrence Erlbaum.

Harrington, A. (Ed.), (1997). The Placebo Effect. Cambridge: Harvard University Press.

Harris, R., Porges, S., Clemenson Carpenter, M. & Vincenz, L. (1993), Hypnotic Susceptibility, mood state, and cardiovascular reactivity, American Journal of Clinical Hypnosis, 31, 15-25.

Hautkappe, H. & Bongartz, W. (1992). Heart-Rate Variability as in Indicator for Post-Hypnotic Amnesia in Real and Simulating Subjects. In Bongartz, E. (Ed.), (1992). Hypnosis: 175 Years after Mesmer: Recent Developments in Theory and Application. Konstanz: Universitatsvergag.

Hilgard, E. (1991). Suggestibility and Suggestions as Related to Hypnosis. In Schumaker, J. (Ed.) Human Suggestibility: Advances in Theory, Research and Application. New York: Routledge.

Hilgard, E. and Hilgard, J, (1983). Hypnosis in the Relief of Pain. Los Altos, California: William Kaufmann, Inc.

Holden, C. (2002). Drugs and placebos look alike in the brain. Science, 292, 947-949.

Hull, C. (1933/1968). Hypnosis and Suggestibility. New York: Appleton-Century-Crofts.

Incyte Corporation, (1999). UniGEMTM V Experiment Analysis: Incyte's Time Course Expression Experiment. http:// gem. incyte.com/ gem/ data/ unigemv/ UniGEMVexperiment.pdf.

Iranmanesh, A., Veldhuis, J., Johnson, M., & Lizarralde, G. (1989). 24-hour pulsatile and circadian patterns of cortisol secretion in alcoholic men. J. Androl., 10, 54-63.

Izquierdo, I., Netto, C., Dalmaz, D., Chaves, M., Pereira, M., & Siegfried, B. (1988). Construction and reconstruction of memories. Brazilian Journal of Medical and Biological Research, 21, 9-25.

Johansson,B. (2002). Environmental Effects on Recovery from Stroke. http://uk.Cambridge.org/resources/0521802547/files/Ch27.pdf

Kitano, H. (2002). Systems Biology: A brief overview. Science, 295, 1662-1664.

Kaufer, D., Friedman, A., Seidman, S. & Soreq, H. (1998). Acute stress facilitates long-lasting changes in cholinergic gene expression. Nature, 393, 373-376.

Kauffman S. (1995). At Home in the Universe: The Search for the Laws of Self-Organization and Complexity. New York: Oxford University Press.

Kempermann, G., Kuhn, G. & Gage, F. (1997). More hippocampal neurons in adult mice living in an enriched environment. Nature, 386, 493-495.

Kiecolt-Glaser, J., Marucha, P., Atkinson, C. & Glaser, R. (2001). Hypnosis as a modulator of cellular immune dysregulation during acute stress. Journal of Consulting and Clinical Psychology, 69, 674-682.

Kirsch, I. & Lynn, S. (1995). The Altered State of Hypnosis: Changes in the Theoretical Landscape. American Psychologist, 50:10, pp. 846-858.

Kleitman, N. (1969). Basic rest-activity cycle in relation to sleep and wakefulness. In

A. Kales (Ed.), Sleep: Physiology & Pathology (pp. 33-38). Philadelphia: Lippincott.

Kleitman, N. & Rossi, E. (1992). The basic rest-activity cycle -- 32 years later: An interview with Nathaniel Kleitman at 96. Interviewed by E. Rossi. In D. Lloyd & E. Rossi (Eds.), Ultradian Rhythms in Life Processes: A Fundamental Inquiry into Chronobiology and Psychobiology. New York: Springer-Verlag, 303-306.

Kronauer, R. (1984). Modeling principles for human circadian rhythms. In Moore-Ede, M. & Czeisler, C. (Eds.) Mathematical Models of the Circadian Sleep-Wake Cycle (pp. 105-128). New York: Raven Press.

Lazarus, A. & Mayne, T. (1991). Relaxation: Some limitations, side effects, and proposed solutions. Psychotherapy, 27, 261-266.

Lippincott, B. (1992). Owls and Larks in Hypnosis: Individual differences in hypnotic susceptibility relating to biological rhythms. American Journal of Clinical Hypnosis, 34, 185-192.

Lloyd, D. & Rossi E. (Eds.) (1992). Ultradian Rhythms in Life Processes: A Fundamental Inquiry into Chronobiology and Psychobiology. New York: Springer.

Lloyd, D. & Rossi, E.(1993). Biological Rhythms as organization and information. Biological Reviews, 68, 563-577.

Malcolm, S. and Goodship, J. (2001). (Eds.) Genotype to Phenotype, 2ed edition. Oxford: Oxford University Press.

Mann, B. & Sanders, S. (1995). The effects of light, temperature, trance length and time of day on hypnotic depth. American Journal of Clinical Hypnosis. 37, 3, p. 43-53.

McGaugh, J., (1989). Involvement of hormonal and neuromodulatory systems in the regulation of memory storage. Annual Reviews Neuroscience, 12: 255-287.

McLaren, A. (2000). Cloning: Pathways to a pluripotent future. Science, 288, 1775-1780.

McKernan, M. & Shinnick-Gallagher, P. (1997). Fear conditioning induces a lasting potentiation of synaptic currents in vitro. Nature, 390, 607-611.

Merchant, K. (1996). Pharmacological Regulation of Gene Expression in the CNS. Boca Raton: CRC Press.

Moller, H. & Volz, H. (1996). Drug treatment of depression in the 1990's. Drugs, 52, 625-638.

Morimoto, R. & Jacob, S. 1998. Stress-Inducible Gene Expression Conference: Book of Abstracts. New York: Cognizant Communication Corp.

Naish, P. (Ed.) (1986). What is Hypnosis? Current Theories and Research. Open University Press, Milton Keynes: Philadelphia.

Newtson, D. (1994). The perception and coupling of behavior waves. In Vallacher, R. & Nowak, J., (Eds.), Dynamical Systems in Social Psychology. New York: Academic Press.

Nishimura, E. Jordan,S., Oshima,H., et al. (2002). Dominant role of the niche in melanocyte stem-cell fate determination. Nature, 416, 854-860.

Nobel, D. (2002). Modeling the heart - From genes to cells to the whole organ. Science, 295, 1678-1682.

Osowiec, D. (1992). Ultradian rhythms in self-actualization, anxiety, and stress-related somatic symptoms. Ph.D. Dissertation, California Institute of Integral Studies.

Otto, R. (1923/1950). The Idea of the Holy. NY: Oxford University Press.

Papez, J. A proposed mechanism of emotion. Archives of Neurology & Psychiatry 1937;38:725-744.

Pardue, M., Feramisco, J. & Lindquist, S., (1989). Stress Induced Proteins. NY: Alan Liss.

Pert, C., Ruff, M., Weber, R. & Herkenham, M. (1985). Neuropeptides and their receptors: A psychosomatic network. The Journal of Immunology, 135(2), 820s-826s.

Pert, C., Ruff, M., Spencer, D. & Rossi, E. (1989). Self-reflective molecular psychology. Psychological Perspectives , 20, 213-221.

Petrovic, P., Kalso, E., Petersson, K. and Ingvar, M. (2002). Placebo and opioid analgesia - imaging a shared neuronal network. Science, 295, 1737-1740.

Pluchino, S. et al. (2003). Injection of adult neurospheres induces recovery in a chronic model of multiple sclerosis. Nature, 422, 688-694.

Price, D. (1998). Mechanisms of pain reduction by hypnotic and placebo suggestions. The Science and Practice of Mindbody Interactions. Sedona Conference.

Quitkin, F., McGrath, P., & Stewart, J. (1996). Chronological milestones to guide drug change: When should clinicians switch antidepressants? Archive of General Psychiatry, 53, 785-792.

Rainville, P., Ducan, G., Price, D., Carrier, B. & Bushnell, C. (1997). Pain affect encoded in human anterior cingulate but not somatosensory cortex. Science, 277, 968-871.

Rainville, P., Hofbauer, R., Paus, T., Duncan, G., Bushnell, M., and Price, D. (1999). Cerebral Mechanisms of Hypnotic Induction and Suggestion. J. Cognitive Neuroscience. 11 (1): 110-125.

Rosa, L., Rosa, E., Sarner, L. & Barrett, S. (1998). A close look at therapeutic touch. Journal of the American Medical Association, 279, 1005-1010.

Rosenberg, S. & Barry, J. The Transformed Cell: Unlocking the Mysteries of Cancer. New York: Putnam/Chapmans, 1992.

Rossi, E. (1972/1985/2000). Dreams, Consciousness, Spirit. The Quantum Experience of Self-Reflection and Co-Creation. Palisades Gateway Publishing: 560 Baywood Way, Los Osos, CA., 93402. Fax 805-528-5977.

Rossi, E. (1973). Psychological shocks and creative moments in psychotherapy. The American Journal of Clinical Hypnosis, 16, 9-22.

Rossi, E. (1982). Hypnosis and ultradian cycles: A new state(s) theory of hypnosis? The American Journal of Clinical Hypnosis, 25 (1), 21-32.

Rossi, E. (1986). The Indirect Trance Assessment Scale (ITAS): A Preliminary Outline and Learning Tool. In Hypnotic and Strategic Interventions: Principles and Practice. Yapko, M. (Ed.), New York: Irvington, pp. 1-29.

Rossi, E. (1986/1993). The Psychobiology of Mind-Body Healing, Revised Edition. New York: Norton.

Rossi, E. (1988). The Psychobiology of Mind-Body Healing: The Vision and State of the Art. In Zeig, J. and Lankton, S. (Eds.), Developing Ericksonian Therapy: State of the Art, 127-148. New York: Brunner/Mazel.

Rossi, E. (1990 a). From mind to molecule: More than a metaphor. In Zeig, J. & Gilligan, S. (Eds.) Brief Therapy: Myths, Methods and Metaphors, (pp. 445-472). New York: Brunner/Mazel.

Rossi, E. (1990 b). Mind-Molecular Communication: Can We Really Talk to Our Genes? Hypnos, 17(1), 3-14.

Rossi, E. (1992). Periodicity in self-hypnosis and the ultradian healing response: A pilot study. Hypnos, 19. 4-13.

Rossi, E. (1996). The Symptom Path to Enlightenment: The New Dynamics of Self-Organization in Hypnotherapy. Phoenix, Arizona: Zeig, Tucker, Theisen, Inc. 1-877-850-0443

Rossi, E. (1997). The symptom path to enlightenment: The psychobiology of Jung's constructive method. Psychological Perspectives, 36, 68-84.

Rossi, E., (1999). Exploring gene expression in sleep, dreams, and Hypnosis with the new DMA microarray technology: A call for clinical-experimental research. Sleep and Hypnosis: An InternationalJournal of Sleep, Dream, and Hypnosis, 2, 40-46.

Rossi, E. (2002). The Psychobiology of Gene Expression: Neuroscience and Neurogenesis in Hypnosis and the Healing Arts. New York: W. W. Norton Professional Books.

Rossi, E., & Cheek, D. (1988). Mind-Body Therapy: Ideodynamic Healing in Hypnosis. New York: W. W. Norton.

Rossi, E., & Nimmons, D. (1991). The Twenty Minute Break: Using The New Science of Ultradian Rhythms. Pacific Palisades, CA.: Palisades Gateway Press.

Rossi, E., & Ryan, M. (Eds.) (1986). Mindbody Communication in Hypnosis. Vol. 3. The Seminars, Workshops, and Lectures of Milton H. Erickson. New York: Irvington.

Routtenberg, A. & Meberg, P. (1998). A novel signaling system from the synapse to the nucleus. Trends in Neurosciences, 21, 106.

Saito, T. & Kano, T. (1992). The diurnal fluctuation of hypnotic susceptibility. The Japanese Journal of Hypnosis. 37, 1, 6-12.

Schäfer, C., Rosenblum, M. & Kurths, J. (1998). Heartbeat synchronized with ventilation. Nature, 392, 239-240.

Schanberg, S. (1995). The genetic basis for touch effects. Touch in Early Development, In Field, T. (Ed.), N.Y.: Lawrence Earlbaum, pp. 67-79.

Scharrer E. & Scharrer B. (1940). Secretory cells within the hypothalamus. Research Publications of the association of Nervous and Mental Diseases. New York: Hafner.

Schlingensiepen, K., Kunst, M., Gerdes, W. & Brysch, W. (1995). Complementary expression patterns of c-jun and jun B in rat brain and analysis of their function with antisense oligionucleotides. In Tölle TR, Schadrack J, Zieglgansberger, W. Immediate Early Genes in the CNS. New York: Springer-Verlag.

Schmitt, F. (1984). Molecular regulators of brain function: A new view. Neuroscience, 13, 991-1001.

Sommer, C. (1993). Ultradian Rhythms and the Common Everyday Trance. Hypnos, 20, 135-140.

Soreq, H. & Friedman, A. (1997). Images of stress in the brain. Discovery, May, 19-20.

Spiegel, D. (1998). Hypnosis and implicit memory: Automatic processing of explicit content. American Journal of Clinical Hypnosis, 40, 231-240.

Spiegel, D. & Barabasz, A. (1988). Effects of hypnotic instructions on P300 event-related-potential amplitudes: research and clinical implications. American Journal of Clinical Hypnosis, 31, 11-16.

Steinman, L. (2003). Collateral damage repaired. Nature, 422, 671-672.

Sturgis, L. & Coe, W. (1990). Physiological responsiveness during hypnosis. The International J. of Clin. Hypnosis, 38, 3, 196-207.

Temple, S. (2001). Stem cell plasticity-building the brain of our dreams. Nature Reviews: Neuroscience, 2, 513-520.

Tinterow, M. (1970). Foundations of Hypnosis. Springfield, Ill: Thomas.

Todorov, I. (1990). How cells maintain stability. Scientific American, 263, 66-75.

Tölle, T.R., Schadrack, J., & Zieglgansberger, W. (1995). Immediate Early Genes in the CNS. New York: Springer-Verlag.

Tully, T. (1996). Discovery of genes involved with learning and memory: An experimental synthesis of Hirschian and Benzerian perspectives. Proc. Natl. Acad. Sci. USA, 93:13460-13467.

Unterwegner, E., Lamas, J., and Bongartz, W. (1992). Heart-Rate Variability of High and Low Susceptible Subjects During Administration of the Stanford Scale, Form C. In Bongartz, E. (Ed.), (1992). Hypnosis: 175 Years after Mesmer: Recent Developments in Theory and Application. Konstanz: Universitatsvergag.

Vallacher, R. & Nowak, J. (1994). Dynamical Systems in Social Psychology. New York: Academic Press.

Van Praag, H., Kempermann, G., & Gage, F. (1999). Running increases cell proliferation and neurogenesis in the adult mouse dentate gyrus. Nature Neuroscience, 2, 266-270.

Vogel, G. (2000). Stem cells: New excitement, Persistent Questions. Science, 290, 1672-1674.

Wagstaff, G. (1986). Hypnosis as compliance and belief: a socio-cognitive view. pp. 57-84 In P. Naish (Ed.) What is hypnosis? Current Theories and Research. Philadelphia: Open University Press, Milton Keynes.

Wallace, B. (1993). Day persons, night persons, and variability in hypnotic susceptibility. Journal of Personality and Social Psychology, 64, 827-833.

Wallace, B. & Kokoszka, A. (1995). Fluctuations in Hypnotic Susceptibility and Imaging Ability over a 16-hour period. The International Journal of Clinical and Experimental Hypnosis, 16, 1, 7-19.

Weinstein, E. & Au, P. (1991). Use of hypnosis before and during angioplasty. American Journal of Clinical Hypnosis, 34, 29-37.

Whitney, A., Diehn, M., Popper, S., Alizadeh, A., Boldrick, J., Relman, D., & Brown, P. (2003). Individuality and variation in gene expression patterns in human blood. Proceedings of the National Academy of Sciences, 100, 1896-1901.

Woody, E., & Farvolden, P. (1998). Dissociation in hypnosis and frontal executive function. American Journal of Clinical Hypnosis, 40, 206-216.

Yerkes, R. & Dodson, J. (1908). The relationship of strength of stimulus to rapidity of habit formation. Journal of Comparative Neurology and Psychology, 18, 459-482.

Ziegelgaensgerger, W. (1998). Signalling in the Pain Matrix. Abstracts of ICMART '98 International Medical Acupuncture Congress

## CHAPTER 4

Aboody, K., Brown, A., Rainov, N., Bower, K., Liu, S., Yang, W., Small, J., Herrlinger, U., Ourednik, V., Black,?., Breakefield, X., & Snyder, E. (2000). Neural stem cells display extensive tropism for pathology in adult brain: Evidence from intracranial gliomas. Proceedings of the National Academy of Science, Vol. 97, Issue 23, 12846-12851.

Barabasz, A. (2002). The reality of trance. Paper presented at The Society for Clinical and Experimental Hypnosis: 53rd Annual Scientific Meeting November 9-10, Boston, Massachusetts.

Bentivoglio, M. & Grassi-Zucconi, G. (1999). Immediate early gene expression in sleep and wakefulness. In Lydic, R. and Baghdoyan, H. (1999). Handbook of behavioral state control: Cellular and molecular mechanisms. New York: CRC Press, 235-253.

Braid, J. (1855/1970). The physiology of fascination and the critics criticized. In Tinterow, M., Foundations of hypnosis. Springfield, IL: C. C. Thomas, Pp. 369-372

Castes, M., Hagel, I., Palenque, M., Canelones, P., Corano, A., & Lynch, N. (1999).

Immunological changes associated with clinical improvement of asthmatic children subjected to psychosocial intervention. Brain & Behavioral Immunology, 13(1), 1-13.

Chin, H. & Moldin, S. (2001). Methods in genomic neuroscience. Boca Raton, FL: CRC Press.

Cirelli, C., & Pompeiano, M., & Tononi, G. (1996). Neuronal gene expression in the waking state: A role for the Locus Coeruleus. Science, 272, 1211-1215.

Cirelli, C., Pompeiano, M. & Tononi, G. (1998). Immediate early genes as a tool to understand the regulation of the sleep-waking cycle. In Lydic, R. (Ed.), Molecular Regulation of Arousal States. New York: CRC Press.

Clayton, D. (2000). The genomic action potential. Neurobiology of Memory and Learning, 74, 185-216.

Collins, F. (2000). In Siri, C. Human genome director says psychologists will play a critical role in the initiatives success. Monitor on Psychology, 31, 14-15

Dunlap, J. (1999). The molecular basis for circadian clocks. Cell, 96,271-290.

Edmonston, W. (1986). The induction of hypnosis. New York: Wiley

Erickson, M., (1948/1980). Hypnotic psychotherapy. In E. Rossi (Ed.), The collected papers of Milton H. Erickson on hypnosis. Vol. 4. Innovative hypnotherapy (Pp. 35-48). New York: Irvington.

Erickson, M. (1958/1980). Naturalistic techniques of hypnosis. In E. Rossi (Ed.), The collected papers of Milton H. Erickson on hypnosis. Vol. I. The nature of hypnosis and suggestion (pp. 168-176). New York: Irvington.

Erickson, M. (1959/1980). Further clinical techniques of hypnosis: Utilization techniques. In E. Rossi (Ed. j, The collected papers of Milton H. Erickson on hypnosis. Vol. I. The nature of hypnosis and suggestion (pp. 177-205). New York: Irvington.

Erickson, M. (1964/1980). The "Surprise" and "My-Friend-John" techniques of hypnosis: Minimal cues and natural field experimentation. In E. Rossi (Ed. j, The collected papers of Milton H. Erickson on hypnosis. I. The nature of hypnosis and suggestion (pp. 340-359). New York: Irvington.

Eriksson, P., Perfilieva, E., Bjork-Ericksson, T., Alborn, A-M., Nordborg, C., Peterson, D., & Gage, F. (1998). Neurogenesis in the adult human hippocampus. Nature Medicine, 4, 1313-1317.

Gage, F. (2000a). Mammalian neural stem cells. Science, 287, 1433-1438.

Gage, F. (2000b). Growing new brain cells. Life Extension, 6,49-54.

Galli, R., Borello, U., Gritti, A., Minasi, M., Bjornson, C., Coletta, M., Mora, M., DeAngelis, M., Fiocco, R., Cossu, G. & Vescovi, A. (2000). Skeletal myogenic potential of human and mouse neural stem cells. Nature Neuroscience, 3, 986 - 991.

Glaser, R., Kennedy, S., Lafuse, W., Bonneau, R., Speicher, C., Hillhouse, J. & Kiecolt-Glaser, J, (1990), Psychological stress-induced modulation of interleukin 2 receptor gene expression and interleukin 2 production in peripheral blood leukocytes. Archives of General Psychiatry, 47, 707-712.

Glaser, R., Lafuse, W., Bonneau, R., Atkinson, C., & Kiecolt-Glaser, J, (1993). Stress-associated modulation of proto-oncogene expression in human peripheral blood leukocytes. Behavioral Neuroscience, 107, 525-529.

Gorton, B., (1949). The physiology of hypnosis: A review of the literature. The

Psychiatric Quarterly, 23, 317-343, 457-485. Gorton, B. (1957). The physiology of hypnosis, I. Journal of the Society of Psychosomatic Dentistry, 4(3}, 86-103. Gorton, B. (1958). The physiology of hypnosis: Vasomotor activity in hypnosis. Journal of the American Society of Psychosomatic Dentistry, 5(1), 20-28. Gould,

E., Alison, J., Reeves, A., Michael, S., Graziano, S., & Gross, C. (1999a).
Neurogenesis in the neocortex of adult primates. Science, 286, 548-552. Gould, E. & Gross, C. (2000). New neurons: Extraordinary evidence or extraordinary
conclusion? Science, 288, 77la. Gould, E., Tanapat, P., Reeves, A., & Shors, T. (1999b). Learning enhances adult
neurogenesis in the hippocampal formation. Nature Neuroscience, 2(3), 260-
265. Gross, C. (2000). Neurogenesis in the adult brain: Death of a dogma. Nature • Reviews
Neuroscience, 1,67-73. Gruzelier, J., Smith, E, Nagy, A., & Henderson, D. (2001). Cellular and humoral immunity,
mood and exam stress: The influences of self-hypnosis and personality
predictors. International Journal of Psychophysiology, 42(1), 55-71. Hilgard, E. & Hilgard, J, (1983). Hypnosis in the relief of pain. Los Altos, CA: William
Kaufmann, Inc. Incyte Corporation (1999). UniGEM™ V Experiment Analysis: Time course gene
expression experiments,• (http://gem.incyte.com/gem/data/unigemv/ index.shtml). Janet, P. (1919/1925). Psychological healing: A historical and clinical study. Two
Volumes. New York: Arno Press. Johnston, D. & Maio-Sin Wu, S. (1995). Foundations of cellular neurophysiology.
Cambridge, MA: MIT Press. Kandel, E. (1983). From metapsychology to molecular biology: Explorations into the nature of anxiety. American Journal of Psychiatry, 140(10), 1277-1293.

Kandel, E. (1989). Genes, nerve cells, and the remembrance of things past. Journal of Neuropsychiatry 1(2), 103-125.

Kandel, E. (1998). A new intellectual framework for psychiatry. The American Journal of Psychiatry, 155,457-469.

Kandel, E. (2000). Cellular mechanisms of learning and the biological basis of individuality. In Kandel, E., Schwartz, J., & Jessell, T. (Eds.) Principles of neural science. 4th Edition. P 1247-1279.

Kaufer, D., Friedman, A., Seidman, S. & Soreq, H. (1998). Acute stress facilitates long-lasting changes in cholinergic gene expression. Nature, 393, 373-377.

Kempermann, G. & Gage, F. (1999). New nerve cells for the adult brain. Scientific American, 280,48-53.

Kempermann, G, Kuhn, G. & Gage, F. (1997). More hippocampal neurons in adult mice living in an enriched environment. Nature, 386, 493-495.

Kiecolt-Glaser, J., Marucha, P., Atkinson, C. & Glaser, R. (2001). Hypnosis as a modulator of cellular immune dysregulation during acute stress. Journal of Consulting and Clinical Psychology, 69, 674-682.

Kleitman, N. & Rossi, E. (1992). The basic rest-activity cycle - 32 years later. An interview with Nethaniel Kleitman at 96. In Lloyd, D. & Rossi, E., Eds. (1992). Ultradian and Hypnosis: An International Journal of Sleep, Dream, and Hypnosis, 2:1, 40-46.

Rossi, E. (2000c). In search of a deep psychobiology of hypnosis: Visionary hypotheses for a new millennium. American Journal of Clinical Hypnosis, 42:3/42:4, 178-207.

Rossi, E. (2001 a). Updating Milton Erickson's neuro-psycho-physiological dynamics of therapeutic hypnosis and psychotherapy. The Milton H. Erickson Foundation Newsletter. Phoenix, AZ, pp. 10-13.

Rossi, E. (2001b). The deep psychobiology of psychotherapy. In Corsini, R. (Ed.) (2001). Handbook of Innovative Therapy, 2nd edition, New York: Wiley, pp. 155-165.

Rossi, E. (2002a). The psychobiology of gene expression: Neuroscience andneurogenesis in therapeutic hypnosis and the healing arts. New York: Norton Professional Books.

Rossi, E. (2002b). A conceptual review of the psychosocial genomics of expectancy and surprise: Neuroscience perspectives about the deep psychobiology of therapeutic hypnosis. American Journal of Clinical Hypnosis, 45:2,103-118.

Rossi, E., & Cheek, D. (1988). Mindbody Therapy: Ideodynamic healing in hypnosis. New York: W. W. Norton.

Sillman, B. (1999). Signaling and Gene Expression in the Immune System. New York: Cold Spring Harbor Press.

Stahl, S. (2000). Essential Psychopharmacology: Neuroscientific Basis and Practical Applications. New York: Cambridge University Press.

Stahl, S. (2000). Essential Psychopharmacology. 2ed Edition. Cambridge: Cambridge University Press.

Tinterow, M. (1970). Foundations of hypnosis. Springfield, IL: C. C. Thomas. Uchida,N., Buck,D., He, D., Reitsma,M., Masek, M., Phan-T.,Tsukamoto-A., Gage, F. & Weissman, I. (2000). Direct isolation of human central nervous system stem cells. Proceedings of the National Academy of Science, Vol. 97, 26, 14720-14725.

Van Praag, H., Kempermann, G., & Gage, F. (1999). Running increases cell proliferation and neurogenesis in the adult mouse dentate gyrus. Nature Neuroscience, 2, 266-270.

Van Praag, H., Kempermann, G, & Gage, F. (2000). Neural consequences of environmental enrichment. Nature Reviews: Nature Neuroscience, 1, 191-198.

Van Praag, H., Schinder, A., Christie, B., Toni, N., Palmer, T., & Gage, F. (2002). Functional neurogenesis in the adult hippocampus. Nature, 415, 1030-1034.

**CHAPTER 5**

Barnier, A., Bryant, R., & Briscoe, S. (2001). Posthypnotic amnesia for material learned before or during hypnosis. The International Journal of Clinical and Experimental Hypnosis, 49:4, 286-304.

Blagosklonny, M., & Pardee, A. (2002). Conceptual biology: Unearthing the gems. Nature, 416, 373.

Braid, J. (1855/1970). The physiology of fascination and the critics criticized. In Tinterow, M., (Ed.). Foundations of hypnosis. Springfield, IL: C. C. Thomas, pp. 376-389.

Castillo, P., Schoch, S., Schmitz, E, Siidhof, T, & Malenka, R. (2002). RIM la is required for presy nap tic long-term potentiation. Nature, 415, 327-330.

David, D., King, B. & Borkardt, J. (2001). Is a capacity for negative priming correlated with hypnotizability? The International Journal of Clinical and Experimental Hypnosis, 49:1,30-37.

David, D. & Brown, R. (2002). Suggestibility and negative priming: Two replication studies. The International Journal of Clinical and Experimental Hypnosis, 50:3, 215-228.

Dobrunz, L. & Garner, C. (2002). Priming plasticity. Nature, 415, 277-278.

Dorfman, J. & Kihlstrom, J. (1994). Semantic priming in posthypnotic amnesia. Paper presented at the annual meeting of the Psychonomic Society, St. Louis, MO. Erickson, M. (1985). In Rossi, E., & Ryan, M. (Eds.) Life reframing in hypnosis: Vol. 2. The seminars, workshops, and lectures of Milton H. Erickson. New York: Irvington

Erickson, M. (1992). In Rossi, E. & Ryan, M. (Eds.) Creative choice in hypnosis: Vol. 3. The seminars, workshops, and lectures of Milton H. Erickson. New York: Irvington

Erickson, M., Rossi, E. (1976). Two-level communication and the microdynamics of trance and suggestion. The American Journal of Clinical Hypnosis, 18, 153-171.

Erickson, M., & Rossi, E. (1919). Hypnotherapy: An exploratory casebook. New York: Irvington.

Erickson, M., & Rossi, E. (1981). Experiencing hypnosis: Therapeutic approaches to altered states. New York: Irvington.

Erickson, M., & Rossi, E. (1989). The February man: Evolving consciousness and identity in hypnotherapy. New York: Brunner/Mazel.

Erickson, M., Rossi, E., & Rossi, S. (1976). Hypnotic realities. New York: Irvington.

Gruzelier, J., Smith, F., Nagy, A., & Henderson (2001). Cellular and humoral immunity, mood and exam stress: The influences of self-hypnosis and personality predictors. International Journal of Psychophysiology, 42(1), 55-71.

Hebb, D. (1949). The organization of behavior: A neuropsychological theory. New York: Wiley.

Kandel, E. (1983). From metapsychology to molecular biology: Explorations into the nature of anxiety. American Journal of Psychiatry, 140(10), 1277-1293.

Kandel, E. (1989). Genes, nerve cells, and the remembrance of things past. Journal of Neuropsychiatry 7(2), 103-125.

Kandel, E. (1998). A new intellectual framework for psychiatry. The American Journal of Psychiatry, 155,457-469.

Kandel, E. (2000). Cellular mechanisms of learning and the biological basis of individuality. In Kandel, E., Schwartz, J., & Jessell, T. (Eds.) Principles of neural science. 4th Edition, pp. 1247-1279.

Kaufer, D., Friedman, A., Seidman, S. & Soreq, H. (1998). Acute stress facilitates long-lasting changes in cholinergic gene expression. Nature, 393, 373-377.

Kiecolt-Glaser, J., Marucha, P., Atkinson, C. & Glaser, R. (2001). Hypnosis as a modulator of cellular immune dysregulation during acute stress. Journal of Consulting & Clinical Psychology, 69(4), 674-682.

Kihlstrom, J. (1998). Hypnosis and the psychological unconscious. In The encyclopedia of mental health, Vol. 2. New York: Academic Press, pp. 467-477.

Kirsch, I. (2000). The response set theory of hypnosis. The American Journal of Clinical Hypnosis, 42:3/4, 274-292.

Kirsch, I. & Lynn, S. (1997). Hypnotic involuntariness and automaticity of everyday life. The American Journal of Clinical Hypnosis, 40, 329-348.

Lodish, H., Berk, A., Zipursky, S., Matsudaria, P., Baltimore, D. & Darnell, J. (Eds.,2000). Molecular cell biology, 4th edition, New York: Freeman.

Meshorer, E., Erb, C., Gazit, R., Pavlovsky, L., Kaufer, D., Friedman, A., Glick, D., Ben-Arie, N. & Soreq, H. (2002). Alternative splicing and Neuritic mRNA translocation under long-term neuronal hypersensitivity. Science, 295, 508- 512.

Moore, D. (2001). The dependent gene: The fallacy of "nature vs. nurture". New York: Freeman.

Otto,R. (1926/1950). The idea of the holy. London: Oxford University Press. Ridley, Matt. (1999). Genome: The autobiography of a species in 23 chapters. New York: HarperCollins.

Rossi, E. (1973). Psychological shocks and creative moments in psychotherapy. The American Journal of Clinical Hypnosis, 16, 9-22.

Rossi, E. (1990). Mind-molecular communication: Can we really talk to our genes? Hypnos, 7 7(1), 3-14.

Rossi, E. (1994). The emergence of mind-gene communication. European Journal of Clinical Hypnosis, 3,4-17.

Rossi, E. (1996). The symptom path to enlightenment: The new dynamics of self - organization in hypnotherapy. New York: Zeig, Tucker , Theisen, Inc.

Rossi, E. (1999). Exploring gene expression in sleep, dreams, and hypnosis with the new DNA microarray technology: A call for clinical-experimental research. Sleep and Hypnosis: An International}ournal of Sleep, Dream, and Hypnosis, 2:1,40-46.

Rossi, E., (2000a). Dreams, consciousness, spirit: The quantum experience of self - reflection and co-creation. (Third edition of Dreams and the Growth of Personality, 1972/1986). New York: Zeig, Tucker, Theisen, Inc.

Rossi, E. (2000b). Exploring gene expression in sleep, dreams and hypnosis with the new DNA microarray technology: A call for clinical-experimental research. Sleep and Hypnosis: An International Journal of Sleep, Dream, and Hypnosis, 2:1, 40-46.

Rossi, E. (2000c). In search of a deep psychobiology of hypnosis: Visionary hypotheses for a new millennium. American Journal of Clinical Hypnosis, 42:3/4, 178-207.

Rossi, E. (2001). The deep psychobiology of psychotherapy. In Corsini, R. (Ed.) (2001). Handbook of Innovative Therapy, 2nd edition, New York: Wiley. Pp. 155-165.

Rossi, E. (2002). The Psychobiology of Gene Expression: Neuroscience and Neurogenesis in Therapeutic Hypnosis and the Healing Arts. New York: Norton Professional Books.

Schoch, S., Castillo, P., Jo,T., Mukherjee, K., Geppert, M., Wang, Y, Schmitz, E, Malenka, R. & Siidhof, T. (2002). RIMla forms a protein scaffold for regulating neurotransmitter release at the active zone. Nature, 415, 321-326.

Waelti, P., Dickinsen, A., & Schultz, W. (2001). Dopamine responses comply with basic assumptions of formal learning theory. Nature, 412, 43-48. Zeig, J. (1990). Seeding. In Zeig, J. and Gilligan, S. (Eds.) Brief therapy: Myths, methods, and metaphors. New York: Brunner/Mazel, Pp. 221-246

Zeig, J. (1999). The virtues of our faults: A key concept of Ericksonian therapy. Sleep and Hypnosis: An International Journal of Sleep, Dream, and Hypnosis, 1:2, 129-138.

Zeki, S. (1999). Inner vision: An exploration of art and the brain. Oxford: Oxford University Press. Zeki, S. (2001). Artistic creativity and the brain. Science, 293, 51-52.

## CHAPTER 6

Aczel, A. (1997). *Fermat's Last Theorem.* New York: Dell.

Barinaga, M. (2003). Studying the well-trained mind. *Science.* 302, 44-46.

Birikh, K., Sklan, E., Shoham, S., & Soreq, H. (2003). Interaction of "readthrough" acetylcholinesterase with RACKl and PKCßll correlates with intensified fear-induced conflict behavior. *Proceedings of the National Academy of Sciences,* 100, 283-288.

Blank, T., Nijholt, I., & Spiess, J. (In Press). Molecular determinants of acute stress on hippocampus-dependent learning and memory and synaptic plasticity.

Brenner, T., Hamra-Amitay, Y., Evron, T., Boneva, N., Seidman, S., & Soreq, H. (2003). The role of readthrough acetylcholinesterase in the pathophysiology of myasthenia gravis. *Federation of American Societies for Experimental Biology, The*

*FASEB Journal,* 17, 214-221.

Cohen, O., Reichenberg, A., Perry, C., Ginzberg, D., Pollmacher, T., Soreq, H., & Yirmiya, R. (In Press). Endotoxin-induced changes in human working and decarative memory associated with cleavage of plasma "readthrough" acetlycholineserase.

Cohen-Cory, S. (2002). The Developing Synapse: Construction and Modulation of Synaptic Structures and Circuits. *Science,* 298: 770-776.

Davidson, R., Kabat-Zinn, J., Schumacher, J., Rosenkranz, M., Muller, D., Santorelli, S., Urbanowski, F., Harrington, A., Bonus, K., & Sheridan, J. (2003). Alterations in brain and immune function produced by mindfulness meditation. *Psychosomatic Medicine, 65,* 564-570.

Ellenberger, E. (1970). *The Discovery of the Unconscious.* N.Y.: Basic Books.

Gibson, G. (2003). *Microarray Analysis: Genome-scale hypothesis scanning. PLOS Biology: Public Library of Science.* 1, 28-29.

Grisaru, D., Sternfeld, M., Eldor, A., Glick, D., & Soreq, H. (1999). Structural roles of acetylcholinesterase variants in biology and pathology. *European Journal of Biochemistry,* 264, 672-686.

Gurwitz, D., Razon, N., Sokolovsky, M. & Soreq, H. (1984). Expression of muscarinic binding sites in primary human brain tumors. *Brain Research,* 316, 61-70.

Hadamard, J., (1954). *The Psychology of Invention in the Mathematical Field.* New York: Dover.

Hobson, J., & Leonard, J. (2001). *Out of its Mind, Psychiatry in Crisis.* Cambridge, Massachusetts: Persus Publishing.

Jung, C. (1923). *Psychological Types or The Psychology of Individuation.* New York: Pantheon Books.

Jung. C. (1943/1996). The Synthetic or Constructive Method. In Jung, C. (1966). *Two essays on analytical psychology. Vol. 7. The Collected Works of C. G. Jung.* (R. F. C. Hull, Trans.). Bollingen Series XX. Princeton, New Jersey: Princeton University Press.

Kahneman, D. (2003). A Perspective on Judgment and Choice. *The American Psychologist,* 58, 697-720.

Kaufer, D., Friedman, A., Seidman, S. & Soreq, H. (1998). Acute stress facilitates long-lasting changes in cholinergic gene expression. *Nature, 393,* 373-377.

Kaufer, D., Friedman, A., & Soreq, H. (1999). The vicious circle of stress and anticholinesterase responses. *The Neuroscientist,* 5, 1-9.

Kempermann, G. & Neumann, H. (2003). Microglia: The enemy within? *Science, 1689,* 302, 1689-1690.

Kleitman, N. (1969). Basic rest-activity cycle in relation to sleep and wakefulness. In A. Kales (Ed.), *Sleep: Physiology & Pathology* (pp. 33-38). Philadelphia: Lippincott.

Kleitman, N. & Rossi, E. (1992). The basic rest-activity cycle — 32 years later: An interview with Nathaniel Kleitman at 96. Interviewed by E. Rossi. In D. Lloyd & E. Rossi (Eds.), *Ultradian Rhythms in Life Processes: A Fundamental Inquiry into Chronobiology and Psychobiology.* New York: Springer-Verlag, 303-306.

Levsky, J., Shenoy, S., Pezo, C., & Singer, R. (2002). Single-Cell Gene Expression Profiling. *Science,* 297, 836-840.

Lewis, B., Green, R. & Brenner, S., (2003). Evidence for the widespread coupling of alternative splicing and nonsense-mediated mRNA decay in humans. *Proceedings of the National Academy of Sciences,* 100, 189-192.

Lloyd, D. & Rossi, E., (Eds.), (1992). *Ultradian Rhythms in Life Processes: A Fundamental Inquiry into Chronobiology and Psychobiology.* New York: Springer-Verlag.

Meshorer, E., Erb, C., Gazit, R., Pavlovsky, L., Kaufer, D., Friedman, A., Glick, D.,

Ben-Arie, N. & Soreq, H. (2002). Alternative splicing and Neuritic mRNA translocation under long-term neuronal hypersensitivity. *Science, 295,* 508-512.

Meshorer, E. & Soreq, H. (2002). Pre-mRNA splicing modulations in senescence. *Aging Cell,* 1, 10-16..

Mor, I., Grisaru, D., Titelbaum, L., Evron, T., Richler, C., Wahrman, J., Sternfeld, M., Yogev, L., Meiri, N., Seidman, S., & Soreq, H. (2001). Modified testicular expression of stress-associated "readthrough" acetylcholinesterase predicts male infertility. *Federation of American Societies for Experimental Biology, The FASEB Journal,* 10, 1096-1119.

Razon, N., Soreq, H., Roth, E., Bartal, A. & Silman, I. (1984). Characterization of levels and forms of cholinesterases in human primary brain tumors. *Experimental Neurology,* 84, 681-695)

Rossi, E. (1967). Game and growth: Two dimensions of our psychotherapeutic zeitgeist. *Journal of Humanistic Psychology*, 7, 136-154.

Rossi, E. (1968). The Breakout Heuristic: A Phenomenology of Growth Therapy with College Students. *Journal of Humanistic Psychology*, 8, 16-28.

Rossi, E. (1972/2000). *Dreams and the Growth of Personality: Expanding Awareness in Psychotherapy.* New York: Pergamon Press. Updated in the 3ed edition as *Dreams, Consciousness, and Spirit.* Phoenix, Arizona: Zeig, Tucker, Theisen.

Rossi, E., (1996). *The Symptom Path to Enlightenment: The New Dynamics of Hypnotherapeutic Work.* Phoenix, Arizona. Zeig, Tucker Theisen Inc.

Rossi, E. (2002a). *The Psychobiology of Gene Expression: Neuroscience and Neurogenesis in Therapeutic Hypnosis and the Healing Arts.* NY: W. W. Norton Professional Books.

Rossi, E. (2002b). Psychosocial Genomics: Gene Expression, Neurogenesis, and Human Experience in Mind-Body Medicine. *Advances: in Mind-Body Medicine,* 22-30.

Rossi, E. (2003a). Gene Expression, Neurogenesis, and Healing: Psychosocial Genomics of Therapeutic Hypnosis. *American Journal of Clinical Hypnosis.* 45:3, 197-216.

Rossi, E. (2003b). Can We Really Talk to Our Genes? Psychosocial Genomics. *Hypnos: The Journal of European Society of Hypnosis in Psychotherapy and Psychosomatic Medicine*, Vol. 30, 1, 6-15.

Rossi, E. (2003c). The Bioinformatics of Psychosocial Genomics in Alternative and Complementary Medicine. *Forschende Komplementarmedizine und Klassische Naturheilkunde,* 10, 143-150.

Rossi, E. & Nimmons, D., (1991). *The Twenty-Minute Break: The Ultradian Healing Response.* Phoenix Arizona. Zeig, Tucker, Theisen Inc.

Smith,.S. (1995). Getting into and out of mental ruts: A theory of fixation, incubation, and insight. In Sternberg, R. & Davidson, J. (Eds.) *The Nature of Insight.* Cambridge, Massachusetts: MIT Press. 229-251.

Soreq, H. & Glick, D. (2000). Novel roles for cholinesterases in stress and inhibitor responses. In Giacobini, E., (ed.), *Cholinesterases and Cholinesterase Inhibitors: Basic, Preclinical and Clinical Aspects.* London: Martin Dunitz, pp. 47-61.

Soreq, H. & Seidman, S. (2001). Acetylcholinesterase – new roles for an old actor. *Nature Reviews: Neuroscience,* 2, 294-302.

Sternfeld, M., Shoham, S., Klein, O., Flores-Flores, C., Evron, T., Idelson, G., Kitsberg, D., Patrick, J., & Soreq, H. (2000). Excess "read-through" acetylcholinesterase attenuates but the "synaptic" variant intensifies neurodeterioration correlates. *Proceedings of the National Academy of Sciences,* 97, 8647-8652.

Wallas, G. (1926). *The Art of Thought.* New York: Harcourt.

Whitney, A., Diehn, M., Popper, S., Alizadeh, A., Boldrick, J., Relman, D. & Brown,

P. (2003) Individuality and variation in gene expression patterns in human blood. *Proceedings of the National Academy of Scientists*, 100, 1896-1901.

Zeki, S. (1999). *Inner Vision: An Exploration of Art and the Brain.* Oxford: Oxford University Press.

Zeki, S. (2001). Artistic creativity and the brain. *Science,* 293, 51-52.

Zilboorg, G., & Henry, G. (1941). *A History of Medical Psychology.* New York: Norton.

**CHAPTER 7**

Aldrich, K., & Bernstein, D. (1987). The effect of time of day on hypnotizability. *International Journal of Clinical & Experimental Hypnosis, 35,* 141-145.

Barabasz, A. & Barabasz, M. (1996). Neurotherapy and alert hypnosis in the treatment of attention deficit hyperactivity disorder. In Lynn, S., Kirsch, J. & Rhue, J., *Clinical Hypnosis Handbook*, APA, pp. 271-291.

Carrer, L. (2002). Ambroise-Auguste Liébeault: The Hypnological Legacy of a Secular Saint. College Station, TX: Virtualbookworm.com Publishing Inc.

Erickson, M., (1948/1980). Hypnotic Psychotherapy. In E. Rossi (Ed.), *The Collected Papers of Milton H. Erickson on Hypnosis. Vol. 4. Innovative Hypnotherapy* (pp. 35-48). New York: Irvington.

Erickson, M. (1958/1980). The hypnotic alteration of blood flow. In E. Rossi (Ed.), *The Collected Papers of Milton E. Erickson on Hypnosis. Vol. 2. Hypnotic Alteration of Sensory, Perceptual, and Psychophysical Processes* (pp. 51-80). New York: Irvington pp. 192-195.

Erickson, M., & Rossi, E. (1979*). Hypnotherapy: An Exploratory Casebook.* New York: Irvington.

Goho, A. (2004). Gene Screen: Ultrasensitive nanowires catch mutations. *Science News,* 165, p. 6.

Hautkappe, H. & Bongartz, W. (1992). *Heart-Rate Variability as in Indicator for Post-Hypnotic Amnesia in Real and Simulating Subjects.* In Bongartz, E. (Ed.), (1992). *Hypnosis: 175 Years after Mesmer: Recent Developments in Theory and Application.* Konstanz: Universitatsvergag.

Hood, L. & Galas, D. (2003). The digital code of DNA. *Nature,*421, 444 – 448.

Hull, C. (1933/1968). *Hypnosis and suggestibility*. New York: Appleton-Century-Crofts.

Johnson, C. (1999). Forty years of PRCs – What have we learned? *Chronobiology International,* 16, 711-743.

Panda, S., Hogenesch, J., & Kay, S. (2002 a). Circadian rhythms from flies to humans. *Nature,* 417, 329-335.

Panda, S., Antoch, M., Miller, B., Su, A., Schook, A., Straume, M., Schultz, P., Kay, S., Takahashi, J., & Hogenesch, J. (2002 b). Coordinated Transcription of key pathways in the mouse by the circadian clock. *Cell*, 109, 307-320.

Pavlov, I. (1927). *Conditioned reflexes: an investigation of the physiological activity of the cerebral cortex.* Anrep, G. (Trans.). Oxford: Oxford University Press.

Rosbash, M. & Takahashi, J. (2002). The cancer connection. *Nature,* 420, 373-374.

Rossi, E. (1990). Mind-molecular communication: Can we really talk to our genes? *Hypnos, 17*(1), 3-14.

Rossi, E. (1992). Periodicity in self-hypnosis and the ultradian healing response: A pilot study. *Hypnos,* 19. 4-13.

Rossi, E. (1996). *The symptom path to enlightenment: The new dynamics of self-organization in hypnotherapy.* N.Y.: Zeig, Tucker , Theisen, Inc.

Rossi, E., (2000 a). *Dreams, consciousness, spirit: The quantum experience of self-reflection and co-creation.* (Third edition of Dreams and the growth of personality,

1972/1986). N. Y.: Zeig, Tucker, Theisen, Inc.

Rossi, E. (2000 b). Exploring gene expression in sleep, dreams and hypnosis with the new DNA microarray technology: A call for clinical-experimental research. *Sleep and Hypnosis: An International Journal of Sleep, Dream, and Hypnosis,* 2:1, 40-46.

Rossi, E. (2000c). In search of a deep psychobiology of hypnosis: Visionary hypotheses for a new millennium. *American Journal of Clinical Hypnosis*, 42:3/42:4, 178-207.

Rossi, E. (2002). *The Psychobiology Of Gene Expression: Neuroscience And Neurogenesis In Hypnosis And The Healing Arts.* New York: W. W. Norton Professional Books.

Rossi, E. (2003a). Gene expression, neurogenesis, and healing: Psychosocial genomics of therapeutic hypnosis. *American Journal of Clinical Hypnosis,* 45:3, 197-216.

Rossi, E. (2003b). Can we really talk to our genes? Psychosocial genomics. *Hypnos,* 30, 6-15.

Rossi, E. (2004). *Discorso Tra Geni (A Discourse with our Genes).* Rome: ELRF Publications.

Rossi, E., & Cheek, D. (1988*). Mind-Body Therapy: Ideodynamic Healing in Hypnosis.* New York: W. W. Norton.

Sanders, S. & Mann, B. (1995). The effects of light, temperature, trance length and time of day on hypnotic depth. *American J. Clinical Hypnosis*. 37, 43-53.

Storch, K., Lipan, O., Leykin, I., Viswanathan, N., Davis, F., Wong, W., & Weitz, C. (2002). Extensive and divergent circadian gene expression in liver and heart *Nature*, 417, 78 – 83.

Ueda, H., Chen, W., Adachi, A., Wakamatsu, H., Hayashi, S., Takasugi, T., Nagano, M., Nakahama, K., Suzuki, Y., Sugano, S., Iino, M., Shigeyoshi, Y., Hashimoto, S. (2002). A transcription factor response element for gene expression during circadian night. *Nature*, 418, 534 – 539.

Unterweger, E., Lamas, J., & Bongartz, W. (1992). Heart-Rate Variability of High and Low Susceptible Subjects During Administration of the Stanford Scale, Form C. In Bongartz, E. (Ed.), (1992). *Hypnosis: 175 Years after Mesmer: Recent Developments in Theory and Application*. Konstanz: Universitatsvergag.

Watson, J. & Crick, F. (1953). Genetical implications of the structure of deoxyribonucleic acid. *Nature,* 171, 964.

Wilber, K. (1993). *The spectrum of consciousness.* 2nd Edition. Wheaton, Illinois: Quest Books.

**CHAPTER 8**

Alejel, T., Beimesche, S., Cierny, I., Krause, D. & Knepel, W. (2002). Effect of antidepressant drugs and psychosocial stress on CRE-dependent reporter gene expression in the brain of transgenic mice. *Experimental and Clinical Endocrinology & Diabetes: Poster Neuro/Psychoendocrinology.*

Alfonso, J., Pollevick, G., van der Hart, G., Flugge, G., Fuchs, E. & Frasch, A. Identification of genes regulated by chronic psychosocial stress and antidepressant treatment in the hippocampus. *European Journal of Neuroscience,* 19, 3, 659-666.

Aserinsky, E. & Kleitman, N. (1953). Regularly occurring periods of eye motility and concomitant phenomena during sleep. *Science*, 118, 273-274.

Bentivoglio, M., & Grassi-Zucconi, G. (1999) Immediate early gene expression in sleep and wakefulness. In Lydic, R. & Baghdoyan, H. (Eds.) 1999. *Handbook of Behavioral State Control: Cellular and Molecular Mechanisms.* New York, CRC Press.

Berman, D. and Dudai, Y. (2001). Memory extinction, learning anew, and learning the new: Dissociations in the molecular machinery of learning in the cortex.

*Science*, 291, 2417-2419.

Bernheim, H. (1886/1957). *Suggestive Therapeutics: A Treatise on the Nature and Uses of Hypnotism*. Westport: Associated Booksellers.

Bongartz, E. (Ed.), (1992) *Hypnosis: 175 Years after Mesmer: Recent Developments in Theory and Application*. Konstanz: Universitatsvergag.

Castes, M., Hagel, I., Palenque, M., Canelones, P., Corano, A., & Lynch, N. (1999). Immunological changes associated with clinical improvement of asthmatic children subjected to psychosocial intervention. *Brain & Behavioral Immunology, 13(1)*, 1-13.

Cirelli, C., Gutierrez, C., & Tononi, G. (2004). Extensive and divergent effects of sleep and wakefulness on brain gene expression. *Neuron*, 41, 35-43.

Cohen-Cory, S. (2002). The Developing Synapse: Construction and Modulation of Synaptic Structures and Circuits. *Science* 298: 770-776.

Dudai, Y. (2000). The shaky trace. *Nature, 406*, 686-687.

Erickson, M., (1948/1980). Hypnotic Psychotherapy. In E. Rossi (Ed.), *The Collected Papers of Milton H. Erickson on Hypnosis. Vol. 4. Innovative Hypnotherapy* (pp. 35-48). New York: Irvington.

Erickson, M. (1970/1980). In E. Rossi (Ed.*), The Collected Papers of Milton H. Erickson on Hypnosis. Vol. 4, Innovative Hypnotherapy.* Hypnosis: Its Renascence as a treatment modality. New York: Irvington. Pp. 52-75.

Erickson, M. (1980). In E. Rossi (Ed.*), The Collected Papers of Milton H. Erickson on Hypnosis. Vol. 3, The Hypnotic Investigation of Psychodynamic processes*. Section 1: Amnesia. New York: Irvington. Pp. 35-90.

Frankland, P., Bontempi, B., Talton, L. Kaczmarek, L., Silva, A. (2004). The Involvement of the Anterior Cingulate Cortex in Remote Contextual Fear Memory. Science, 304, 881-883.

Glaser, R., Lafuse, W., Bonneau, R., Atkinson, C., & Kiecolt-Glaser, J, (1993). Stress-associated modulation of proto-oncogene expression in human peripheral blood leukocytes. *Behavioral Neuroscience*, 107, 525-529.

Greenfield, S. (1994). A model explaining Brazilian spiritist surgeries and other unusual religious-based healing. *Subtle Energies*, 5, 109-141.

Greenfield, S. (2000). Religious altered states and cultural-biological transduction in healing. Paper presented for the *Congresso Brasilerio de Etnopsiciatria*, Fortaleza CE, September 28-30. To Appear in Portuguese in the *Revista Brasileria de Etnopsiquiatria*.

Hoffman, K. & McNaughton, B. (2002). Coordinated reactivation of distributed memory traces in the primate neocortex. *Science, 297*, 2070-2073.

Izquierdo, I., Netto, C., Dalmaz, D., Chaves, M., Pereira, M., & Siegfried, B. (1988). Construction and reconstruction of memories. *Brazilian Journal of Medical and Biological Research, 21*, 9-25.

Janet, P. (1925/1976). *Psychological Healing: A historical and clinical study* (2 volumes). (Paul, E. & Paul, C., Trans.). N.Y.: Arno Press.

Kandel, E. (2000). Cellular mechanisms of learning and the biological basis of individuality. In Kandel, E., Schwartz, J., & Jessell, T. (Eds.) *Principles of Neural Science. 4th Edition*. P 1247-1279.

Kaufer, D., Friedman, A., Seidman, S. & Soreq, H. (1998). Acute stress facilitates long-lasting changes in cholinergic gene expression. *Nature, 393*, 373-377.

Kleitman, N. & Rossi, E. (1992). The basic rest-activity cycle – 32 years later. An interview with Nathaniel Kleitman at 96. In Lloyd, D. & Rossi, E., Eds. (1992). *Ultradian Rhythms in Life Processes: A Fundamental Inquiry into Chronobiology and Psychobiology*. New York: Springer-Verlag, 303-306.

Laughlin, S. & Sejnowski, J. (2003). Communication in neuronal networks. *Science, 301,* 1870-1874.

***Lee, J., Everitt, B., and Thomas, K. (2004).*** Independent Cellular Processes for Hippocampal Memory Consolidation and Reconsolidation. **Science**, 304, 5672, 839-843.

Levsky, J., Shenoy, S., Pezo, R. and Singer, R. (2002). Single-cell gene expression profiling. *Science*, 297, 836-840.

Lloyd, D. & Rossi, E., Eds. (1992). *Ultradian Rhythms in Life Processes: A Fundamental Inquiry into Chronobiology and Psychobiology*. New York: Springer-Verlag.

Lloyd, D. & Rossi, E., (1993). Biological Rhythms as organization and information. *Biological Reviews*, 68, 563-577.

Lüscher, C., Nicoll, R., Malenka, C., & Muller, D. (2000). Synaptic plasticity and dynamic modulation of the postsynaptic membrane. *Nature Neuroscience* 3, 545-567.

Matus, A. (2000). Actin-based plasticity in dendritic spines. *Science,* 290, 754-758.

Milad, M. and Quirk, G. (2002). Neurons in medial prefrontal cortex signal memory for fear extinction. *Nature,* 420, 70-74.

Nader, K., Schafe, G., & Le Doux, J. (2000 a). Fear memories require protein synthesis in the amygdala for reconsolidation after retrieval. *Nature, 406,* 722-726.

Nader, K., Schafe, G., & Le Doux, J. (2000 b). The labile nature of consolidation theory. *Nature Reviews: Neuroscience,* 1, 216-219.

Netz, R. (2002). Proof, amazement, and the unexpected. *Science,* 298, 967-968.

Pavlov, I. (1927). *Conditioned reflexes, An investigation of the physiological activity of the cerebral cortex.* London: Oxford University Press.

Rainville, P., Ducan, G., Price, D., Carrier, B. & Bushnell, C. (1997). Pain affect encoded in human anterior cingulate but not somatosensory cortex. *Science,* 277, 968-871.

Rainville, P., Hofbauer, R., Paus, T., Duncan, G., Busnell, M., & Price, D. (1999). Cerebral mechanisms of hypnotic induction and suggestion. *Journal of Cognitive Neuroscience,* 11, (1), 110-125.

Ribeiro, S., Goyal, V., Mello, C. & Pavlides, C. (1999). Brain gene expression during REM sleep depends on prior waking experience. *Learning & Memory, 6:* 500-508.

Ribeiro, S. (2003). Dream, memory and Freud's reconciliation with the brain. *Revista Brasileira de Psiquiatria (Review of Brazilian Psychiatry),* 25, 1-13.

Ribeiro, S., Mello, C., Velho, T., Gardner, T., Jarvis, E., & Pavlides, C. (2002). Induction of hippocampal long-term potentiation during waking leads to increased extrahippocampal zif-268 expression during ensuing rapid-eye-movement sleep. *Journal of neuroscience,* 22(24), 10914-10923.

Ribeiro, S., Gervasoni, D., Soares, E., Zhou, Y., Lin, S., Pantoja, J., Lavine, M., & Nicolelis, M. (2004). Long-lasting novelty-induced neuronal reverberation during slow-wave sleep in multiple forebrain areas. *Public Library of Science, Biology. (PLoS),* 2 (1), 126-137.

Ridley, M. (1999). *Genome: The Autobiography of a Species in 23 Chapters.* N.Y.: HarperCollins.

Rosbash, M., & Takakshi, J. (2003). Circadian rhythms: the cancer connection. *Nature,* 420, 373-374.

Rossi, E. (1967). Game and growth: Two dimensions of our psychotherapeutic zeitgeist. *Journal of Humanistic Psychology*, 7, 136-154.

Rossi, E. (1968). The Breakout Heuristic: A Phenomenology of Growth Therapy with College Students. *Journal of Humanistic Psychology*, 8, 16-28.

Rossi, E. (1972/1985/2000). *Dreams, consciousness, spirit* (3ed edition of *Dreams and the growth of personality*) Phoenix: Zeig, Tucker, Theisen.

Rossi, E. (1973 a). Psychological shocks and creative moments in psychotherapy. *The American Journal of Clinical Hypnosis, 16,* 9-22.

Rossi, E. (1973 b). The dream-protein hypothesis, *American Journal of Psychiatry,* 130, 1094-1097.

Rossi, E. (1982). Hypnosis and ultradian cycles: A new state(s) theory of hypnosis? *The American Journal of Clinical Hypnosis,* 25 (1), 21-32.

Rossi, E. (1986). Altered states of consciousness in everyday life: The ultradian rhythms. In B. Wolman & M. Ullman (Eds.), *Handbook of Altered States of Consciousness* (pp. 97-132). New York: Van Nostrand.

Rossi, E. (1986/1993). *The Psychobiology of Mind-Body Healing*, Revised Edition. New York: Norton.

Rossi, E. (1990). Mind-Molecular Communication: Can We Really Talk to Our Genes? *Hypnos, 17*(1), 3-14.

Rossi, E. (1992). The wave nature of consciousness: A new direction for the evolution of psychotherapy. In Zeig, J. (Ed.) The Evolution of Psychotherapy: The Second Conference. New York: Brunner-Mazel, pp 216-235.

Rossi, E. (1994). The emergence of mind-gene communication. *European Journal of Clinical Hypnosis,* 3, 4-17.

Rossi, E. (1996). *The Symptom Path to Enlightenment: The New Dynamics of Self-Organization in Hypnotherapy.* Pheonix, Arizona.: Zeig, Tucker, Theisen, Inc.

Rossi, E. (1999). Sleep, dream, hypnosis and healing: behavioral state-related gene expression and psychotherapy. *Sleep and Hypnosis: An International Journal of Sleep, Dream, and Hypnosis,* 1:3, 141-157.

Rossi, E. (2000 a). In search of a deep psychobiology of hypnosis: Visionary hypotheses for a new millennium. *American Journal of Clinical Hypnosis*, 42:3/42:4, 178-207.

Rossi, E. (2000 b). Exploring gene expression in sleep, dreams and hypnosis with the new DNA microarray technology: A call for clinical-experimental research. *Sleep and Hypnosis: An International Journal of Sleep, Dream, and Hypnosis,* 2:1, 40-46.

Rossi, E. (2001a). Updating Milton Erickson's neuro-psycho-physiological dynamics of therapeutic hypnosis and psychotherapy. *The Milton H. Erickson Foundation Newsletter.* Phoenix, Arizona, pp. 10-13.

Rossi, E. (2001b). The deep psychobiology of psychotherapy. In Corsini, R. (Ed.) (2001). *Handbook of Innovative Therapy,* 2nd edition, N.Y.: Wiley. P. 155-165.

Rossi, E. (2002a). *The Psychobiology of Gene Expression: Neuroscience and Neurogenesis in Hypnosis and the Healing Arts.* NY: Norton Professional Books.

Rossi, E. (2002b). A conceptual review of the psychosocial genomics of expectancy and surprise: Neuroscience perspectives about the deep psychobiology of therapeutic hypnosis. *American Journal of Clinical Hypnosis*, 45:2,103-118.

Rossi, E. (2003). Gene Expression, Neurogenesis, and Healing: Psychosocial Genomics of Therapeutic Hypnosis. *American Journal of Clinical Hypnosis*, 45, 3, 197-216.

Rossi, E. (2004). A Bioinformatics Approach to the Psychosocial Genomics of Therapeutic Hypnosis. *Hypnos, 31,* 15-21.

Rossi, E. (2004, In Press). *A Discourse with Our Genes: The Neuroscience of Therapeutic Hypnosis and Psychotherapy.* Benevento, Italy: Editris.

Rossi, E., & Cheek, D. (1988). *Mindbody Therapy: Ideodynamic Healing in Hypnosis*. New York: W. W. Norton.

Rossi, E., & Nimmons, D. (1991). *The Twenty Minute Break: The Ultradian Healing Response*. N.Y.: Zeig, Tucker, Theisen, Inc.

Schanberg, S. (1995). The genetic basis for touch effects. Field, T. (Ed.) *Touch in Early*

*Development.* N.Y.: Lawrence Erlbaum, 67-79.

Selkoe, D. (2002). Alzheimer's Disease Is a Synaptic Failure. *Science*, 298, 789-791.

Sheng, M. and Kim, M. (2002). Postsynaptic signaling and plasticity mechanisms. *Science*, 298, 776-780.

Shimizu, E, Tang, Y., Rampon, C., & Tsien, J. (2000). NMDA receptor-dependent synaptic reinforcement as a crucial process for memory consolidation. *Science*, 290, 1170-1174.

Smith, S. (1995). Getting into and out of mental ruts: A theory of fixation, incubation, and insight. In Sternberg, R. & Davidson, J. (Eds.) *The Nature of Insight.* Cambridge, Massachusetts: MIT Press. 229-251.

Soreq, H. & Seidman, S. (2001). Acetylcholinesterase – new roles for an old actor. *Nature Reviews: Neuroscience*, 2, 294-302.

Spedding, M., Neau, I., & Harsing, L. (2003). Brain plasticity and pathology in psychiatric disease: sites of action for potential therapy. *Current Opinion in Pharmacology*, 3, 33-40.

Sutherland, G. & McNaughton, B. (2000). Memory trace reactivation in hippocampal and neocortical neuronal ensembles. *Current Opinion in Neurobiology, 10,* 180-186.

Van Praag, H., Schinder, A., Christie, B., Toni, N., Palmer, T. and Gage, F. (2002). Functional neurogenesis in the adult hippocampus. *Nature*, 415, 1030-1034.

Wallas, G. (1926). *The Art of Thought.* New York: Harcourt.

Weitzenhoffer, A. (2000). *The Practice of Hypnotism. 2ed Edition.* New York: Wiley.

Weitzenhoffer, A. (2001). For the record: A commentary on the role of suggestion in hypnosis. *American Journal of Clinical Hypnosis, 44,* 155-157.

Weitzenhoffer, A. (2002). Symptom removal: The nineteenth century experience. *American Journal of Clinical Hypnosis*, 45, 129 -136.

Whitney, A., Diehn, M., Popper, S., Alizadeh, A., Boldrick, J., Relman, D. & Brown, P. (2003) Individuality and variation in gene expression patterns in human blood. *Proceedings of the National Academy of Sciences,*100, 1896-1901.

Wittenberg, G., Sullivan, M., & Tsien, J. (2003). Synaptic reentry reinforcement based network model for long-term memory consolidation. *Hippocampus, 12,* 637-647.

**CHAPTER 9**

Autelitano, D.J. (1998). Stress-induced stimulation of pituitary POMC gene expression is associated with activation of transcription factor AP-1 in hypothalamus and pituitary. *Brain Research Bulletin*, 45(1), 75-82.

Azar, B. (1999). New pieces filling in addiction puzzle. *APA Monitor*, 30, 1-15.

Bentivoglio, M. and Grassi-Zucconi, G. (1999). Immediate-early gene expression in sleep and wakefulness. In Lydic, R. and Baghdoyan, H. (1999). *Handbook of Behavioral State Control: Cellular and Molecular Mechanisms.* New York: CRC Press, 235-253.

Berman, D. and Dudai, Y. (2001). Memory extinction, learning anew, and learning the new: Dissociations in the molecular machinery of learning in the cortex. *Science*, 291, 2417-2419.

Chin, H. & Moldin, S. (2001). *Methods in Genomic Neuroscience.* Boca Raton: CRC Press.

Dudai, Y. (2000). The shaky trace. *Nature, 406*, 686-687.

Erickson, M. (1958/1980). Naturalistic techniques of hypnosis. In E. Rossi (Eds.), *The Collected Papers of Milton H. Erickson on Hypnosis. Vol. I. The Nature of Hypnosis*

*and Suggestion* (pp. 168-176). New York: Irvington.
Erickson, M. (1959/1980). Further clinical techniques of hypnosis: Utilization techniques. In E. Rossi (Eds.), *The Collected Papers of Milton H. Erickson on Hypnosis. Vol. I. The Nature of Hypnosis and Suggestion* (pp. 177-205). New York: Irvington.
Erickson, M. and Erickson, E. (1941/1980). Concerning the nature and character of post-hypnotic behavior. In E. Rossi (Eds.), *The Collected Papers of Milton H. Erickson on Hypnosis. I. The Nature of Hypnosis and Suggestion* (pp. 360-365). New York: Irvington.
Erickson, M., & Rossi, E. (1974/1980). Varieties of Hypnotic Amnesia. In Rossi, E. (Ed.), *The Collected Papers of Milton H. Erickson on Hypnosis. Vol. Hypnotic Investigation of Psychodynamic Processes.* (pp. 71-90). New York: Irvington.
Eriksson, P., Perfilieva, E., Björk-Ericksson, T., Alborn, A., Nordborg, C., Peterson, D., & Gage, F. (1998). Neurogenesis in the adult human hippocampus. *Nature Medicine*, 4, 1313-1317.
Gelb, M. (1998). *How to Think Like Leonardo da Vinci: Seven Steps to Genius Every Day.* New York: Dell.
Gould E., Tanapat, P., McEwen, B., Flügge, G., & Fuchs, E. (1998). Proliferation of granule cell precursors in the dentate gyrus of adult monkeys is diminished by stress. *Proceeding of the National Academy of Sciences,* 95, 3168-3171.
Gould E., Tanapat, P., Reeves, A., & Shors, T. (1999). Learning enhances adult neurogenesis in the hippocampal formation. *Nature Neuroscience*, 2(3), 260-265.
Janet, P. (1925/1976). *Psychological Healing: A historical and clinical study* (2 volumes). (Paul, E. & Paul, C., Trans.). N.Y.: Macmillan/Arno Press Reprint.
Kandel, E. (1983). From metapsychology to molecular biology: Explorations into the nature of anxiety. *American Journal of Psychiatry, 140*(10), 1277-1293.
Kandel, E. (1989). Genes, nerve cells, and the remembrance of things past. *Journal of Neuropsychiatry 1*(2), 103-125.
Kandel, E. (2000). Cellular mechanisms of learning and the biological basis of individuality. In Kandel, E., Schwartz, J., & Jessell, T. (Eds.) *Principles of Neural Science. 4th Edition.* P 1247-1279.
Kandel, E. (2001). The molecular biology of memory storage: A dialogue between genes and synapses. *Science, 294,* 1030-1038.
Kelner, K. & Bloom, F. (2000). *The Best of Science: Neuroscience.* Washington: American Association for the Advancement of Science.
Kempermann, G. & Gage, F. (1999). New nerve cells for the adult brain. *Scientific American,* 280, 48-53.
Kempermann, G., Kuhn, G. & Gage, F. (1997). More hippocampal neurons in adult mice living in an enriched environment. *Nature,* 386, 493-495.
Kleitman, N. & Rossi, E. (1992). The basic rest-activity cycle – 32 years later. An interview with Nethaniel Kleitman at 96. In Lloyd, D. & Rossi, E., Eds. (1992). *Ultradian Rhythms in Life Processes: A Fundamental Inquiry into Chronobiology and Psychobiology*. New York: Springer-Verlag, 303-306.
Lloyd, D. & Rossi, E., Eds. (1992). *Ultradian Rhythms in Life Processes: A Fundamental Inquiry into Chronobiology and Psychobiology*. New York: Springer-Verlag.
Lloyd, D. & Rossi, E., (1993). Biological Rhythms as organization and information. *Biological Reviews*, 68, 563-577.
Lydic, R. (Ed.) (1998). Molecular Regulation of Arousal States. New York: CRC Press.
Lydic, R. and Baghdoyan, H. (1999). *Handbook of Behavioral State Control: Cellular and Molecular Mechanisms.* New York: CRC Press.
Nader, K., Schafe, G., & Le Doux, J. (2000a). Fear memories require protein synthe-

sis in the amygdala for reconsolidation after retrieval. *Nature, 406,* 722-726.

Nader, K., Schafe, G., & Le Doux, J. (2000b). The labile nature of consolidation theory. *Nature Reviews: Neuroscience,* 1, 216-219.

O'Neill, M., Hicks, C., Shaw, G., Parameswaran, T., Cardwell, G., & O'Neill, M. (1998). Effects of 5-hydroxytryptamine2 receptor antagonism on the behavioral activation and immediate-early gene expression induced by dizocilpine. *Journal of Pharmacological Experimental Therapy,* 287(3), 839-846.

Ribeiro, S., Goyal, V., Mello, C. & Pavlides, C. (1999). Brain gene expression during REM sleep depends on prior waking experience. *Learning & Memory, 6:* 500-508.

Ridley, Matt. (1999). *Genome: The Autobiography of a Species in 23 Chapters.* N.Y.: HarperCollins.

Rossi, E. (1967). Game and growth: Two dimensions of our psychotherapeutic zeitgeist. *Journal of Humanistic Psychology*, 7, 136-154.

Rossi, E. (1968). The Breakout Heuristic: A Phenomenology of Growth Therapy with College Students. *Journal of Humanistic Psychology*, 8, 16-28.

Rossi, E. (1986/1993). *The Psychobiology of Mindbody Healing: New Concepts of Therapeutic Hypnosis.* 2nd Edition. New York: Norton.

Rossi, E. (1992). Videotape (Code E 297-V8) available to licensed professionals and researchers from the Erickson Foundation, 3606 North 24th street, Phoenix, Arizona, 85016; www.erickson-foundation.org; Tel. 602-956-6196.

Rossi, E. (1996). *The Symptom Path to Enlightenment: The New Dynamics of Self-Organization in Hypnotherapy*. N.Y.: Zeig,Tucker, Theisen, Inc.

Rossi, E. (1999). Sleep, dream, hypnosis and healing: behavioral state-related gene expression and psychotherapy. *Sleep and Hypnosis: An International Journal of Sleep, Dream, and Hypnosis,* 1:3, 141-157.

Rossi, E. (2000a). Exploring gene expression in sleep, dreams, and hypnosis with the new DNA microarray technology: A call for clinical-experimental research. *Sleep and Hypnosis: An International Journal of Sleep, Dream, and Hypnosis,* 2:1, 40-46.

Rossi, E. (2000b). In search of a deep psychobiology of hypnosis: Visionary hypotheses for a new millennium. *American Journal of Clinical Hypnosis*, 42:3/42:4, 178-207.

Rossi, E. (2000c). *Dreams, Consciousness, Spirit: The Quantum Experience of Self-Reflection and Co-Creation*. (Third edition of *Dreams and the Growth of Personality,* 1972/1986). N. Y.: Zeig, Tucker, Theisen, Inc.

Rossi, E. (2000d). The numinosum and the brain: The weaving thread of consciousness. *Psychological Perspectives,* 40, 94-103.

Rossi, E. (2001a). The deep psychobiology of psychotherapy. In Corsini, R. (Ed.) *Handbook of Innovative Therapy,* 2nd edition, N.Y.: Wiley, 155-165.

Rossi, E. (2001b). Updating Milton Erickson's neuro-psycho-physiological dynamics of therapeutic hypnosis and psychotherapy. *The Milton H. Erickson Foundation Newsletter.* Phoenix, Arizona, pp. 10-13.

Rossi, E. (2002). *The Psychobiology of Gene Expression: Neuroscience and Neurogenesis in Hypnosis and the Healing Arts.* New York: W. W. Norton Professional Books.

Rossi, E., & Cheek, D. (1988). *Mindbody Therapy: Ideodynamic Healing in Hypnosis*. New York: W. W. Norton.

Rossi, E., Lippincott, B., & Bessette, A. (1994). Ultradian Dynamics in Hypnotherapy: Part One. *The European Journal of Clinical Hypnosis, 2, 1, p. 10-20.*

Rossi, E., Lippincott, B., & Bessette, A. (1995). Ultradian Dynamics in Hypnotherapy: Part Two. *The European Journal of Clinical Hypnosis, 2, 2, p. 6-14.*

Schanberg, S. (1995). The genetic basis for touch effects. Field, T. (Ed.) *Touch in Early Development.* N.Y.: Lawrence Erlbaum, 67-79.

Schrödinger, E. (1944). *What is Life? & Mind and Matter.* New York: Cambridge University Press.

Senba, E., & Ueyama, T. (1997). Stress-induced expression of immediate-early genes in the brain and peripheral organs of the rat. *Neuroscience Research,* 29(3), 183-207.

Shimizu, E, Tang, Y., Rampon, C., & Tsien, J. (2000). NMDA receptor-dependent synaptic reinforcement as a crucial process for memory consolidation. *Science,* 290, 1170-1174.

Snowdon, D. (2001). *Aging with Grace.* New York: Bantam.

Sternberg, E. (2000). *The Balance Within: The Science of Connecting Health and Emotions.* N.Y.: Freeman.

Sternberg, R. (2001). What is the common thread of creativity? *American Psychologist,* 56, 360-362.

Sternberg, R. & Davidson, J. (1995). *The Nature of Insight.* Cambridge, Mass.: The MIT Press.

Tölle TR, Schadrack J, Zieglgansberger, W. (1995). *Immediate-early Genes in the CNS.* New York: Springer-Verlag.

Toppila, J. Stenberg D, Alanko L, Asikainen M, Urban J., Turek F., Porkka- Heiskanen T. (1995). REM sleep deprivation induces galanin gene expression in the rat brain. *Neuroscience Letters,* 183, 171-174.

Van Praag, H., Kempermann, G., and Gage, F. (1999). Running increases cell proliferation and neurogenesis in the adult mouse dentate gyrus. *Nature Neuroscience,* 2, 266-270.

Van Praag, H., Kempermann, G., and Gage, F. (2000). Neural consequences of environmental enrichment. *Nature Reviews: Neuroscience,* 1, 191-198.

Whitney, A., Diehn, M., Popper, S., Alizadeh, A., Boldrick, J., Relman, D. & Brown, P. (2003) Individuality and variation in gene expression patterns in human blood. *Proceedings of the National Academy of Sciences,* 100, 1896-1901.

Zeig, J. (1997). Experiential approaches to clinician development. In Zeig, J. (Ed.). *The Evolution of Psychotherapy: The Third Conference.* N.Y.: Brunner/Mazel.

**CHAPTER 10**

Ahmadias, N., Alaei, H., and Hänninen, O. (2003). Effect of exercise on learning, memory and Levels of epinephrine in rats. Hippocampus. Journal of Sports Science and Medicine, 2, 106-109

Berdanier, C. (Ed.) (1996). *Nutrients and gene expression.* New York: CRC Press.

Ding, X. et al. (2002). Resveratrol inhibits proliferation and induces apoposis in human pancreatic cancer cells. *Pancreas,* 25, e71-e76.

Dinges, D. & Broughton, R. (1989). *Sleep and Alertness.* New York: Raven Press.

Erickson, M. (1965/1980). A special inquiry with Aldous Huxley into the nature and character of the various states of consciousness. In E. Rossi (Eds.), *The Collected Papers of Milton H. Erickson on Hypnosis. I. The Nature of Hypnosis and Suggestion* (pp. 83-107). New York: Irvington.

Erickson, M. (1980). An introduction to the study and application of hypnosis for pain control. In E. Rossi (Ed.), *The Collected Papers of Milton H. Erickson on Hypnosis. IV. Innovative Hypnotherapy; Section nine.* (pp. 237-245). New York: Irvington.

Finneran, R. (Ed.) (1997). *The Collected Works of W. B. Yeats. Vol. 1. The Poems.* Vascillation, New York: Schribner, Pp. 255.

Gstaiger, M. et al. (2003). Control of nutrient-sensitive transcription programs by the unconventional prefoldin URI. *Science,* 302, 1208-1212

Hobson, J., (1997). Consciousness as a state-dependent phenomenon. In *Scientific Approaches to the Question of Consciousness,* Cohen, J.& Schooler, J., eds. Mahwah, N.J.: Lawrence Erlbaum Associates, pp. 379-396.

Hobson, J. (1999). *Consciousness.* New York: Scientific American Library.

Jang, J. et al. (2003). Protective effect of resveratrol on beta-amyloid-induced oxidative PC12 cell death. *Free Radical Biological Medicine,* 34, 1100-1110.

Jenkins, D. (1989). Nibbling versus gorging: Metabolic advantages of increased meal frequency. *The New England Journal of Medicine*, 321, 929-935.

Jung. C. (1943/1966). The Synthetic or Constructive Method. In Jung, C. (1966). *Two essays on analytical psychology. Vol. 7. The Collected Works of C. G. Jung.* (R. F. C. Hull, Trans.). Bollingen Series XX. Princeton, New Jersey: Princeton University Press.

Kempermann, G. & Neumann, H. (2003). Microglia: The enemy within? *Science,* 302, 1689-1690.

Kleitman, N. (1963). *Sleep and Wakefulness as Alternating Phases in the Cycle of Existence.* Chicago, Ill: University of Chicago.

Kleitman, N. (1969). Basic rest-activity cycle in relation to sleep and wakefulness. In A. Kales (Ed.), *Sleep: Physiology & Pathology* (pp. 33-38). Philadelphia: Lippincott.

Kleitman, N. (1970). Implications of the rest-activity cycle: Implications for organizing activity. In E. Hartmann (Ed.), *Sleep and Dreaming*. Boston: Little, Brown.

Kleitman, N. & Rossi, E. (1992). The basic rest-activity cycle – 32 years later. An interview with Nethaniel Kleitman at 96. In Lloyd, D. & Rossi, E., Eds. (1992). *Ultradian Rhythms in Life Processes: A Fundamental Inquiry into Chronobiology and Psychobiology*. New York: Springer-Verlag, 303-306.

Ko, J., Na, D., Lee, Y., Shin, S., Kim, J., Hwang, B., Min, B., Park, D. (2002). cDNA microarray analysis of the differential gene expression in the neuropathic pain and electroacupuncture treatment models. *Journal of Biochemistry & Molecular Biology,* 35(4), 420-427.

Lloyd, D. & Rossi, E., (Eds.), (1992). *Ultradian Rhythms in Life Processes: A Fundamental Inquiry into Chronobiology and Psychobiology*. New York: Springer-Verlag.

Lloyd, D. & Rossi, E., (1993). Biological Rhythms as organization and information. *Biological Reviews*, 68, 563-577.

Lu, R. et al., (1999). Resveratrol, a natural product derived from grape, exhibits antiestrogenic activity and inhibits the growth of human breast cancer cells. *Journal of Cell Physiology,* 179, 297-304.

Monje, M., Toda, H., & Palmer, T. (2003). Inflammatory Blockade Restores Adult Hippocampal Neurogenesis, *Science, 302,* 1760-1765.

Moustaïd-Moussa, N. and Berdanier, C. (2001). *Nutrient-Gene Interactions in Health and Disease: The Effect of Macronutrients and Micronutrients on Gene Expression.* Boca Raton: CRC Press.

Narayanan, B. et al. (2002). Int4ractive gene expression pattern in prostrate cancer cells exposed to phenolic antioxidants. *Life Sciences,* 70, 1821-1839.

Odum, H. (1988). Self-Organization, Transformity, and Information. *Science,* 242, 1132-1139.

Pace-Asciak, C. et al. (1995). The red wine phenolics trans-resveratrol and quercetin block human platlet aggregation and eiosanoid synthesis: implications for protection against coronary heart disease, *Clinical Chemistry Acta,* 235, 207-219.

Pennebaker, J. (1997). *Opening Up: The Healing Power of Expressing Emotions.* (Revised edition). New York: Guilford Press.

Rhodes, J., van Pragg, H., Jeffrey, S., Girard, I., Mitchell, G. & Garland, T. (2003).

Exercise increases hippocampal Neurogenesis to high levels but does not improve spatial learning in mice bread for increased voluntary wheel turning. *Behavioral Neuroscience, 117,* 1005-1016.

Ribeiro, S., Goyal, V., Mello, C., Pavlides, C. (1999) Brain gene expression during REM sleep depends on prior waking experience. *Learn & Memory*, 6: 500–508.

Ribeiro, S., Mello, C., Velho, T., Gardner, T., Jarvis, E., Pavlides, C. (2002) Induction of hippocampal long-term potentiation during waking leads to increased extra hippocampal zif-268 expression during ensuing rapid-eye-movement sleep. *Journal of Neuroscience*, 22 (24): 10914–10923.

Ribeiro, S., Gervasoni, D., Soares, E., Zhou, Y., Lin, S., Pantoja, J., Lavine, M., Nicolelis, M. (2004). Long-lasting novelty-induced neuronal reverberation during slow-wave sleep in multiple forebrain areas. *Public Library of Science, Biology (PLoS),* Vol.2, 1, 126-137.

Ridley, Matt. (1999). *Genome: The Autobiography of a Species in 23 Chapters.* N.Y.: HarperCollins.

Rossi, E. (2002). *The Psychobiology of Gene Expression: Neuroscience and Neurogenesis in Hypnosis and the Healing Arts.* New York: W. W. Norton Professional Books.

Rossi, E., & Nimmons, D. (1991). *The Twenty Minute Break: The Ultradian Healing Response.* N.Y.: Zeig, Tucker, Theisen, Inc.

Sinha, K. et al. (2002). Protective effects of resveratrol against oxidative stress in middle cerebral artery occlusion model of stroke in rats. *Life Sciences,* 71, 655-665.

Smith, A. and Tart, C. (1998). Cosmic consciousness experience and psychedelic experiences: A first person comparison. *Journal of Consciousness Studies,* 5, 97-107.

Strickgold, R., Malia, A., Maguire, D., Roddenberry, D., & O'Connor, M. (2000a). Replaying the game: Hypnogogic images in normals and amnesiacs. *Science,* 290, 350-353.

Strickgold, R., Whidbee, D., Schirmer, B., Patel, V. & Hobson, J. (2000b). Visual discrimination task improvement: A multi-step process occurring during sleep. *Journal of Cognitive Neuroscience,* 12:2, In Press.

Van Gelder, R. (2003). Oscillatory mechanisms underlying the mammalian circadian clock. *Science,* 302, www.stke.org

Yen, G. et al. (2003). Effects of resveratrol and 4-hexylresorcinol on hydrogen peroxide induced oxidative damage in human lymphocytes. *Free Radical Research,* 37, 509-14.

## APPENDIX 1

BOWERS, K. Has the sun set on the Stanford Scales? *American Journal of Clinical Hypnosis,* 1982, *2,* 79-88.

BYKOV, V. L., & KATINAS, G. S. Temporal organization of the thyroid in A/He mice (morphometric investigation). *BioL Bull. Acad. Sci. USSR,* 1979, 6, 247-249.

DELGADO-GARCIA, J. M., GRAU, C., DEFEUDCS, P., DEL Pozo. F., JIMENEZ, J. M., & DELGADO, J. M. Ultradian rhythms in the mobility and behavior of rhesus monkeys. *Experimental Brain Research,* 1976,25,79-91.

DEMENT, W., & KLEITMAN, N. Cyclic variations in EEC during sleep and their relation to eye movements, body motility, and dreaming. *Elec-troencephalography & Clinical Neurophvsiologv,* 1957,9,673-690.

ERICKSON, M. H., & Rossi, E. L. *Hypnotherapy: An Exploratory Casebook.* New York: Irvington, 1979.

ERICKSON, M. H., Rossi, E. L., & Rossi, S. I. *Hypnotic Realities.* New York: Irvington,

1976.

ERIKSSON, E., ED'EN, MODTCH, K., & H"AGGENDAL, J. Ultradian rhythm in rat hypothalamic dopamine levels. *Journal of Neural Transmission,* 1980, *48,* 305-310.

FILICORI. M., BOLELLI, G., FRANCESCHETT1, F., & LAFISCA, S. The ultradian pulsatile release of gona-dotropins in normal female subject. *Acta Europaea Fertilitatis.* 1979, *10,* 29-33.

FISCHER, R. A cartography of ecstatic and meditative states. *Science.* 1971, *174,* 897-904.

FRIEDMAN, A. H., & PIEPHO, R. W. Effect of photo period reversal on patterns for GABA levels in rat brain. *International Journal of Chronobiology,* 1978, 5, 445-458.

FRIEDMAN, S., & FISCHER, C. On the presence of a rhythmic, diurnal, oral instinctual drive cycle in man: a preliminary report. *Journal of the American Psychoanalytic Association,* 1967, *15,* 317-343.

GLOBUS, G. G. Rapid eye movement cycle in real

time. *Archive* of General Psychophysiology,* 1966, /5, 654-69.

GLOBUS, G. G. Observations on sub-circadian rhythms. *Psychophysiology.* 1968, *4.* 366.

GLOBUS, G. G., PHOEBUS, E., & MOORE, C. REM "sleep" manifestations during waking. *Psychophysiology,* 1970, 7, 308.

GOPHER. D.. & LAVIE P. Short-term rhythms in the performance of a simple motor task. *Journal of Motor Behavior.* 1980, *12,* 207-221.

HIATT, J. F., & KRIPKE, D. F. Ultradian rhythms in waking gastric activity. *Psychosomatic Medicine,* 1975, *37,* 320-325.

HILGARD, E. The eyeroll sign and other scores of the Hypnotic Induction Profile (HIP) as related to the Stanford Hypnotic Susceptibility Scales, Form C (SHSS:C). *American Journal of Clinical Hypnosis,* 1982a, 2, 89-97.

HILGARD, E. Further discussion of the HIP and the Stanford Form C: A reply to reply by Frischolz, Spiegel,Tryon & Fischer. *American Journal of Clinical Hypnosis,* 1982b, 2, 106-107.

HORNE, J., & WHITEHEAD, M. Ultradian and other rhythms in human respiration rate. *Experientia, 1976,32,* 1165-1167.

HUNSAKER, W. G., REISER, B., & WOLYNETZ, M. Vaginal temperature rhythms in sheep. *International Journal of Chronobiology,* 1977,4, 151-162.

KATZ, R. J. The temporal structure of motivation. III. Identification and ecological significance of ultra-dian rhythms of intracranial reinforcement. *Behavioral & Neural Biology,* 1980, *30,* 148-159.

KLEIN, R., & ARMITAGE, R. Rhythms in human performance: l-!/:-hour oscillations in cognitive style. *Science.* 1979, *204.* 1326-1328.

KLEITMAN, N. *Sleep and Wakefulness* (2nd Ed.). Chicago: University of Chicago Press, 1963.

KLEITMAN, N. Basic rest-activity cycle in relation to sleep and wakefulness. In A. Kales (Ed.). *Sleep: Physiology & Pathology.* Philadelphia: Lippincott, 1969, 33-38.

KRIPKE, D. F. An ultradian biological rhythm associated with perceptual deprivation and REM sleep. *Psychosomatic Medicine,* 1972, *34,* 221-234.

KRIPKE. D. F. Ultradian rhythms in sleep and wakeful-ness. In E. D. Weitzman (Ed.), *Advances in Sleep Research.* Vol. 1. New York: Spectrum, 1974, 305-325,

KRIPKE. D. F. Ultradian rhythms in behavior and physiology. In F. M. Brown and R. C. Graeber (Eds.), *Rhythmic Aspects of Behavior.* Hillsdale, New Jersey: Erlbaum Associates, 1982a.*

* Projected publication date of April, 1982. according to *Forthcoming Books Index.*

KRIPKE. D. F. Telephone conversation with Ernest Rossi. March, 1982b.

KRIPKE, D. F., MULLANEY. D. G.. WYBORNEY, V. G.. & MESSIN, S. There's no basic rest-activity cycle. In F. D. Stott et al., (Eds.) ISAM 1977: *Proceedings of the Second International Symposium on Ambulatory Monitoring.* London: Academic Press. 1978. 105-113.

KRIPKE. D. F., & SONNENSCHEIN, D. A biologic rhythm in waking fantasy. In K. Pope & J. Stringer (Eds.), *The Stream of Consciousness.* New York: Plenum, 1978. 321-332.

KRYNICKI, V. Time trends and periodic cycles in REM sleep eye movements. *Electroencephatography & Clinical Neurophysiology,* 1975, *39,* 507-513.

LAVIE, P. Ultradian rhythms in the perception of two apparent motions. *Chronobiologia.* 1976, *3,* 214-218.

LAVIE, P. Nonstationarity in human perceptual ultradian rhythms. *Chronobiologia,* 1977, *4,* 38-48.

LAVIE, P., & KRIPKE. D. F. Ultradian rhythms in urine flow in waking humans. *Nature,* 1977,269, 142-144.

LAVIE. P., & KRIPKE, D. F. Ultradian circa I[1]/; hour rhythms: A multioscillatory system. *Life Sciences,* 29, New York: Pergamon, 1981, 2445-2450.

LAVIE, P., LORD, J. W., & FRANK, R. A. Basic rest-activity cycle in the perception of the spiral aftereffect: a sensitive detector of a basic biological rhythm. *Behavioral Biology,* 1974, *11.* 373-379.

LAVIE, P., & SCHERSON, A. Ultrashort sleep-walking schedule. I. Evidence of ultradian rhythmicity in "sleepability." *Electroencephalography & Clinical Neurophysiology,* 1981,52, 163-174.

LAVIE, P., & SCHULZ, H. Ultradian rhythms in the pupil. *Sleep Research,* 1978, 7, 307.

LEVIN, B. E., GOLDSTEIN, A., & NATELSON, B. H. Ultradian rhythm of plasma noradrenaline in rhesus monkeys. *Nature,* 1978, 272, 164-166.

LEWIS, B. D., KRIPKE, D. F., & BOWDEN, D. M. Ultradian rhythms in hand-mouth behavior of the rhesus monkey. *Physiology & Behavior,* 1977,*18.* 283-286.

LOVETT, J. W. Two biological rhythms of perception distinguishing between intact and relatively damaged brain function in man. *International Journal of Chronobiology.* 1976, *4,* 39-49.

LOVETT, J. W. Sinus tachycardia and abnormal cardiac rate variation in schizophrenia. *Neuropsycho-biology.* 1980, *6,* 305-312.

LOVETT, J. W., PAYNE, W. D., & PODNIEKS, 1. An ultradian rhythm of reaction time measurements in man. *Neuropsychobiology,* 1978, *4,* 93-98.

LOVETT, J. W., & PODNIEKS, I. Comparison between some biological clocks regulating sensory and psychomotor aspects of perception in man. *Neuropsychobiology, 1975, 1,* 261-266.

LUBOSHITZKY. R.. LAVIE, P., SoiK, Y., CLICK. S. M., LEROITH, D.. SHENN-ORR, Z., & BARZILAI, D. Anti-diuretic hormone secretion and urine flow in aged catheterized patients. *T.l.T. Journal of Life Sciences.* 1978, *8,* 99-103.

MAXIM, P. E.. BOWDEN, D. M., & SACKETT, G. P. Ultradian rhythms of solitary and social behavior in rhesus monkeys. *Physiology & Behavior,* 1976, *17,* 337-344.

MEIER-KOLL, A., POHL. P., SCHAFF, C, & STANK-IEWITZ, C. Ein chronobiologischer aspekt stereotypen vehaltens. *Archiv fuer Psychiatric und Nervenkrankheiten (Archives of Psychiatry & Neurological Sciences},* 1978,225, 179-191.

MILLARD, W. J., REPPERT, S. M., SAGAR. S. M., & MARTIN, J. B. Light-dark entrapment of the growth hormone ultradian rhythm in the rat is mediated by the arcuate nucleus. *Endocrinology,* 1981, *108,* 2394-2396.

NABER, D., WIRZ-JUSTICE, A., KAFKA, M. S., & WEHR, T. A. Dopamine receptor binding in rat striatum: ultradian rhythm and its modification by chronic imipramine. *Psychopharmacology* (Berlin), 1980, *68,* 1-5.

OHLMEYER, P., BRILMAYER, H., & HUELLSTRUNG, H. *Pflu. Arch. Ges. Physiol.,* 1944, *248,*

599-660.

ORR, W. D., HOFFMAN, H. J., & HEGGE, F. W. Ultradian rhythms in extended performance. *Aerospace Medicine,* 1974, *45,* 995-1000.

ORR, W. C., HOFFMAN, H. J., & HEGGE. F. W. The assessment of time-dependent changes in human performance. *Chronobiologia,* 1976, *3,* 293—305.

OSWALD, L. MERRINGTON, J., & LEWIS, H. Cyclical "on demand" oral intake by adults. *Nature,* 1970, 225, 959-960.

PODNIEKS, I., & LOVETT, J. W. Spontaneous rhythms of perceptual motor performance in intact and damaged brain of man. *Biological Psychology, 1915, 3,* 201-212.

QUABBEE, H. J., GREGOR, M., BUMKE-VOGT, C., ECK-HOF, A., & WITT. I. Twenty-four-hour pattern of growth hormone secretion in the rhesus monkey: studies including alterations of the sleep/wake and sleep stage cycles. *Endocrinology,* 1981, *109,* 513-522.

RASMUSSEN, D. D.. & MALVEN, P. V. Relationship between rhythmic motor activity and plasma luteinizing hormone in ovariectomized sheep. *Neuroendocrinology,* 1981, *32,* 364-369.

REINBERG, A., MIGRAINE, C., APFELBAUM, M., BRI-GANT, L., GHATA, J., VIEUX, N., LAPORTE, A., & NICOLAi. Orcadian and ultradian rhythms in the feeding behavior and nutrient intakes of oil refinery

operators with shift-work every 3-4 days. *Diabete and Metabolisme.* 1979, 5. 33-41.

ROMANO, S.. & GIZDULICH, P. Suggestion of ultradian rhythm in peripheral blood flow. *Chronobiologia,* 1980, 7, 259-261.

Rossi, E. L. Hypnotist describes natural rhythm of trance readiness. *Brain/Mind Bulletin,* 1981, *6,* No. 7.

SHIOTSUKA, R., JOVONOVICH, J., & JOVONOVICH, J. A. In vitro data on drug sensitivity: circadian and ultradian corticosterone rhythms in adrenal organ cultures. In J. Aschoff et al. (Eds.), *Chronobiological Aspects of Endocrinology.* Stuttgart: Schattaucr, 1974, 255-267.

SfMON, M. L.. & GEORGE, R. Diurnal variations in plasma corticosterone and growth hormone as correlated with regional variations in norepine-phrine, dopamine and serotonin content of rat brain. *Neuroendocrinology,* 1975, *17,* 125-138.

STEINER, R. A., PETERSON, A. P., Yu, J. Y., CONNER, H., GILBERT, M., TERPENNING, B., & BREMNER, W. J. Ultradian luteinizing hormone and testosterone rhythms in the adult male monkey, macaca fas-cicularis. *Endocrinology,* 1980, *107,* 1489-1493.

TANNENBAUM, G. S., & MARTIN, J. B. Evidence for an endogenous ultradian rhythm governing growth hormone secretion in the rat. *Endocrinology,* 1976, 95, 562-570.

TANNENBAUM, G. S.. MARTIN, J. B., & COLLE, E, Ultradian growth hormone rhythm in the rat: effects of feeding, hyperglycetnia, and insulin-induced hypoglycemia. *Endocrinology,* 1976, *99,* 720-727.

TIERNEY,, L. R., MCGUIRE, R. J., & WALTON, H. J. Distributions of body-rocking manifested by severely mentally deficient adults in ward environments. *Journal of Mental Deficiency Research,* 1978, 22, 243-254.

ULLNER, R. E. On the development of ultradian rhythms: the rapid eye movement activity in premature children. In L. E. Scheving et al. (Eds.), *Chro-nobiology.* Tokyo: Igaku Shoin, 1974, 478-481.

WEDA, T. *Archives of Psychological Monographs,* 1922, *8.*

WEITZMAN, E. D. Temporal organization of neuroen-docrinal function in relation to the sleep-waking cycle in man. In *Recent Studies of Hypothalamic Function,* International Symposium, Calgary, 1973. Basel: Karger, 1974, 26-38.

YEN, S. S. C., VANDENBERG, G., TSAI, C. C.. & PARKER, D. C. Ultradian fluctuations of gonadotropins. In M. Ferin et al. (Eds.), *Biorhythms and Human Reproduction.* New York: Wiley & Sons, 1974, 203-218.

## APPENDIX 2

Barabasz, A. & Barabasz, M. (1996). Neurotherapy and alert hypnosis in the treatment of attention deficit hyperactivity disorder. In Lynn, S., Kirsch, J. & Rhue, J. Clinical Hypnosis Handbook, APA 271-291.

Barabasz, A., Barabasz, M., Jensen, S., Calvin, S., Trevisan, M. & Warner. D. (1999). Cortical event related potentials show the structure of hypnotic suggestions is crucial. International Journal of Clinical and Experimental Hypnosis, 47, 1, 5-22.

Bradbury, J. (2003). Human epigenome project - up and running. Public Library of Science - Biology (PLoS), vol. 1, (3), 316-319.

Draganski, B., Gasert, C., Busch, V., Schuierer, G., Bogdahn, U., & May, A. (2004). Changes in the grey matter induced by training. Nature, 427, 311-312.

Erickson, M. (1958/1980). Naturalistic techniques of hypnosis. In E. Rossi (Eds.), The Collected Papers of Milton H. Erickson on Hypnosis. Vol. I. The Nature of Hypnosis and Suggestion (pp. 168-176). New York: Irvington.

Erickson, M. (1959/1980). Further clinical techniques of hypnosis: Utilization techniques. In E. Rossi (Eds.), The Collected Papers of Milton H. Erickson on Hypnosis. Vol. I. The Nature of Hypnosis and Suggestion (pp. 177-205). New York: Irvington.

Erickson, M., & Rossi, E. (1981). Experiencing Hypnosis: Therapeutic Approaches to Altered States. New York: Irvington.

Erickson, M., Rossi, E. & Rossi, S. (1976). Hypnotic Realities: The Induction of Clinical Hypnosis and Forms of Indirect Suggestion. New York: Irvington.

Gelb, M. (1998). How to Think Like Leonardo da Vinci: Seven Steps to Genius Every Day. New York: Dell Trade Paperback.

Goho, A. (2004). Gene Screen: Ultrasensitive nanowires catch mutations. Science News, 165, 6.

Hahm, J. and Lieber, C. (2004). Direct ultrasensitive electrical detection of DNA and DNA sequence variations using nanowire nanosensors. Nano Letters, 4(1), 51 - 54; DOI: 10.1021/nl034853b

Hilgard, E. (1965). Hypnotic Susceptibility. N.Y.: Harcourt, Brace & World.

Kandel, E. (1998). A new intellectual framework for psychiatry? The American Journal of Psychiatry, 155, 457-469.

Kandel, E. (2000). Cellular mechanisms of learning and the biological basis of individuality. In Kandel, E., Schwartz, J., & Jessell, T. (Eds.) Principles of Neural Science. 4th Edition. P 1247-1279.

Kandel, E. (2001). The molecular biology of memory storage: A dialogue between genes and synapses. Science, 294, 1030-1038.

Ko, J., Na, D., Lee, Y., Shin, S., Kim, J., Hwang, B., Min, B., Park, D. (2002). cDNA microarray analysis of the differential gene expression in the neuropathic pain and electroacupuncture treatment models. Journal of Biochemistry & Molecular Biology, 35(4), 420-427.

Kripke, D., (1974). Ultradian rhythms in sleep and wakefulness. In E. Weitzman (Ed.), Advances in sleep research, (Vol. 1). New York: Spectrum.

Kripke, D., & Sonnenschein, D. (1978). A biological rhythm in waking fantasy. In K. Pope & J. Singer (Eds.), The stream of consciousness. New York: Plenum.

Levsky, J., Shenoy, S., Pezo, C., & Singer, R. (2002). Single-Cell Gene Expression Profiling. Science, 297, 836-840.

Lippincott, B. (1992). Owls and Larks in Hypnosis: Individual differences in hypnotic susceptibility relating to biological rhythms. American Journal of Clinical Hypnosis, 34, 185-192.

Lippincott, B., (1993). The temperature rhythm and hypnotizability: A brief report. Contemporary Hypnosis, 10,155-158.

Lloyd, D. & Rossi, E., (Eds.), (1992). Ultradian Rhythms in Life Processes: A Fundamental Inquiry into Chronobiology and Psychobiology. New York: Springer-Verlag.

Maquet, P. and Ruby, P. (2004). Insight and the sleep committee. Nature, 427, 304-305.

Powell, K. (2003). Economy of the Mind. Public Library of Science, Biology (PLoS), Vol.1, 213-215.

Rainville, P. (2002). Brain mechanisms of pain affect and pain modulation. Cognitive Neuroscience, 195-203.

Ribeiro, S., Goyal, V., Mello, C., Pavlides, C. (1999) Brain gene expression during REM sleep depends on prior waking experience. Learn & Memory, 6: 500-508.

Ribeiro, S., Mello, C., Velho, T., Gardner, T., Jarvis, E., Pavlides, C. (2002) Induction of hippocampal long-term potentiation during waking leads to increased extra hippocampal zif-268 expression during ensuing rapid-eye-movement sleep. Journal of Neuroscience, 22 (24): 10914-10923.

Ribeiro, S., Gervasoni, D., Soares, E., Zhou, Y., Lin, S., Pantoja, J., Lavine, M., Nicolelis, M. (2004). Long-lasting novelty-induced neuronal reverberation during slow-wave sleep in multiple forebrain areas. Public Library of Science, Biology (PLoS), Vol.2, 1, 126-137.

Rossi, E. (1981). Hypnotist describes natural rhythm of trance readiness. Brain Mind Bulletin, 6, 1.

Rossi, E. (1982). Hypnosis and ultradian cycles: A new state(s) theory of hypnosis? The American Journal of Clinical Hypnosis, 25 (1), 21-32.

Rossi, E. (1985). Unity and diversity in Ericksonian approaches: Now and the future. In J. Zeig (Ed.), Ericksonian Psychotherapy. Vol. 1. Structures. New York: Brunner/Mazel, pp. 15-30.

Rossi, E. (1986a). Altered states of consciousness in everyday life: The ultradian rhythms. In B. Wolman & M. Ullman (Eds.), Handbook of Altered States of Consciousness (pp. 97-132). New York: Van Nostrand.

Rossi, E. (1986b). The Indirect Trance Assessment Scale (ITAS). In Yapko, M. (Ed.). Hypnotic and Strategic Interventions: Principles and Practice. New York: Irvington.

Rossi, E. (1986c/1993). The Psychobiology of Mind-Body Healing, 2ed Edition. New York: Norton.

Rossi, E. (1988). The Feigenbaum scenario as a model of the limits of conscious information processing. Biosystems, 46, 113-122.

Rossi, E. (1990). Mind-Molecular Communication: Can We Really Talk to our Genes? Hypnos, 17, 3-14.

Rossi, E. (1992). Periodicity in self-hypnosis and the ultradian healing response: A pilot study. Hypnos, 19, 4-13.

Rossi, E. (1994). The emergence of mind-gene communication. European Journal of Clinical Hypnosis, 3, 1-14.

Rossi, E. (1996). The Symptom Path to Enlightenment: The New Dynamics of Self-Organization in Hypnotherapy; An Advanced Manual for Beginners. New York: Zeig, Tucker, Theisen.

Rossi, E. (2002). The Psychobiology of Gene Expression: Neuroscience and Neurogenesis in Hypnosis and the Healing Arts. New York: W. W. Norton Professional Books.

Rossi, E., (2004). Gene expression and brain plasticity in stroke rehabilitation: A personal memoir of mind-body healing dreams. American Journal of Clinical Hypnosis, 46:3, 215-227.

Rossi, E., & Cheek, D. (1988). Mind-Body Therapy: Ideodynamic Healing in Hypnosis. New York: W. W. Norton.

Rossi, E., & Lippincott. B. (1993). A clinical-experimental exploration of Erickson's naturalistic approach: Ultradian time and trance phenomena. Hypnos, 20, 10-20.

Rossi, E., Lippincott, B., & Bessette, A. (1994). Ultradian Dynamics in Hypnotherapy: Part One. The European Journal of Clinical Hypnosis, 2, 1, p. 10-20.

Rossi, E., Lippincott, B., & Bessette, A. (1995). Ultradian Dynamics in Hypnotherapy: Part Two. The European Journal of Clinical Hypnosis, 2, 2, p. 6-14.

Sanders, S. & Mann, B. (1995). The effects of light, temperature, trance length and time of day on hypnotic depth. American J. Clinical Hypnosis. 37, 43-53.

Storch, K., Lipan, O., Leykin, I., Viswanathan, N., Davis, F., Wong, W., & Weitz, C. (2002). Extensive and divergent circadian gene expression in liver and heart Nature, 417, 78 - 83.

Si, K., Lindquist, S., & Kandel, E. (2003 a). A neuronal isoform of the Aplysia CPEB has prion-like properties. Cell, 115, 879-891.

Si, K., Giustetto, M., Etkin, A., Hsu, R., Janisiewicz, A., Miniaci, M., Kim, J., Zhu, H. & Kandel, E. (2003 b). A neuronal isoform of CPEB regulates local protein synthesis and stabilizes synapse-specific long-term facilitation in Aplysia, Cell, 115, 893-904.

Sjöberg, P. (2002). Numerical solution of the master equation in molecular biology. Master's Degree Project (MSc Thesis): Uppsala University. http://www.icm.uu.se/molbio/paul/masterreport.pdf

Wagner, U., Gais, S., Haider, H., Verleger, R., & Born, J. (2004). Sleep inspires insight. Nature, 427, 352-355.

Weitzenhoffer, A. (2000). The Practice of Hypnotism. 2ed Edition. New York: Wiley.

Weitzenhoffer, A. (2001). For the record: A commentary on the role of suggestion in hypnosis. American Journal of Clinical Hypnosis, 44, 155-157.

# Index

Ernest Lawrence Rossi, Ph.D.
A DISCOURSE WITH OUR GENES
*The Psychosocial and Cultural Genomics*
*of Therapeutic Hypnosis and Psychotherapy*

Edizioni Editris sas

*Printed in Italy*